Themes in Contemporary Archaeology

The Themes in Contemporary Archaeology series provides cutting edge perspectives on key areas of debate in current archaeological enquiry, with a particular emphasis on European archaeology. The series has a broad coverage, encompassing all archaeological periods and all approaches. Examples of topics welcome in the series include, but are not limited to:

- from theoretical debate to archaeological practice
- landscape studies
- bioarchaeology
- issues of cultural heritage

The volumes are based on research presented at the Annual Meetings of the European Association of Archaeologists. They include proceedings of individual sessions, which can be enlarged if necessary to provide coherence for publication. Each volume undergoes a strict peer-review process, ensuring volumes of high quality that capture current debates in the field.

SERIES EDITORS
The Series editors are Peter Attema, Agathe Reingruber and Robin Skeates (themes@e-a-a.org)

DISCOUNT FOR MEMBERS
All EAA members can buy print and e-copies of all THEMES volumes at 30% discount - please email administrator@e-a-a.org who will assist you.

Please refer the following link for the EAA guidelines

https://www.e-a-a.org/EAA/Navigation_Publications/THEMES.aspx?WebsiteKey=59c4f6fb-bee0-4278-9c2f-9529288deb09&THEMES=4#THEMES

Christina Tsouparopoulou • Lærke Recht
Editors

Human and Aquatic Beings: Interactions in and beyond the Eastern Mediterranean (3rd–1st Millennia BCE)

Springer

EAA
European Association of Archaeologists

Editors
Christina Tsouparopoulou
Institute of Archaeology
UKSW Warsaw
Warszawa, Poland

Lærke Recht
Institut für Antike
University of Graz
Graz, Austria

ISSN 2730-7441 ISSN 2730-745X (electronic)
Themes in Contemporary Archaeology
ISBN 978-3-031-73645-2 ISBN 978-3-031-73643-8 (eBook)
https://doi.org/10.1007/978-3-031-73643-8

Cover image: Original pencil drawing by M. Yamasaki, modified and digitised by L. Recht

This Springer imprint is published by the registered company Springer Nature Switzerland AG
The registered company address is: Gewerbestrasse 11, 6330 Cham, Switzerland

Funding

Christina Tsouparopoulou has been supported by the EU Horizon 2020 Research and Innovation Programme under the Marie Skłodowska-Curie grant agreements no. 748293 and 951328; by the Swedish Research Council, under Grant No. VR 2016-02028; and by Poland's National Science Center (NCN) under the OPUS grant agreement No. 2021/41/B/HS3/04062.

Contents

About the Editors

Christina Tsouparopoulou Assistant Professor in Mesopotamian Archaeology, Institute of Archaeology, UKSW Warsaw and Honorary Fellow, Department of Archaeology, Durham University. Main research interests: Bronze Age Mesopotamia and Eastern Mediterranean, religion, popular material culture, text and object, human-animal relations, digital humanities.

Lærke Recht Professor of Early Eastern Mediterranean Archaeology, Department of Ancient Eastern Mediterranean Studies, Institute of Classics, University of Graz and International Institute for Mesopotamian Area Studies Research Fellow. Main research interests: Bronze Age Aegean, Cyprus and Southwest Asia, human-animal relations, iconography, religion, digital archaeology.

List of Figures

List of Tables

List of Appendices (online supplementary material)

1 Creatures of the Water and Their Impact on Human Lives

Lærke Recht and Christina Tsouparopoulou

Fish carry stories in their bones
(Todd 2018: 66)

Abstract

In studies of human-animal relations there is often a focus on mammals and other large animals, while creatures of the sea are ignored or underrepresented. This is despite their enormous importance both in the island communities of Crete and Cyprus, and in the river-dependent areas of Egypt and Western Asia. Fish and molluscs were part of subsistence strategies, and thus central to ancient economies, but they also feature prominently in religious practices and as powerful symbols in iconography. Further, various types of water (oceans, rivers, lakes and so on) may give rise to a range of relations. We here present and discuss some of the ways in which this topic can be approached and what we might learn from employing a variety of methods and data—including zooarchaeological, stable isotope, iconographic, functional and archaeological studies, contextualising the chapters that appear in this volume. These themes are brought together to examine how animals of the water impacted human lives and are placed in the broader contexts of human-environment relations in the ancient Eastern Mediterranean and beyond.

L. Recht (✉)
Institut für Antike, University of Graz, Graz, Austria
e-mail: rechtl@tcd.ie

C. Tsouparopoulou
Institute of Archaeology, UKSW Warsaw, Warszawa, Poland

Department of Archaeology, Durham University, Durham, UK

1.1 Human-Aquatic Relations

This volume is a collection of chapters that explore relations between humans and aquatic beings in the past. To the quotation from Todd that fish carry stories in their bones, we may add that molluscs also carry stories in their shells. This is the central idea of the volume, to extract and explore some of the information that can be gained from aquatic beings, with a series of case studies of the many ways in which fish and molluscs have interacted with humans and impacted human lives. Geographically, the chapters cover the Eastern Mediterranean and adjacent areas: the Aegean, Cyprus, Egypt, West Asia and Iran. The focus is primarily on the third to first millennia BCE, although examples and discussion sometimes include earlier and later periods. The methods and data used include the physical remains of aquatic beings and zooarchaeological analysis but go much beyond that: the stories are not only in the physical remains, but also in images, human remains, tools and installations, and in written records and ancient knowledge of aquatic spaces and their inhabitants. The chapters here are case studies: the volume does not aim to cover every aspect or type of interaction in the Eastern Mediterranean and beyond over these millennia, nor all available methodologies. Instead, the studies illustrate a part of the existing material record and a range of possible approaches. We take the opportunity here to point to some themes and methods that appear in several of the chapters more or less explicitly, as well as to discuss themes that we consider worth delving into further, even if they only appear to a limited extent in the studies that follow.

The study of fish and molluscs has been part of zooarchaeology and zoological studies of archaeological assemblages since the nineteenth century, but only more

C. Tsouparopoulou, L. Recht (eds.), *Human and Aquatic Beings: Interactions in and beyond the Eastern Mediterranean (3rd–1st Millennia BCE)*, Themes in Contemporary Archaeology, https://doi.org/10.1007/978-3-031-73643-8_1

consistently so from the later part of the twentieth century (Wheeler and Jones 1989: 1–3). Approaches inspired by the field of Human-Animal Studies and multispecies ethnography have become increasingly frequent in zooarchaeology and archaeology more broadly in the last decades (e.g. Russell 2012; Sykes 2015; Stépanoff and Vigne 2019). While there is much variety within such studies, they generally attempt to de-centre humans and present a more balanced discussion of relations between humans and other animals, for example by acknowledging active engagement of all players and species entanglement. The interactions may be more or less symmetrical, and studies also include more detailed analyses of how nonhuman animals resist contact or certain types of relations with humans (e.g. Ingold 1994; Recht 2019, 2023; see also Hribal 2007 for a historical approach). Research in Human-Animal Studies has mostly explored interactions with medium to large-sized terrestrial species, probably due to their more immediately evocative nature and sometimes more intimate relations with humans. Aquatic species have received less attention, with important exceptions especially coming from ethnographic studies (e.g. Todd 2014, 2018; papers in Goyal et al. 2023).

Some of these ethnographic studies bear particular relevance to the aim of this volume, contributing to a greater balance, in particular with more specific archaeological perspectives from the early Eastern Mediterranean world. For example, in her work on human-fish relations among Indigenous peoples in Canada, Todd speaks of 'fish pluralities' and 'active sites of engagement' as a way of expressing the many ways of interacting with and knowing fish (Todd 2014, 2018). These concepts, extended to molluscs, find resonance in the context of past communities and the very different parts of the world discussed here, where the meshwork of aquatic beings, human beings, landscape, knowledge, memory, practice, subsistence, material and craft is evident in every chapter presented.

While the chapters here do not all explicitly draw on the field of Human-Animal Studies, they do all concern interactions between humans and aquatic animals, and explore various types of relations, their expression in the archaeological record, and means of extrapolating qualities of the relations. Below, we highlight themes present within the chapters, as well as additional themes we find pertinent for more detailed discussion. Thus, our aim extends beyond merely introducing the volume; we seek to offer a substantive contribution to the dialogue it initiates.

1.2 Underwater Spaces

Aquatic spaces were central to all the cultures of the eastern Mediterranean and beyond. The Mediterranean itself was not simply a big open space: for island communities, it was both a means of connecting and a source of isolation (perhaps even protection). Without it, the great 'international' age of the Late Bronze Age and its extensive trade networks would have looked very different. Equally important were the riverine landscapes of Mesopotamia and Egypt, which fostered the first urban centres. The dependence on the Nile, Tigris and Euphrates (and their tributaries) in everyday life is reflected in the importance of irrigation and seasonal flooding in official matters, with, for example, the digging and maintenance of canals being one of the key responsibilities of Mesopotamian rulers. Rivers and oceans were the most important and efficient means of transport, both locally and long-distance. Thus, humans frequently travelled on and fished in fresh and sea water, and some humans may even have spent the majority of their lives on or near water. All of these activities mostly relate to water as a surface, a space to travel upon and to fish from.

However, seas, oceans and lakes all have depth—they are three-dimensional spaces, and most aquatic beings live *within* these spaces, not on their surfaces or at their edges. Although human beings could only ever visit the underwater realm for limited periods of time, the range of species, iconographic evidence and written sources provide evidence that such visits did indeed take place. Minoan so-called Marine style vases depict submarine landscapes of octopuses, sea anemones, star fish, molluscs, argonauts, fish, dolphins and the rocky seabed. The floating asymmetrical composition of some specimens suggests a space conceptualised not as a terrestrial one with a fixed reference point in the form of a groundline (e.g. Popham 1967: pl. 79b), perhaps also reflecting a human experience of being underwater. A Mesopotamian understanding of fish as being in and of the water (more specifically, the rivers) can be seen as represented already in early engravings on seals, and in later periods, on Neo-Assyrian palace reliefs. For example, on reliefs from Khorsabad depicting transport of timber on the river, fish swim in the carefully carved water along with crabs, turtles, eel-like creatures, goat-fish and other unidentifiable creatures (Botta 1849: pl. 32–34; see also BM 102981). Whether this extended to substantial underwater experience is not clear. The occasional presence of fantastic creatures such as fish-goats, the generic or somewhat comical rendering of fish, and the absence of any other underwater features may indicate a lack of familiarity with or interest in the landscape below the surface, at least on the part of the engraver.

How underwater spaces were conceptualised and experienced is thus a difficult question and one that remains to be addressed in detail. The demographic extent of diving, frequency and expertise are also currently mostly unknown. The chapter by Yamasaki is therefore particularly relevant in starting to explore this much underrepresented aspect of relations between (deep) water, humans and aquatic beings.

1.3 Fish and Mollusc Remains and Human Diet

Like all faunal remains, the remains of fish and molluscs on archaeological sites are subject to preservation, excavation strategies and taphonomic processes. They have the added disadvantage of usually consisting of very small and fragile elements, and thus tend to be underrepresented in faunal assemblages, in particular when dry or wet sieving is not carried out.[1] Identification of fish and molluscs to species also requires special expertise, and these are therefore not always recorded in detail in faunal reports, making it difficult to reconstruct exactly which taxa or species were present and exploited at specific sites, along with questions of seasonality, trade and fishing technologies that otherwise have the potential to be inferred from faunal assemblages.[2] These are also the challenges faced by Michael in her longue durée study of human-fish relations on Cyprus. She is nevertheless able, through systematically mapping the archaeological sites where finds of fishing gear and fish remains have been recovered, to demonstrate regional and chronological patterns in fishing activity. In her chapter, she discusses, among others, that the Late Bronze Age site of Hala Sultan Tekke has revealed shell gathering and the exploitation of coastal and pelagic resources caught in an area ranging from 1 to 300 m depth (also in Michael 2022: 351–53), indicating that fishing was practised within the areas of the continental shelf and the lagoon, which acted as the harbour of the site during the second millennium BCE. Moreover, the retrieval of imported fish species, such as *Lates niloticus* and *Clarias* sp. at Hala Sultan Tekke attests to the presence of exchange networks also evident in the material culture of the site.

Consumption constituted one of the main spheres of interaction between humans and aquatic beings in the regions covered in this volume. We take it that a large amount of the remains of fish recovered from archaeological sites is the result of human consumption of these aquatic species, and that much of the evidence of fishing relates to food practices. This can also be detected through the application of stable isotope analysis, where aquatic species of different trophic levels can be distinguished from terrestrial species. Such analyses have been carried out on human skeletons from a number of sites in the Aegean, and in their chapter, Mylona and Allshouse carefully integrate these results with those of faunal assemblages to demonstrate a range of patterns in the consumption of aquatic species at different sites in the region. This complements earlier research that indicates a surprisingly low level of reliance on marine resources in the Neolithic and Bronze Age of Greece (Papathanasiou 2015). Similar studies from northern Africa and southwest Asia have so far either failed to identify aquatic species in human dietary practices or have done so to a limited extent (e.g. Herrscher et al. 2021; Sołtysiak and Fernandes 2021; Stantis et al. 2021).

Molluscs were certainly also consumed by humans, but the hard shells which survive in archaeological contexts in the eastern Mediterranean could indicate a wide variety of additional types of interaction. Even large deposits could reflect production, such as the great amounts of murex shells recorded at some Levantine and Cypriot sites used in the production of purple dye, as discussed in the chapter by Yamasaki.

1.4 Fishing and Collecting

Some of the methods used for fishing and collecting in the areas covered here are known or inferred from fishing equipment, the species present in faunal assemblages (which provide information concerning typical habitats and thus indirectly possible means of catching), visual evidence, written records and ethnographic studies. For example, fishing and fishing technology have been discussed for the Aegean (e.g. Rose 1994; Mylona 2014, 2021; Shapland 2022: ch. 6), Classical antiquity (e.g. Bekker-Nielsen and Casasola 2010; Bekker-Nielsen 2020), Mesopotamia (e.g. Englund 1990; Sahrhage 1999, 2008b; Potts 2012), Egypt (Aleem 1972; Brewer and Friedman 1989; Sahrhage 2008a; Brewer 2012), the Levant (e.g. Galili et al. 2002) and the Indus Valley (Sahrhage 2008c; Belcher 2023). Some of the known fishing methods include the use of hook and line, harpoon and spear, nets, traps and ruses; ponds may also have been used to keep or breed fish, as is known from ancient Egypt (Aleem 1972).

In addition to the diverse methods used for fishing and collecting, the social organisation of fishing emerges as a notable theme for further consideration. Papers in Bekker-Nielsen's edited volume (2010) reveal how fishing practices were not only a means of subsistence but also a reflection of the social and economic structures within communities. These inferences shed light on community hierarchies, trade practices, and the roles of individuals within the fishing process, offering a nuanced understanding of ancient

[1] Wheeler and Jones provide a test example of collection by hand compared to collection by sieve, which resulted in the presence of two and six different species, respectively (1989: 38–39, with more references to various experiments with similar results).

[2] We may assume, however, that the range of species would have been substantial. For example, a list published in 1969 mentions about 150 different species of fish present in the waters of Iraq (Mahdi and Georg 1969). Correct identification is of course key in order to address such questions, as is knowledge of potential taphonomic processes and the properties of the skeletal remains of aquatic species (discussion of this and examples of incorrect inferences can also be found in Wheeler and Jones 1989). Unfortunately, legacy data do not always offer detailed information about taxa or species identification, as illustrated in several chapters here.

societies' reliance on aquatic resources. Moreover, the implications of fishing techniques, discerned from the remains of gear and fish, allow us to further understand the technological advancements and environmental adaptations of ancient civilisations. The types of gear used, coupled with the species identified in faunal assemblages, suggest sophisticated knowledge of marine and freshwater ecosystems, fishing seasons, and the behaviour of target species. This reciprocal relationship between the remains of fishing gear and fish provides a window into the past, revealing not only the techniques employed but also the cultural significance of fishing in shaping social practices and dietary preferences. Such insights underscore the intricate link between human ingenuity in exploiting aquatic resources and the broader socio-economic and environmental contexts of ancient fishing communities.

Michael, in her chapter, highlights fishing both as an activity and lifestyle on Cyprus, involving interdependent economic, cultural, technological and environmental aspects. She examines the archaeological finds of fishing technology (e.g. harpoons or spears, hooks, traps, stone, clay and lead weights for the net or line weirs, fish-tanks) and the fishbone assemblages recovered at a variety of archaeological sites on Cyprus. Stamatopoulou closely examines the fishing gear excavated at Vergina in northern Greece, which includes more than 20 fishing-net sinkers and a line sinker, made of lead, and a bronze fishing hook, and concludes that fishing at Vergina aimed at supplementing the diet and was not professionally or recreationally performed.

Implicit but rarely discussed in the archaeological literature is the meshwork of actors, material and knowledge. We can return to Todd's work for her description of what 'fishing' entails:

> fishing is the constellation of activities that go into a fishing trip, including the following: preparation (packing, cooking, inviting people and spreading the word about a trip); thinking about fish and determining where the best place to try and get fish will be (based on time of year, who one is travelling with, available resources, preferred fishing locales); travelling in and navigating sentient landscapes (Anderson 2000; Ingold 1993); skills required to get *at* fish (using an auger to drill a hole through the ice, preparing and setting a net, knowing how to select the right rod and line and lure); and skills and jobs that are employed to support fishing *at* a fishing site (starting a fire to make tea and lunch, making sure children and "greenhorns" [unskilled beginners, both Inuvialuit and non-Inuvialuit] are looked after). When fishing, one is thus not focused only on the "catch," but also on entering into a series of relationships with fish, the environment, and other people in order to *try* to get a fish. (Todd 2014: 224)

All types of fishing would have constituted embodied practices that involved knowledge of the local environment, the behaviour of the water and its beings, and fishing technology. Learning these skills and living in and with the landscape likely occurred through communities of practice and more or less formal apprenticeships. Belcher's ethnoarchaeological study of fishing practices in Pakistani Punjab, aimed at providing a better understanding of third millennium BCE practices in the Indus Valley, reveal fascinating insights into fishing as a family activity, gendered distribution of different tasks (varying based on household or more commercial level), and how ownership of traps and nets can affect access (Belcher 2023). For example, not everyone can afford the necessary nets, or to maintain the nets already owned, and therefore occasionally borrow from a fellow community member, with that person then receiving a cut of the catch or profits. Fishing itself seems to be almost exclusively a male activity, although at the household level, women might be in charge of preparing and cooking. These social conditions and dynamics are generally lost in the archaeological record, but studies like this do offer both possibilities and potential means of identifying similar contexts based on faunal remains and fishing equipment.

Similar communities of practice apply to harvesting or collecting of molluscs (see e.g. Burgos et al. 2019), which also requires knowledge of habitat, where and when to find molluscs, and how to process them for the desired purposes. The harvesting itself, however, may need very little skill. An ethnoarchaeological study by ElMali of recent practices on the coasts of Oman demonstrates that women, men and children all participate in the collection of molluscs with no special training (ElMali 1999), although it can be assumed that even the simple act of locating and picking up the animals involves a level of imitation and observation of other community members. Just as fascinating are the post-collection processes recorded, where the methods for preparing the molluscs for consumption reveal subsistence strategies and long term preservation. In addition, the shells were discarded in a non-systematic manner that means they would not be recognised archaeologically.

1.5 Trade

Aquatic beings were part of short and long-distance trade from at least as early as the third millennium BCE (for a broad diachronic and geographical overview, see also Van Neer et al. 2004). Some fish were traded whole, as exemplified by the presence of Nile perch on Bronze Age Cyprus (see also Michael's chapter). Egyptian sources tell us that fish could be preserved by drying, salting and pickling, or as fish cakes or fillets (Aleem 1972); it is therefore likely that exported fish were dried. In the case of molluscs, it is harder to determine in what form they travelled, and whether they were traded for their meat or shells, since only the shells are preserved in the archaeological record.

Some of the trade networks involving aquatic resources that can be reconstructed from texts and archaeological evidence are discussed in the chapter by Yamasaki, who notes exchange between Egypt and Cyprus, Egypt and the Levant (including Ugarit and Byblos), Ugarit and Tyre; the trade between Egypt and Cyprus is also part of Michael's chapter. Meanwhile, Schlegel's survey of the find spots of engraved *Tridacna* in the first millennium BCE reveals an extensive and broad exchange network, with the *Tridacna* likely originating in the Gulf and widely distributed, from Susa in the east to Italy in the west, as well as in Egypt and northern and southern Mesopotamia. Even further east, Naseri, Motarjem and Reese offer a similarly broad geographical distribution of shell discs made from species originating in the Gulf and Oman region. The shell material from the third millennium BCE cemetery at Ur equally demonstrates long-distance trade between Mesopotamia, the Gulf and the Indus Valley, as noted in Recht and Tsouparopoulou's chapter (see also Crawford 1973).

There may be a correlation between butchery patterns and trade, at least at a local level. Interestingly, and perhaps somewhat counterintuitively, Belcher's ethnoarchaeological study also demonstrated a correlation between species diversity, butchery patterns and the mechanisms of distribution (Belcher 2023). A greater species diversity was associated with more direct exchange between fishers and customers, while indirect trade through specialist merchants or fishmongers resulted in a more limited selection of species, possibly with an emphasis on those that have more meat and can be preserved for longer. Similarly, a greater diversity in butchery patterns was typically related to more direct trade, while indirect trade had more specialisation and standardisation of butchery methods.

1.6 Shells and Fish as Raw Material: Transformations and Materiality

The entangled relations of material, quality, embodiment, humans, animals and plants are recognised in studies in materiality or new materialism. For example, the capacity of clay to be shaped, to act as a body, to transform itself, its contents and its 'users' has been discussed for various ceramic containers from Greece and Cyprus (Recht 2014; Steel 2020; Voulgari 2021). Aquatic beings can be part of such entanglements in a number of ways. As we see in the papers of Eleftheriou, Panagiotaki and Reese, Recht and Tsouparopoulou, Schlegel, Ftaimi, Patrevita, and Naseri, Motarjem and Reese, the shells of various aquatic species were transformed into containers, seals, inlays, jewellery, decorative and symbolic objects. To achieve this, they were in some cases simply used as they were, but in other cases their shape was modified, incised or painted, sometimes to the extent that they would no longer have been recognisable as once having been the shells of living animals. In the case of the latter, objects are likely to have maintained the light hue and sheen characteristic of many shells, but their origin was otherwise obliterated, and may not have been known by anyone not involved in the process of manufacture.

This process ranges from the fairly modest decoration of cockle shells with paint and/or small piercings, as in those recovered from the Temple Repositories at Knossos (Eleftheriou, Panagiotaki and Reese) to those requiring highly skilled artisans, as for example the cylinder seals of Mesopotamia (Ftaimi; Recht and Tsouparopoulou) or the gold imitation cowrie-shells of Egypt (Patrevita).

Organic material such as fish skin does not preserve well in the areas discussed here, and we do not have explicit evidence of how this may have been used as a raw material, but there are some more indirect indications, and practices from elsewhere in the world hint at possible uses of this material. Fish skin and fish leather have been, and some places still are, used for clothes, bags, parchment and musical instruments (Vávra 2020; Rahme 2021). Fish leather has many of the same qualities as other types of leather, but may be more durable, and can be made waterproof with certain treatments.

It is possible that the images of fish-cloaked figures discussed by Ftaimi are a reference to real priests dressed in fish-cloaks for designated occasions. If so, such cloaks may have been partially or wholly made from actual fish skins (see also *RLA*, 'Maske'). Other parts of fish and fish oil are also known to have been used in medicinal recipes in Mesopotamia and Egypt (Thompson 1937; Brewer 2012; Arbøll 2021).

1.7 Iconography and Symbolling of Aquatic Beings

The aquatic world also found its way into visual representations in different media and types of material culture: jewellery, seals, ceramics, figurines, wall reliefs and paintings, plaques, statues and so on (see chapters by Lekka, Yamasaki, Schlegel, Ftaimi, Patrevita, and Recht and Tsouparopoulou). Not only the raw material, but also images travelled. While the *Tridacna* of Schlegel's chapter may have come to the Levant to be engraved there, many of the vessels with marine imagery in Lekka's chapter were produced in Greece and exported to Cyprus, where in turn they were re-produced on locally made ceramics. The precise species is often not recognisable. Fish in particular are often rendered in a generic and standardised manner, the equivalent of a modern-day stick-man drawing. Outside three-dimensional skeuomorphs, molluscs are rarely depicted in the iconographic repertoire, with the notable exception of Minoan Crete. Nevertheless, a great range of aquatic beings occurs in the images of the Eastern Mediterranean and beyond, including tuna fish, dolphins, octopuses, squids, cuttlefish, sea anemones, triton shells, eels, turtles, crabs and cockle shells, along with hybrid and mythological creatures. Without assuming ancient adherence to a Linnaean system and to absolute zoological accuracy, careful analysis of the iconography may also reveal details that provide further information or identification, as for example done in a collaboration between archaeologist Alberti and marine biologist Bello on octopuses in Bronze Age Aegean iconography (Alberti and Bello 2021).

Sometimes aquatic beings were used to represent environments—the spaces related to water, including those underwater noted above. In these instances, they act as a sort of shorthand for the seascape (or index, as Shapland [2022: 163] calls it), rather than refer to specific animals or animal species. Elsewhere, the reference seems to be to concepts related to abundance, such as scenes of feasting or offerings, as that found on the Standard of Ur, discussed by Recht and Tsouparopoulou. These are not mutually exclusive, and many of the scenes on the pottery examined by Lekka may be understood to both indicate a space and the wealth of the sea.

Aquatic beings were part of mythology and religion. There is no universality to their meaning or what specific species symbolised, but some commonalities must have occurred. An association with death and rebirth can be detected in Egypt, Mesopotamia and the Aegean. In Egypt, fish could be a symbol of rebirth, and the deceased was believed to change into a fish in order to reach the netherworld (Brewer 2012); in Mesopotamia, shell containers were sometimes placed with the deceased (chapter by Recht and Tsouparopoulou); and in Late Bronze Age Crete, funerary larnakes were frequently decorated with marine motifs (Watrous 1991; Saunders 2008). It is possible that a similar association was present on Cyprus (chapter by Lekka).

In other instances, aquatic beings are associated with individual deities or appear in the mythological repertoire of a region. The mythological realm opens up the possibility of hybrid creatures. The Mesopotamian god Ea/Enki was closely associated with fish: substantial quantities of fish remains have been associated with his temples (especially at his Eridu temple), and in the mid-third millennium BCE, he is depicted with a flowing stream of fish from his shoulders. His daughter Nanshe similarly appears closely associated with fish in literary compositions of the early second millennium BCE, where she is either presented surrounded by fish or as the 'mother of fish' and the 'queen of the fishermen' (Tsouparopoulou 2024, with references therein). In later periods, Enki's creatures included the turtle and the goat-fish. The fish-cloaked figures of the first millennium BCE investigated by Ftaimi may belong in this category, although possibly also represented in rituals by priests in fish-garb.

1.8 Aquatic Beings and Identity

Humans often define themselves and others through relations with other animals and animal bodies—a person may be a cattle farmer, a shepherd, an equestrian or a dog-owner. Groups of people may be identified or self-identify through their foodways (e.g. Sykes 2015: 150–55; Twiss 2019), with perhaps the most famous example being the attitudes towards the consumption of pig. Thus, the Hebrew taboo against pork consumption may have partly developed as a means of self-identification and differentiation (Hesse 1994; Price 2020). Notably, taboos related to fish are also known (e.g. Begossi et al. 2004; Meyer-Rochow 2009), including from ancient Egyptian textual sources (Brewer 2012). Nonetheless, such taboos pertaining to fish consumption have yet to be archaeologically detected.

Meat from different species and parts of the animal body may be associated with social status, which can be studied through faunal assemblages (Crabtree 1990; Ashby 2002; Greenfield and Matney 2021). It is important to note, however, that the specific species and animal parts are highly dependent on contextual factors, precluding a universal correlation between particular species and high or low social status. Similarly, the control and utilisation of fish resources, including access to fishing locations, can serve as leverage in local power dynamics, as exemplified in the case of the American Northwest (Hayden 2021). Social status can be further investigated through stable isotope analyses of human teeth to determine health and diet (e.g. Beck 2020). Mylona and Allshouse combine a survey of evidence from the faunal record with that of results from available stable isotope analyses from the Bronze Age Aegean and demonstrate differential access to fish and seafood within Aegean populations. For example, they show that the people buried at the Gypsades Cemetery in Knossos embraced a diet of low trophic level fish and molluscs (and had poorer burials), while the people buried in Grave Circle A at Mycenae (of elite status) consumed larger fish of higher trophic level.

Other means of expressing identity occur through clothes and other material carried on the body, or personal items used as part of daily activities. As mentioned, such material can be made from or imitate nonhuman animal bodies, including those of aquatic beings. In the chapters presented here, there are examples of jewellery and elements of clothes made from shells (Patrevita; Recht and Tsouparopoulou; Naseri, Motarjem and Reese), jewellery imitating shells (Patrevita), seals made from shells (Recht and Tsouparopoulou; Ftaimi), and personal items such as small containers made of or imitating shells (Recht and Tsouparopoulou; Schlegel). What is more, the intense purple colour extracted from murex shells, and applied to textiles attainable solely by the wealthiest (as discussed in Yamasaki), is likely to have been a highly charged marker of identity.

Connections with wealth is further emphasised through certain species of molluscs as a long-distance good, the craftsmanship involved in their transformation, and their imitations in precious stone and metals. However, aspects of identity here may be intersectional, referring not only to social (economic) status, but also to gender, age and ethnicity, and some aquatic beings may have been more strongly associated with women (Ftaimi; Schlegel; Recht and Tsouparopoulou). The connection with water and aquatic beings varies from direct and explicit, for example with the use of shells in their original form, through to the more commonly transformed items, where the animal itself may only be recognised through prior knowledge of production processes, hue and sheen. Such a transformation and the associated tacit knowledge may have served to add to the social and economic value of the items.

The chapters of this volume thus offer case studies and new insights into interactions between aquatic and human beings in the Eastern Mediterranean and beyond, with an overall aim of further introducing aquatic beings into the research and discussions on human-animal relations in the past.

Bibliography

Alberti, L. & Bello, G. 2021. How Many Tentacles? Octo-pus and X-pus in the Aegean Bronze Age: A New Archaeozoological Approach. In: L. Recht and K. Zeman-Wiśniewska, eds. *Animal Iconography in the Archaeological Record: New Approaches, New Dimensions*. Sheffield: Equinox, pp. 71–90.

Aleem, A.A. 1972. Fishing Industry in Ancient Egypt. *Proceedings of the Royal Society of Edinburgh, Section B: Biological Sciences,* 73: 333–43. https://doi.org/10.1017/S0080455X00002381.

Arbøll, T.P. 2021. A New Look at Eels and Their Use in Mesopotamian Medicine. In: L. Recht & C. Tsouparopoulou, eds. *Fierce Lions, Angry Mice and Fat-tailed Sheep: Animal Encounters in the Ancient Near East.* Cambridge: McDonald Institute for Archaeological Research, pp. 179–91.

Ashby, S.P. 2002. The Role of Zooarchaeology in the Interpretation of Socioeconomic Status: A Discussion with Reference to Medieval Europe. *Archaeological Review from Cambridge,* 18: 37–59.

Beck, J. 2020. The Human Position: The Potential of Bioarchaeology for Studies of Inequality in Iberian Late Prehistory. In: P. Díaz del Río, K. Lillios & I. Sastre Prats, eds. *The Matter of Prehistory: Papers in Honor of Antonio Gilman Guillén*. Madrid: Consejo Superior de Investigaciones Científicas, CSIC, pp. 153–72.

Begossi, A., Hanazaki, N. & Ramos, R.M. 2004. Food Chain and the Reasons for Fish Food Taboos among Amazonian and Atlantic Forest Fishers (Brazil). *Ecological Applications,* 14: 1334–43.

Bekker-Nielsen, T. 2020. Beyond the Mediterranean: Fish and Fishing in the Black Sea. In: J.M. Vargas Gíron, ed. *El instrumental de pesca en el Fretum Gaditanum (siglos V a.C. – VI d.C.)*. Oxford: Archaeopress, pp. 173–75.

Bekker-Nielsen, T. & Casasola, D.B. eds. 2010. *Ancient Nets and Fishing Gear: Proceedings of the International Workshop on "Nets and Fishing Gear in Classical Antiquity: A First Approach".* Cádiz and Aarhus: Publicaciones de la Universidad de Cádiz and Aarhus University Press.

Belcher, W.R. 2023. Ethnoarchaeological Studies of Riverine Fisheries and Butchery in Pakistani Punjab. In: P. Goyal, G.S. Abhayan & S. Channarayapatna, eds. *Animals in Archaeology: Integrating Landscapes, Environment and Humans in South Asia (a Festschrift for Prof. P.P. Joglekar).* Thiruvananthapuram: Department of Archaeology, University of Kerala, pp. 173–95.

Botta, P.E. 1849. *Monument de Ninive, 1. Architecture et sculpture.* Paris: Imprimerie Nationale.

Brewer, D.J. 2012. Fish, Fishing, Pharaonic Egypt. In: R.S. Bagnall, K. Brodersen, C.B. Champion, A. Erskine & S.R. Huebner, eds. *The Encyclopedia of Ancient History*. Chichester: Wiley. https://doi.org/10.1002/9781444338386.wbeah15161.

Brewer, Douglas J. & Friedman, R.F. 1989. *Fish and Fishing in Ancient Egypt.* Cairo: American University in Cairo Press.

Burgos, A., Younger, A.C. & Wolverton, S. 2019. Human Mollusk Interactions in a Changing World. *Journal of Ethnobiology,* 39: 175–81. https://doi.org/10.2993/0278-0771-39.2.175.

Crabtree, P.J. 1990. Zooarchaeology and Complex Societies: Some Uses of Faunal Analysis for the Study of Trade, Social Status, and Ethnicity. *Archaeological Method and Theory,* 2: 155–205.

Crawford, H.E.W. 1973. Mesopotamia's Invisible Exports in the Third Millennium B.C. *World Archaeology,* 5: 232–41.

ElMali, A.T. 1999. Mollusc Harvesting along the Coasts of Oman: A Supplementary Diet. *Proceedings of the Seminar for Arabian Studies,* 29: 45–53.

Englund, R.K. 1990. *Organisation und Verwaltung der Ur III-Fischerei.* Berliner Beiträge zum Vorderen Orient 10. Berlin: Reimer.

Galili E., Rosen, B. & Sharvit, J. 2002. Fishing-gear Sinkers Recovered from an Underwater Wreckage Site, off the Carmel Coast, Israel. *International Journal of Nautical Archaeology,* 31: 182–201.

Goyal, P., Abhayan, G.S. & Channarayapatna, S. eds. 2023. *Animals in Archaeology: Integrating Landscapes, Environment and Humans in South Asia (a Festschrift for Prof. P.P. Joglekar).* Thiruvananthapuram: University of Kerala.

Greenfield, T. & Matney, T. 2021. Stews, Ewes, and Social Cues: Commoner Diets at Neo-Assyrian Tušhan. In: L. Recht & C. Tsouparopoulou, eds. *Fierce Lions, Angry Mice and Fat-tailed Sheep: Animal Encounters in the Ancient Near East.* Cambridge: McDonald Institute for Archaeological Research, pp. 161–78.

Hayden, B. 2021. Transegalitarian Societies on the American Northwest Plateau: Social Dynamics and Cultural/Technological Changes. In: O. Cerasuolo, ed. *The Archaeology of Inequality: Tracing the Archaeological Record.* Albany: SUNY Press, pp. 35–50.

Herrscher, E., Poulmarc'h, M., Palumbi, G., Paz, S., Rova, E., Gogochuri, G., Longford, C. et al. 2021. Dietary Practices, Cultural and Social Identity in the Early Bronze Age Southern Caucasus. *Paléorient,* 47(1): 151–74. https://doi.org/10.4000/paleorient.946.

Hesse, B. 1994. Husbandry, Dietary Taboos and the Bones of the Ancient Near East: Zooarchaeology in the Post-processual World. In: D. Small, ed. *Methods in the Mediterranean. Historical and Archaeological Views on Texts and Archaeology. Mnemosyne, Supplements.* Leiden: Brill, pp. 197–232. https://doi.org/10.1163/9789004329409_009.

Hribal, J. 2007. Animals, Agency, and Class: Writing the History of Animals from Below. *Human Ecology Review,* 14: 101–12.

Ingold, T. 1994. From Trust to Domination: An Alternative History of Human-Animal Relations. In: A. Manning & J. Serpell, eds. *Animals and Human Society: Changing Perspectives.* London & New York: Routledge, pp. 1–22.

Mahdi, N. & Georg, P. V. 1969. *A Systematic List of the Vertebrates of Iraq.* Matḥaf al-Tārīkh al-Ṭabī'ī al-'Irāqī 26. Baghdad: University of Baghdad, Iraq Natural History Museum.

Meyer-Rochow, V.B. 2009. Food Taboos: Their Origins and Purposes. *Journal of Ethnobiology and Ethnomedicine,* 5: paper no. 18. https://doi.org/10.1186/1746-4269-5-18.

Michael, M. 2022. To Fish or Not to Fish? The Case Study of Fishing Activity in Cyprus (PhD dissertation, University of Southampton, Faculty of Arts and Humanities, Department of Archaeology).

Mylona, D. 2014. Aquatic Animal Resources in Prehistoric Aegean, Greece. *Journal of Biological Research-Thessaloniki,* 21(1): paper no. 2. https://doi.org/10.1186/2241-5793-21-2.

Mylona, D. 2021. Processed Fish, Marine Dyes and the Fishing Domaine of the Bronze Age Aegean. In: R. Laffineur & T.G. Palaima, eds. *Zoia: Animal-Human Interactions in the Aegean Middle and Late Bronze Age*. Aegaeum 45. Leuven: Peeters, pp. 113–24.

Papathanasiou, A. 2015. Stable Isotope Analyses in Neolithic and Bronze Age Greece: An Overview. In: A. Papathanasiou, M.P. Richards & S.C. Fox, eds. *Archaeodiet in the Greek World. Dietary Reconstruction from Stable Isotope Analysis.* Hesperia Supplements 49. Princeton, New Jersey: The American School of Classical Studies at Athens, pp. 25–55.

Popham, M. 1967. Late Minoan Pottery, a Summary. *The Annual of the British School at Athens,* 62: 337–51.

Potts, D.T. 2012. Fish and Fishing. In: D.T. Potts, ed. *A Companion to the Archaeology of the Ancient Near East.* Blackwell Companions to the Ancient World. Chichester: Wiley-Blackwell, pp. 220–35. https://doi.org/10.1002/9781444360790.ch12.

Price, M. 2020. *Evolution of a Taboo: Pigs and People in the Ancient Near East.* Oxford: Oxford University Press.

Rahme, L. 2021. Fish Skin, a Sustainable Material Used from Ancient Times to Today's Fashion. *FormAkademisk – forskningstidsskrift for design og designdidaktikk,* 14: 1–16. https://doi.org/10.7577/formakademisk.4183.

Recht, L. 2014. Transformers Energize! Aegean Bronze Age Rhyta in Moments of Transformation. In: A. Bokern & C. Rowan, eds. *Embodying Value? The Transformation of Objects in and from the Ancient World.* Oxford: Archaeopress, pp. 35–51.

Recht, L. 2019. Animals as Social Actors: Cases of Equid Resistance in the Ancient Near East. *Cambridge Archaeological Journal,* 29: 593–606.

Recht, L. 2023. Animals, Violence and Inequality in Ancient Mesopotamia. In: T.P. Leppard & S. Murray, eds. *Violence and Inequality: An Archaeological History.* Denver: University Press of Colorado, pp. 162–92. https://doi.org/10.5876/9781646424979.c006.

Rose, M. J. 1994. With Line and Glittering Bronze Hook: Fishing in the Aegean Bronze Age (PhD dissertation). Indiana University.

Russell, N. 2012. *Social Zooarchaeology: Humans and Animals in Prehistory*. Cambridge: Cambridge University Press.

Sahrhage, D. 1999. *Fischfang und Fischkult im alten Mesopotamien.* Frankfurt am Mein: Peter Lang.

Sahrhage, D. 2008a. Fishing in Ancient Egypt. In: H. Selin, ed. *Encyclopaedia of the History of Science, Technology, and Medicine in Non-Western Cultures*. Dordrecht: Springer, pp. 922–27. https://doi.org/10.1007/978-1-4020-4425-0_8591.

Sahrhage, D. 2008b. Fishing in Mesopotamia. In: H. Selin, ed. *Encyclopaedia of the History of Science, Technology, and Medicine in Non-Western Cultures*. Dordrecht: Springer, pp. 930–35. https://doi.org/10.1007/978-1-4020-4425-0_8592.

Sahrhage, Dietrich. 2008c. Fishing in the Indus Valley. In: H. Selin, ed. *Encyclopaedia of the History of Science, Technology, and Medicine in Non-Western Cultures*. Dordrecht: Springer, 927–930. https://doi.org/10.1007/978-1-4020-4425-0_8593.

Saunders, E. 2008. Pictures from the Sea: The Role of Marine Imagery and Artefacts in the Bronze Age Aegean (PhD dissertation). Trinity College Dublin.

Shapland, A. 2022. *Human-Animal Relations in Bronze Age Crete: A History through Objects*. Cambridge: Cambridge University Press.

Sołtysiak, A. & Fernandes, R. 2021. Much Ado about Nothing: Assessing the Impact of the 4.2 kya Event on Human Subsistence Patterns in Northern Mesopotamia Using Stable Isotope Analysis. *Antiquity,* 95: 1145–60. https://doi.org/10.15184/aqy.2021.117.

Stantis, C., Kharobi, A., Maaranen, N., Macpherson, C., Bietak, M., Prell, S. & Schutkowski, H. 2021. Multi-isotopic Study of Diet and Mobility in the Northeastern Nile Delta. *Archaeological and Anthropological Sciences,* 13: 105. https://doi.org/10.1007/s12520-021-01344-x.

Steel, L. 2020. Feats of Clay: Considering the Materiality of Late Bronze Age Cyprus. *Sustainability,* 12: 1–10. https://doi.org/10.3390/su12176942.

Stépanoff, C. & Vigne, J.D. eds. 2019. *Hybrid Communities: Biosocial Approaches to Domestication and Other Trans-species Relationships*. London: Routledge.

Sykes, N. 2015. *Beastly Questions: Animal Answers to Archaeological Issues*. London: Bloomsbury Academic.

Thompson, R.C. 1937. Assyrian Prescriptions for the Head. *The American Journal of Semitic Languages and Literatures,* 53: 217–38.

Todd, Z. 2014. Fish Pluralities: Human-Animal Relations and Sites of Engagement in Paulatuuq, Arctic Canada, *Études/Inuit/Studies,* 38(1/2): 217–38.

Todd, Z. 2018. Refracting the State through Human-Fish Relations: Fishing, Indigenous Legal Orders and Colonialism in North/Western Canada. *Decolonization: Indigeneity, Education & Society,* 7: 60–75.

Tsouparopoulou, C. 2024. The Early Dynastic 'Maison des Fruits' at Tell K in Tello (Ĝirsu). *Bulletin of the American Schools of Overseas Research,* 391: 191–225. https://doi.org/10.1086/730298.

Twiss, K.C. 2019. *The Archaeology of Food: Identity, Politics, and Ideology in the Prehistoric and Historic Past.* Cambridge: Cambridge University Press.

Van Neer, W., Lernau, O., Friedman, R., Mumford, G., Poblóme, J. & Waelkens, M. 2004. Fish Remains from Archaeological Sites as Indicators of Former Trade Connections in the Eastern Mediterranean. *Paléorient,* 30: 101–47.

Vávra, R. 2020. Fish and Chaps: Some Ethnoarchaeological Thoughts on Fish Skin Use in European Prehistory. *Open Archaeology,* 6: 329–47. https://doi.org/10.1515/opar-2020-0115.

Voulgari, E. 2021. Neolithic Vessels with Animal Characteristics: Modifications of Material as Negotiations of Clay Bodyscapes. In: L. Recht & K. Zeman-Wiśniewska, eds. *Animal Iconography in the Archaeological Record: New Approaches, New Dimensions.* Sheffield: Equinox, pp. 22–41.

Watrous, L. V. 1991. The Origin and Iconography of the Late Minoan Painted Larnax. *Hesperia,* 60: 285–307. https://doi.org/10.2307/148065.

Wheeler, A. & Jones, A.K.G. 1989. *Fishes.* Cambridge: Cambridge University Press.

Isotopes and Fish Bones in the Bronze Age Aegean: Expanding Our Understanding of Aquatic Diets

2

Dimitra Mylona and Aurora Allshouse

Abstract

Isotopic analyses of human bones in a search for ancient individual and community-specific dietary habits are becoming increasingly common in Greek archaeology, highlighting dominant dietary items but also absent or under-consumed ones. Fish and marine molluscs fall in the latter category having been, in this region, largely undetected isotopically. At the same time, accumulated conventional zooarchaeological studies of fish bone and molluscan remains from sites of wide chronological and geographical distribution suggest a constant and, at places, intense consumption of aquatic foods. The two sets of data, isotopic and zooarchaeological, offer an apparently contradictory picture. This paper provides a critical review of the current state of research in the context of Prehistoric, mostly Bronze Age Aegean. Additionally, it explores possible ways to achieve nuanced, in-depth understanding of the exploitation of aquatic, mostly marine, food resources, combining the methodologies and findings of both approaches. It elaborates on certain methodological and interpretational issues, and it suggests possible directions for future research. The combined analysis of the available data suggests that both geographical and social factors may have affected fish consumption in the past. It becomes clear, however, that to be able to offer meaningful and secure evaluations in a time and site-specific manner, we need targeted, high quality excavation data that would provide the context of analysis. We would also need detailed publication of human and fish isotopic comparative data and fish and molluscan archaeological remains.

D. Mylona (✉)
INSTAP Study Center for East Crete, Pacheia Ammos, Ierapetra, Crete, Greece
e-mail: dmylona@hotmail.com

A. Allshouse
Harvard Department of Anthropology, Tozzer Anthropology Building, Cambridge, MA, USA

2.1 Introduction

The intensity of marine consumption in the Prehistoric Aegean has proven to be a contentious issue.[1] On the one hand, at sites where appropriate field methods (e.g. water flotation) are applied during excavation, the remains of marine fish and invertebrates are abundant (Mylona 2003). On the other, biochemical analyses of human remains have failed to find evidence for widespread marine consumption (Triantaphyllou et al. 2008; Richards and Hedges 2008; Papathanasiou and Richards 2015). This has led to contrasting claims about the relative importance of resources from the sea to the diets of people in the Prehistoric Aegean Sea. One interpretation suggests that fish and marine molluscs were commonly consumed along the coasts of the Aegean and that they were important enough to be part of the Neolithic and Bronze Age cultural scene (for a review, see Mylona 2016), while another claims that marine resources were, at most, utilised sporadically (e.g. Papathanasiou and Richards 2015). The two main methods used to explore this issue—zooarchaeological analysis and stable isotope analysis—each have benefits and limitations that should complement each other.

Zooarchaeological studies of fish bones and marine molluscs offer a detailed insight into the presence and articulation of ancient fisheries, clarifying, among other things, the relative importance of different aquatic species in diet in a space and time-specific way (indicatively Rose 1994; Theodoropoulou 2007; Βεροπουλίδου 2011; Reese 2019; Mylona 2020; and for a full bibliography Mylona 2022). However, fish bones are not always collected and studied and, additionally, they often suffer considerably from post-depositional erosion and other taphonomic hazards (Jones 1986; Wheeler and Jones 1989: 61–78; Nicholson 1995; Willis and Boehm 2014). Due to their nature, the amounts of consumed marine foods, at any given site and point in

[1] Both authors contributed equally to this work.

C. Tsouparopoulou, L. Recht (eds.), *Human and Aquatic Beings: Interactions in and beyond the Eastern Mediterranean (3rd–1st Millennia BCE)*, Themes in Contemporary Archaeology, https://doi.org/10.1007/978-3-031-73643-8_2

time, cannot be estimated with any accuracy. Moreover, comparisons with the amounts of other types of edible items (as documented by their remains: bones, carbonised seeds and so on) is far from exact or straightforward.

Stable isotope analysis can somewhat remedy this problem. This method measures the ratio of heavy to light carbon and nitrogen in individual skeletons, thus providing a direct but averaged picture of individual food consumption (for the method, see references below). Isotopic analysis and its quest for a marine foods signature have faced certain challenges. The sensitivity of the method in identifying the isotopic signature of the marine elements in past diets was the first, and one that is constantly improving. Issues of interpretation of data are also a matter of debate (indicatively Hedges 2004), and the method does not provide the taxonomic specificity that zooarchaeological and paleoethnobotanical analysis can. Additionally, the estimated dietary composition drawn from stable isotope analysis is completely reliant on the reference data used. Therefore, if inappropriate isotopic proxies are used for comparison, some food sources are likely to be underestimated or even overlooked completely (Cheung and Szpak 2021).

The traditional interpretation of human isotopic evidence in the circum-Aegean region was that the human values were overwhelmingly consistent with a population consuming a mostly C3 terrestrial diet, with little to no marine component (Triantaphyllou et al. 2008; Richards and Hedges 2008; Papathanasiou and Richards 2015). In this scheme, the roles of fish and seafood were considered, almost always and everywhere, minor. This observation was in stark contrast to the archaeological picture of an abundance of fish bones and seashells in most coastal sites and in several inland ones. This narrative has been repeated even as recently as 2022 (e.g. Dotsika et al. 2022), despite mounting evidence that the standard models used for interpreting stable isotope values in the Aegean basin are insufficient for calculating marine contribution (Vika and Theodoropoulou 2012; Allshouse et al. Forthcoming).

Here, we discuss some of these analytical challenges in the context of the Prehistoric Aegean, with emphasis on the Bronze Age. Neolithic cases are also included in the discussion for comparison. This chapter offers an alternative interpretation based on multiple lines of evidence. By integrating zooarchaeological and stable isotopic data, we counterbalance the limits of each approach to gain a better, fuller understanding of the consumption of marine resources in the Prehistoric Aegean Sea.

2.2 Methods

We compiled zooarchaeological data and stable isotopic data from previously published studies. For the zooarchaeological remains, this paper focuses only on fish bones due to space restrictions. Obviously, marine molluscs, an abundant and well-preserved class of archaeological finds, are also part of the picture and need to be taken into consideration, but they are outside the scope of the present study. Table 2.1 presents marine finds from a selection of sites, namely those that have been collected with appropriate field methods (mostly water-flotation or even dry sieving with a fine mesh) and have been analysed and published in detail. It has been observed that, although the taxonomic variety in these assemblages is quite large, most of the consumed fish fall within a few taxonomic groups. Table 2.1 focuses on the five most commonly eaten fish (or types of fish). Due to the frequency and intensity of their consumption, these are the ones which would likely have affected the isotopic profile of the consumers the most. Also, each of the top five taxonomic groups has been assigned to a trophic level, a value that reflects its position within the marine food chain. This is an important consideration for stable isotope interpretation, as fish low on the food chain will have a different impact on the $\delta^{13}C$ and $\delta^{15}N$ of the humans consuming them than larger, piscivorous fish, due to the cumulative effect of metabolic fractionation on high-trophic level individuals (Vander Zanden and Rasmussen 2001). Other data pertaining the fish bone assemblages is also given, including their geographic position in relation to distinct marine regions (see caption of Table 2.1 for details).

Human isotopic data were collated from eight previous publications (Papathanasiou 2003; Petroutsa and Manolis 2010; Richards and Hedges 2008; Richards and Vika 2008; Triantaphyllou et al. 2008; Kontopoulos and Sampson 2015; Nafplioti 2016) for a total sample of 466 individuals from the Prehistoric Aegean, and date ranges from the Neolithic to the Late Helladic periods (Table 2.2). These data were supplemented by floral and faunal isotopic data from Vika and Theodoropoulou (2012), Bogaard et al. (2013), Papathanasiou (2015), and Price et al. (2017), in order to construct a regionally appropriate isoscape. Plotting and statistical analyses were undertaken using R statistical software (R core team 2020) and RStudio user interface (Rstudio team 2022). In this paper, the discussion will focus on the Bronze Age cases (Fig. 2.1).

2.3 The Zooarchaeological Evidence

Fisheries in the prehistoric Aegean Sea cannot be seen as uniform in terms of their targeted aquatic species. This reflects the marked hydrogeographic difference between the northern and southern Aegean and the specific physical and biological characteristics of different locations within each area (Stergiou et al. 1997; Mylona 2008: 33–37; 2016: 58–60). Water productivity (as a reflection of nutrients availability, salinity, water depth, temperature etc.) is a factor that defines the types of species present and their relative abundance. The currently observed abundance of collective small

Table 2.1 The archaeo-ichthyological record from Neolithic and Bronze Age Greece

Site	Marine region	Date	Collection method	Total number of identifiable bones	Non-identifiable bones	Taxa		Trophic level	Number of other taxa	References
Cyclope cave, Yioura island	North Aegean Sea	Mesolithic	HC, DS	4809	902	*Oblada melanura*	487 (10.12%)	3.10	26	Powell (2003: 80–81); Mylona (2011: 244)
						Scombridae	450 (9.35%)	4.29		
						Serranidae	439 (9.12%)	3.91		
						Scorpaenidae	318 (6.61%)	4.19		
						Diplodus vulgaris	236 (4.90%)	3.31		
Limenaria, Thassos island	Thracian Sea	Neolithic	HC,WF	247	30	*Scomber japonicus*	180 (72.84%)	3.4	4	Theodoropoulou (2007: 325)
						Mugil cephalus	45 (18.21%)	2.5		
						Anguilla anguilla	6 (2.42%)	3.6		
						Euthynnus alletteratus	5 (2.02%)	4.50		
						Argyrosomus regious	4 (1.61%)	4.3		
Mikro Vouni, Samothrace island	Thracian Sea	Neolithic	HC, DS, WF	393	265	*Epinephelus* sp.	69 (17.55%)	4.13	30	Theodoropoulou (2007: 388)
						Chondrychthyes	56 (14.24%)	4.09		
						Sparus aurata	49 (12.46%)	3.7		
						Pagrus pagrus	30 (7.63%)	3.77		
						Diplodus sargus	20 (5.08%)	3.17		
Kryoneri Serres	Strymonic Gulf	Neolithic	HC&WF	1118	468	River fish	878 (78.53%)	—	8	Mylona, pers. analysis
						Clupeidae	80 (7.15%)	3.1		
						Mugilidae	59 (5.27%)	2.62		
						Dicentrarchus labrax	5 (0.44%)	3.5		
						Anguilla anguilla	5 (0.44%)	3.6		
Toumba Thessaloniki	Thermaic Gulf	Bronze Age	HC&WF	966	735	*Sparus aurata*	235 (24.32%)	3.7	50	Theodoropoulou (2007: 439–40)
						Mugilidae	111 (11.49%)	2.62		
						Sparidae indeterminate	90 (9.31%)	—		
						Oblada melanura	72 (7.45%)	3.10		
						Pagellus erythrinus	66 (6.83%)	3.37		
Makrygialos, Pieria	Thermaic Gulf	Neolithic	HC&WF	227	509	Sparidae indeterminate	99 (43.61%)	—	16	Mylona, pers. analysis
						Mugilidae	67 (29.51%)	2.62		
						Sparus aurata	16 (7.04%)	3.7		
						Serranidae indeterminate	13 (5.72%)	3.91		
						Epinephelus sp.	7 (3.08%)			

(continued)

Table 2.1 (continued)

Site	Marine region	Date	Collection method	Total number of identifiable bones	Non-identifiable bones	Taxa		Trophic level	Number of other taxa	References
Cyclope cave, Yioura island	North Aegean Sea	Neolithic	HC, DS	2004	430	*Oblada melanura*	430 (21.45%)	3.10	26	Powell (2003: 80–81); Mylona (2011: 244)
						Diplodus vulgaris	263 (13.12%)	3.31		
						Serranidae indeterminate	231 (11.52%)	3.91		
						Scorpaenidae	104 (5.18%)	4.19		
						Pagellus erythrinus	101 (5.03%)	3.37		
Alepotrypa cave	Mesenian Gulf	Neolithic	HC, WF	316	182	*Thunnus* sp.	135 (42.72%)	4.30	11	Theodoropoulou (2018: 333)
						Auxis rochei	44 (13.92%)	4.4		
						Euthynnus alletteratus	40 (12.65%)	4.50		
						Chondrychthyes	30 (9.49%)	4.09		
						Epinephelus sp.	8 (2.53%)	4.13		
Ftelia, Mykonos island	Central Aegean Sea	Neolithic	HC, DS	28	24	Sparidae indeterminate	8 (28.57%)	—	4	Mylona (2018: 230)
						Pagrus pagrus	8 (28.57%)	3.77		
						Epinephelus sp.	8 (28.57%)	4.13		
						Dentex dentex	3 (10.71%)	4.50		
						Diplodus annularis	1 (0.50%)	3.36		
Akrotiri, Thera island	Aegean Sea	Bronze Age	HC, WF	617	452	Small Sparidae indeterminate	238 (38.57%)	—	20	Mylona (2020: 184)
						Maena smaris	126 (20.42%)	3.00		
						Maena smaris/ *Boops boops*/ Mullidae	114 (18.47%)	3.00/2.92/3.30		
						Pagellus erythrinus	26 (4.21%)	3.37		
						Pagrus pagrus	26 (4.21%)	3.77		
Mochlos, Crete	Cretan Sea	Bronze Age	HC, WF	296	559	Centracanthidae	102 (34.45%)	3.00	20	Mylona (2022: table 16)
						Sparidae indeterminate	79 (26.68%)	—		
						Maena smaris/ *Boops boops*/ Mullidae	33 (11.14%)	3.00/2.92/3.30		
						Boops boops	21 (7.09%)	2.92		
						Chromis chromis	17 (5.74%)	3.70		

Kommos, Crete	Libyan Sea	Bronze Age	HC, WF	562	0	Sparidae/ Centracanthidae	389 (69.21%)	3.00/2.92	8	Rose (1995: 210)
						Sparidae indeterminate	106 (18.61%)	—		
						Serranidae indeterminate	20 (3.55%)	3.91		
						Sparisoma cretense	8 (1.42%)	2.9		
						Labridae	4 (0.71%)	3.34		
Palaikastro, Crete	Cretan Sea	Bronze Age	HC, WF	542	127	*Maena smaris/ Boops boops/* Mullidae	212 (39.11%)	3.00/2.92/3.30	21	Mylona (2019: 374)
						Sparidae indeterminate	97 (17.89%)	—		
						Chromis chromis	44 (8.11%)	3.70		
						Serranus sp.	41 (7.56%)	3.70		
						Boops boops	29 (5.35%)	2.92		

The table includes assemblages that conform to three rules: they are reasonably large; they have been water floated or at least dry sieved; and they are fully analysed. It presents the five most common taxa in each assemblage and their relative abundance. Those taxa may be identified to the level of species, genus, family or groups of the above. This flexible choice of taxonomic categories is unavoidable and mostly reflect the preservation status of each assemblage (that permits a more or less precise identification of fish bones). For each of those cases, the table provides the name of the site, its geographical location in terms of marine regions, the dating of the assemblage, the number of identifiable bones in each and the number of non-identifiable ones. Besides the five most common taxa, the table provides the number of the rest of the taxa in the assemblage. This figure, which corresponds with the part of the assemblage that is not taken up by the five most common taxa, provides a measure of its variability. The table also includes the trophic level of each taxon, based on Stergiou and Karpouzi 2002 and on Froese and Pauly 2022. When the reported taxonomic groups are very broad, the figure of trophic level is an approximation based on the combination of the relevant figures of the most common of that group in the Aegean. For example, Serranidae include both large fish such as groupers (*Epinephelus* sp. - mean fractional trophic level: 4.13) and smaller ones such as the combers (Serranus sp. – mean fractional trophic level: 3.70); in the table, a mean value is presented. The value for Chondrychtyes combines the value for various genera and species commonly occurring in the Aegean, i.e. *Galeus* sp. (4.16), *Isurus oxyrinchus* (4.5), *Carcharodon carcharias* (4.5), *Squatina squatina* (4.1), *Raja clavata* (3.8), *Dasyatis* sp. (4.1), and it must be viewed as indicative only. No value is given for the Sparidae family, because it includes many species that have varied trophic levels (*Salpa salpa*: 2.13 to *Dentex dentex*: 4.50)

Table 2.2 Summary table of human $\delta^{13}C$ and $\delta^{15}N$ data used in this study

Site	Period	μ $\delta^{13}C$	σ $\delta^{13}C$	μ $\delta^{15}N$	σ $\delta^{15}N$	N
Alepotrypa	Neolithic	−19.95	0.45	7.17	1.01	26
Franchthi	Neolithic	−18.65	0.78	9.23	1.83	11
Kephala	Neolithic	−19.09	n/a[a]	9.17	n/a	5
Kouveleiki	Neolithic	−19.84	n/a	8.09	n/a	2
Stavropouli	Neolithic	−19.55	n/a	9.3	n/a	2
Tharounia	Neolithic	−19.94	0.71	8.12	1.22	31
Theopetra	Neolithic	−19.83	0.86	7.4	1.05	13
Spathes	LBA	−19.33	n/a	7.97	n/a	6
Zeli	LBA	−19.69	0.23	8.4	0.61	20
Kalapodi	LBA	−19.66	1.39	8.53	1.04	14
Kalamaki	EBA/LBA	−19.33	0.64	8.2	0.97	32
Agia Triada	LBA	−19.94	0.77	7.22	0.54	70
Almyri	LBA	−19.18	0.25	9.31	0.32	34
Mycenae	MBA/LBA	−19.02	0.7	9.54	1.82	18
Batsorachi	LBA	−19.37	n/a	8.2	n/a	7
Loupouno	LBA	−19.17	n/a	8.17	n/a	3
Lerna	MBA/LBA	−19.55	0.32	8.39	0.75	39
Sykia	LBA	−19.35	n/a	8.28	n/a	6
Armenoi	LBA	−19.77	0.26	7.8	0.49	39
Knossos	MBA/LBA	−19.33	1.13	8.22	0.97	62
All Sites		−19.59	0.71	8.19	1.11	466

Sources: Kontopoulos and Sampson 2015; Nafplioti 2016; Papathanasiou 2003; Petroutsa and Manolis 2010; Richards and Hedges 2008; Richards and Vika 2008; Triantaphyllou et al. 2008
All isotopic data are reported in ‰, as compared to standard. μ = mean/average σ = standard deviation, N = number of individuals
[a]Standard deviation was not calculated for sites with less than 10 individuals

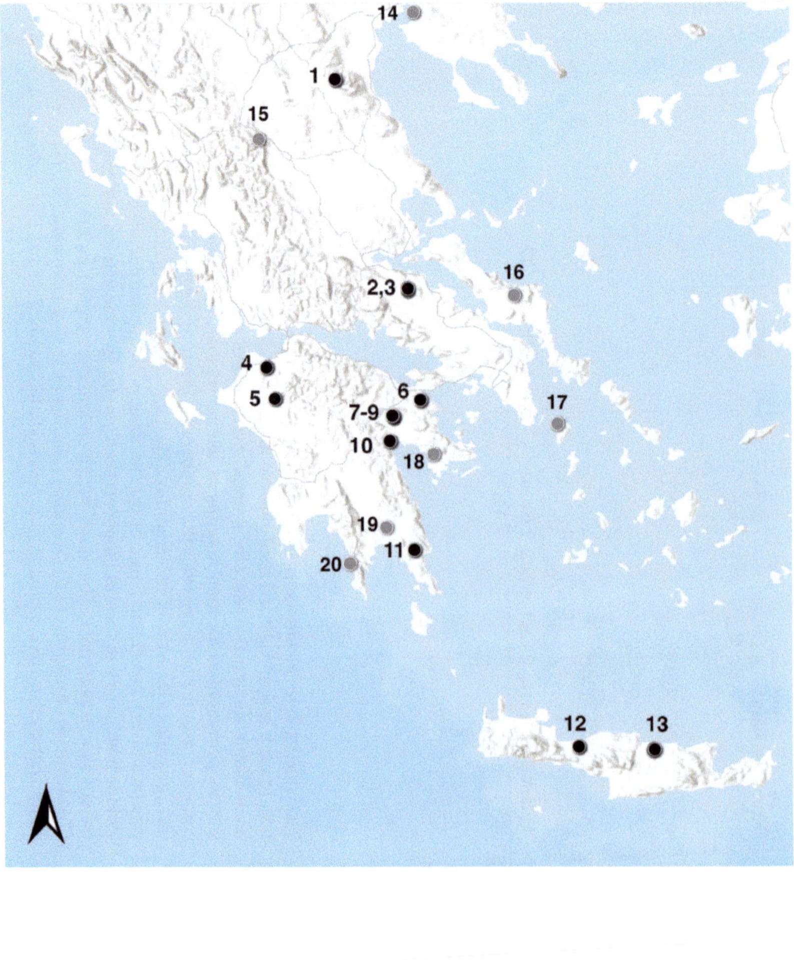

Fig. 2.1 Map showing the approximate location of the Neolithic and Bronze Age sites that produced human isotopic data: 1. Spathes, 2. Zeli, 3. Kalapodi, 4. Kalamaki, 5. Agia Triada, 6. Almyri, 7. Mycenae, 8. Batsorachi, 9. Loupouno, 10. Lerna, 11. Sykia, 12. Armenoi, 13. Knossos, 14. Stavrouopoli, 15. Theopetra, 16. Tharounia, 17. Kephala, 18. Franchthi, 19. Kouveleiki, 20. Alepotrypa. (Map by A. Allshouse)

pelagics, especially sardines and anchovies, in the northern Aegean as opposed to the combined picarel and bogue in its southern part, exemplifies this phenomenon, which is, however, rather complex and fluid. The presence of transitional waters, coastal lagoons and river estuaries has a decisive impact on fisheries, now as in the past. Although the north-south division of the Aegean Sea and the coastal configuration on a local scale were pivotal in the shaping of local prehistoric fisheries, there is a unifying element to them all. Whether in the rich waters of the north Aegean or in the rocky oligotrophic coasts of the south, fishing took place very near the shore, in the relatively shallow waters (a few metres deep at most) (Mylona 2021: 113–15). That feature in itself implies a certain type of selectivity in the types of fish caught and consumed.

Table 2.1, which summarises the aquatic species most commonly consumed in various prehistoric sites around the Aegean, illustrates the above observations. Assemblages from sites on north Aegean coasts (e.g. Limenaria – Theodoropoulou 2007: 325; Mikro Vouni – Theodoropoulou 2007: 388) are dominated by two types of fish. On the one hand there are pelagic predators of high trophic level, such as the migratory Scombridae (various types of tunas and smaller members of the family such as the mackerels), but also several types of sharks and rays, species that prey on pelagic fish and demersal carnivorous fish such as groupers or Scorpion fish. The abundant smaller pelagics like the Clupeidae (*Sardina pilchardus*, *Sardinella* sp. etc) and Engraulidae (*Engraulis encrasicolus*), which form the prey of those migratory and carnivorous fish, are under-represented at those sites. On the other hand, the most common fish in the assemblages of the north Aegean also include species inhabiting, for part of their life cycle, the transitional waters and/or move from marine to fresh water environments (e.g. Limenaria – Theodoropoulou 2007: 325; Mikro Vouni – Theodoropoulou 2007: 388; Toumba Thessaloniki – Theodoropoulou 2007: 439–40; Makrygialos – Mylona pers. analysis). The Mugilidae, mullets of various types, are such fish. Most mullets reproduce in the sea but the young fish retreat to coastal lagoons and river estuaries to feed and grow. They are low in the trophic level, feeding, as adults, mostly on detritus, micro-algae and benthic organisms (Froese and Pauly 2022). Some members of the Sparidae, the sea bream family, e.g. *Sparus aurata*, a top predator (Mariani 2006), do the reverse, spawning in the lagoons and migrating to the sea to feed. Other species, for instance *Argyrosomus regious*, also high in the trophic chain, also prefer the rich waters of coastal lagoons and river estuaries when young to feed and find protection. Finally, the eel (*Anguilla anguilla*) a catadromus, high trophic level species, is also common at those sites.

By contrast, fish bone assemblages in the southern Aegean are totally different in taxonomic composition (with some exceptions, e.g. Neolithic Alepotrypa Cave—see Table 2.1). There the assemblages are dominated by very small fish, less than 15 cm in length and even much smaller, almost all inshore species of a clearly marine character (Fig. 2.2). At certain sites, some brackish water fish (and shellfish) are present, but their numbers are extremely low, probably reflecting the small size and rarity of such ecosystems in their region (e.g. Mugilidae and Ostreidae from Late Minoan IB Mochlos—Mylona 2022). The same scarcity is observed for the migratory fish or the large pelagic predators such as the sharks (e.g. Scombridae at Middle Minoan III Kommos—Rose 1995: 231). On the contrary, those assemblages are dominated by picarel (*Centracanthus cirrus*/*Spicara smaris*/*Spicara maena*/*Spicara flexuosa*),[2] bogue (*Boops boops*), other sparids, mostly small, damsel fish (*Chromis chromis*), combers (*Serranus cabrilla*, *S. hepatus*, *S. scriba*) and groupers (*Epinephelus* sp.). It is interesting that especially the small sparids that were consumed at these sites, e.g. *Pagellus erythrinus*, *Lithognathus mormyris*, *Pagrus pagrus*, *Dentex dentex*, are fish that can grow to quite large sizes, but in the prehistoric southern Aegean they were caught when juvenile, from the nurseries of the shallows. Most of the fish that were consumed in large amounts in the southern Aegean are of rather low trophic level, feeding on zooplankton or having a mixed diet (indicatively: *Spicara maena* and *S. smaris*—Karachle and Stergiou 2014; *Boops boops*, El-Maremie and El-Mor 2015; *Chromis chromis*, Pinnegar 2018; but for carnivorous combers *Serranus cabrilla* and *S. hepatus*, Rachedi et al. 2018 and Bilecenoglu 2009).

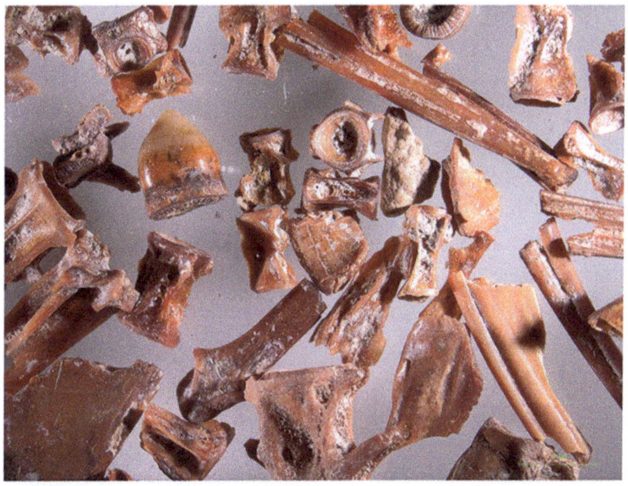

Fig. 2.2 Fish bones collected by water flotation at Bronze Age Akrotiri, Thera. They are from very small fish, mostly picarel and small sea breams. (Photograph: Akrotiri Excavations Archive)

This marked differentiation in the character and taxonomic composition of consumed fish in northern and southern Aegean prehistoric sites forms the background on which a

[2]These taxa previously formed a distinct family known as the Centracanthidae, but now they are considered members of the Sparidae family.

more complex picture is built once we start looking at the specific site by site choices and the social context of fish consumption. For example, a detailed contextual analysis of fish bones and marine mollusks from the Neopalatial horizon at Mochlos, Crete (1750/1700 to 1500/1450 BCE), which is discussed in detail later in this paper—also Mylona (2022), suggests that the consumption of the small inshore fish that were mentioned above, along with rock clinging molluscs of the shoreline such as limpets and top shells, were the most widely consumed marine foods. It appears that these taxa formed the backbone of marine foods consumption. In contrast, larger fish, most of which are top predators and therefore of quite high trophic level (e.g. groupers—*Epinephelus* sp., common dentex-*Dentex dentex*, red porgy-*Pagrus pagrus*, striped seabream- *Lithognanthus mormyris*, common Pandora- *Pagellus erythrinus* and parrot fish- *Sparisoma cretense*), are found concentrated in a few dwellings only, and it appears that their bones are linked to specific special eating events. The Neopalatial Mochlos case study clearly documents that fish consumption was linked to social distinction. So far, this type of analysis is rather rare, and for this reason, the social conditions of fish consumption can mostly be hypothesised rather than documented in specific detail. The example of Mochlos, however, suggests that such differentiation in the consumption of marine foods was a reality even within the confines of single settlements.

2.4 The Isotopic Evidence

Before discussing the details of the isotopic evidence from the Aegean, a brief explanation of the basic principles is required. Stable isotope analysis of human tissue has been used to better understand ancient mobility and migration and dietary composition throughout the human past (Lee-Thorp 2008). While there are a number of elements used for a variety of analytical purposes, those most commonly employed for dietary reconstruction are carbon and nitrogen, both of which are incorporated into human tissue primarily through the metabolism of foods consumed. Because the ratios of heavy to light carbon and nitrogen vary in a predictable manner between food sources, $\delta^{13}C$ and $\delta^{15}N$ from human bone collagen can then be used to reconstruct an approximation of the dietary composition of an individual over their last years of life (Deniro and Epstein 1981; Chisholm et al. 1982; Schoeninger et al. 1983).

Stable isotope analysis measures the ratio of heavy to light isotopes of an element within a sample and compares this against a standard (V-PDB for carbon, AIR for nitrogen). This can be expressed as:

$$\delta X = (R_{sample}/R_{standard} - 1) \times 1000$$

where R represents the ratio of $^{13}C/^{12}C$ or $^{15}N/\ ^{14}N$. The resulting value, expressed as δX ‰, is therefore representative of the difference between the isotopic ratio in the sample and in the standard.

Carbon and nitrogen isotope ratios are significant for understanding dietary composition because these ratios differ across organisms and ecosystems. Because bone collagen is synthesised primarily from the protein fraction of foods we consume, the $\delta^{13}C$ and $\delta^{15}N$ of human bone collagen will be reflective of dietary $\delta^{13}C$ and $\delta^{15}N$ (Hedges and Reynard 2007). When the variation in $\delta^{13}C$ and $\delta^{15}N$ within ecosystems has been well documented, this value can then be used to reconstruct an approximation of the types of foods consumed (Deniro and Epstein 1981).

The most significant factor in $\delta^{15}N$ variation is the trophic position of foods consumed: primary producers like terrestrial plants and marine phytoplankton represent the local baseline and $\delta^{15}N$ will be enriched by 3–5‰ with each step up the food chain due to metabolic fractionation (Deniro and Epstein 1981; Hedges and Reynard 2007). $\delta^{13}C$, on the other hand, is primarily used to distinguish between the consumption of C3 photosynthesisers such as wheat, barley, and rice, which will generally display more negative $\delta^{13}C$, and C4 photosynthesisers such as maize, millet, and sorghum will have higher $\delta^{13}C$. Marine food consumption has been demonstrated to enrich both $\delta^{13}C$ and $\delta^{15}N$, due to a ^{13}C enriched baseline and to longer, more complex food chains than typically found in terrestrial environments (Chisholm et al. 1982; Schoeninger et al. 1983).

Based on decades of observational and experimental studies, these fundamental principles (Fig. 2.3) have been vital in dietary reconstruction to the present day. It has been demonstrated in recent years, however, that these isotopic ratios are sensitive to differences in climate, amount of precipitation, and various anthropogenic forces, such as human induced eutrophication of marine waters (Amundson et al. 2003; Fraser et al. 2011; Isaakidou et al. 2022; Szpak 2014; Graven et al. 2020). They are therefore generalisations that must be refined by isotopic study of the region of interest.

The conclusion reached by the studies that utilise stable isotope analysis to investigate human diet in Prehistoric Aegean has been that people followed a C3 terrestrial diet, in other words a diet based on wheat, barley and so on, with minimal input from marine or freshwater fish and/or C4 plants (Papathanasiou and Richards 2015; Dotsika et al. 2019). This interpretation is rooted in those fundamental isotopic principles discussed above: the human isotopic values for this region are similar to those we would typically expect from humans consuming a mainly terrestrial-based diet. The average $\delta^{15}N$ of human bone collagen in the Aegean region is around 8‰, much lower than what has been observed in other ancient fish-consuming populations,

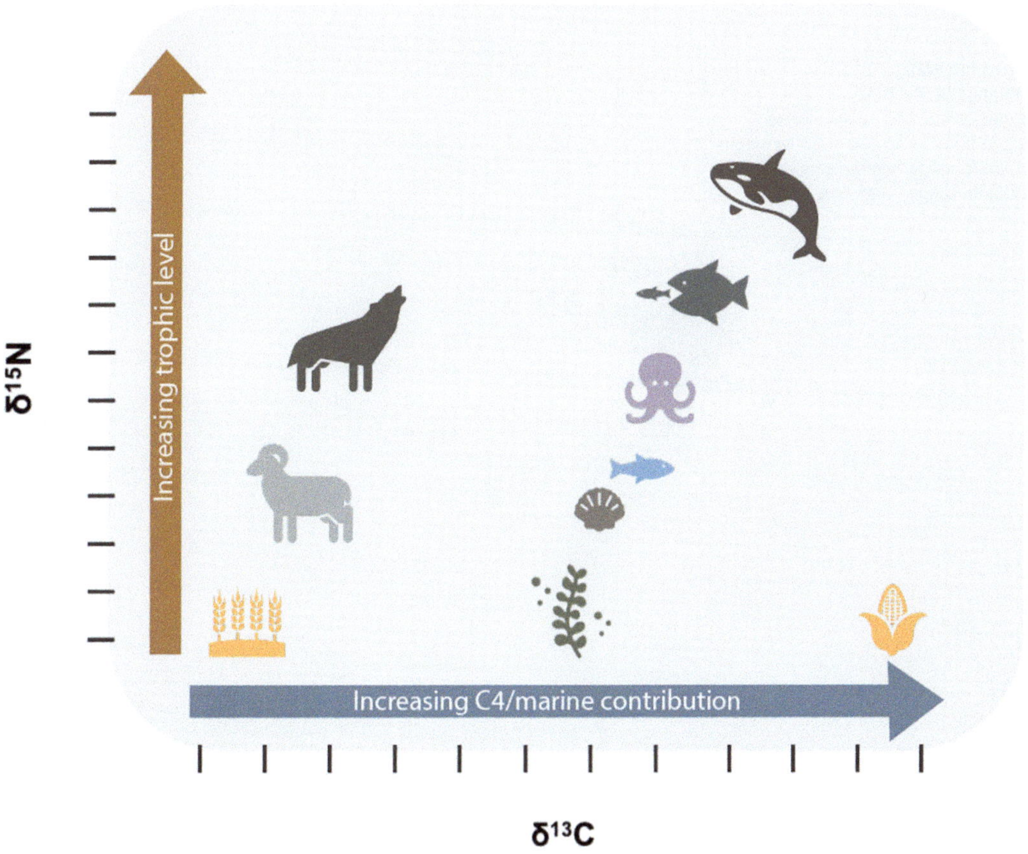

Fig. 2.3 General principles of stable isotope analysis, indicating the isotopic relationship between different groups of organisms. (Graphic by A. Allshouse)

and similarly the average $\delta^{13}C$ is only −19.5‰, far less enriched than what is expected from marine food chains (e.g. Chisholm et al. 1982; Cramp et al. 2014).

To further examine this initial understanding of fish in Aegean prehistoric diets and in view of the contrasting picture emerging from the zooarchaeological record, we collected and plotted all published isotopic $\delta^{15}N$ and $\delta^{13}C$ values from our case region (Table 2.2 and Fig. 2.4). In the same graph and in addition to the human values, we incorporated ranges representing idealised consumer diets (calculated values for a hypothetical human consuming 100% of their diet protein from a single resource class) for a comparison (Fig. 2.4). These windows were calculated using exclusively archaeological Aegean values on the remains of plants, animals, and fish (Bogaard et al. 2013; Papathanasiou 2015; Price et al. 2017; Vika and Theodoropoulou 2012). This approach, while not quantitative, provides a local and contemporary frame of reference for understanding and interpreting the observed human values. By observing how the human isotopic values compare to the various plants and animals of the prehistoric Aegean, we can quickly see that marine foods must have played a significant role in prehistoric human diets, as many humans are enriched in $\delta^{13}C$ beyond the range possible if consuming only terrestrial resources.

2.5 Some Interpretive Challenges

Recent work on the isotopic ratios of Aegean foodstuffs has complicated the interpretation of Aegean human dietary isotopes. Vika and Theodoropoulou (2012) measured bone collagen $\delta^{13}C$ and $\delta^{15}N$ from archaeological fish from Aegean sites, both marine and fluvial. Specifically, they analysed bones from Mugilidae, Serranidae, Scorpaenidae, Sparidae, Thuninidae (which are actually part of the Scombridae family), Cyprinidae, Siluridae, Moronidae and Sciaenidae families without specifying the species, a fact that, as has been discussed above, may cause problems in interpretation. Most members of these families are omnivorous or carnivorous fish, and thus they are not representative of the isotopic values of marine primary consumers (herbivores and filter feeders). Primary consumers are at the base of the food chain, we can expect that their $\delta^{15}N$ and, to a lesser extent $\delta^{13}C$, will be lower than their omnivorous or carnivorous counterparts (Schoeninger et al. 1983). This

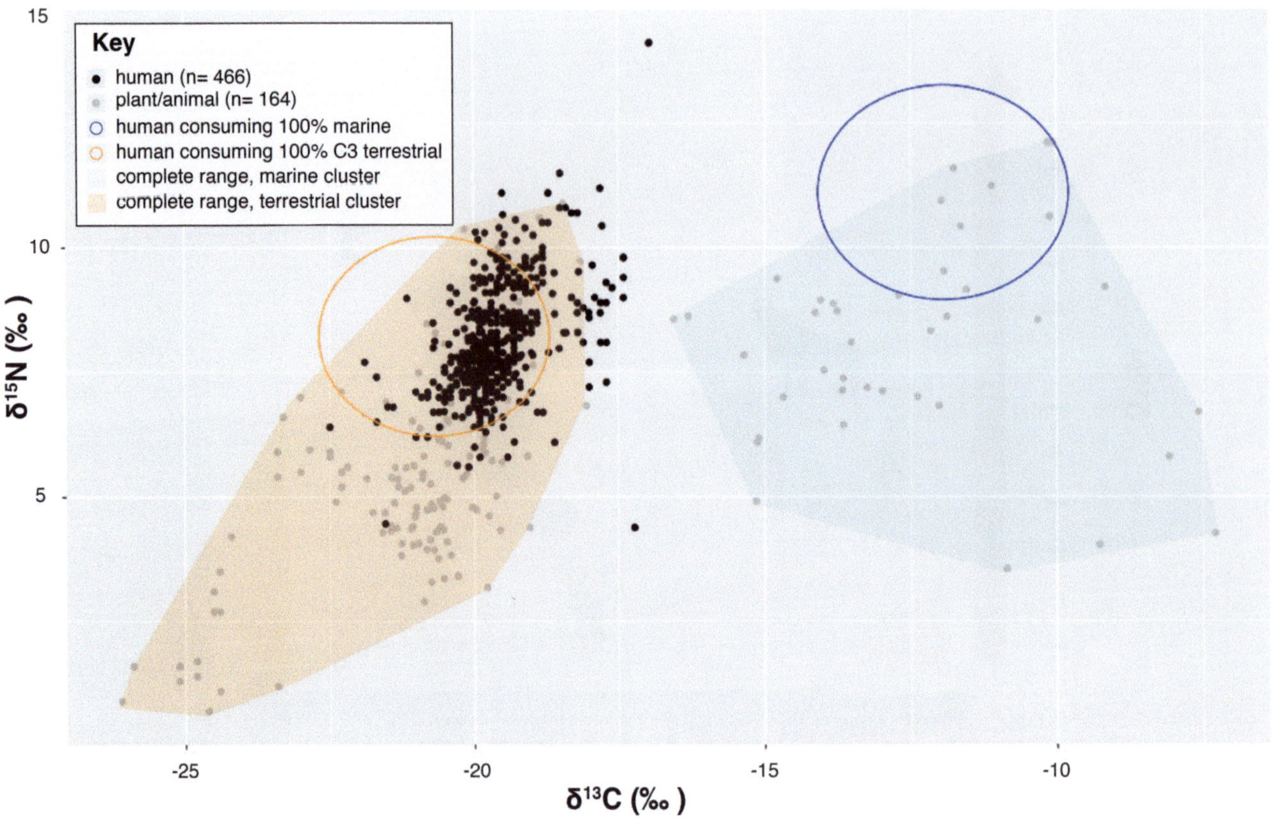

Fig. 2.4 Prehistoric Aegean $\delta^{15}N$ and $\delta^{13}C$ values: humans (black points) compared to plants and animals (grey points). Circles represent where we would expect a hypothetical human consumer eating exclusively C3 terrestrial (orange) and exclusively marine (blue) foods to plot based on existing isotopic values on archaeological foodstuffs. These ranges, drawn from the mean and standard deviation of the two food types, were adjusted for the diet-to-collagen offset (+3‰ $\delta^{15}N$, +0.5‰ $\delta^{13}C$). Shaded areas represent the actual observed range of terrestrial (orange) and marine (blue) plant and animal values (Vika and Theodoropoulou 2012; Bogaard et al. 2013; Papathanasiou 2015; Price et al. 2017). (Graphic by A. Allshouse)

should also factor into any interpretations of human dietary stable isotopes, as the lower isotopic ratios of marine primary consumers would have a lesser impact on the $\delta^{15}N$ and $\delta^{13}C$ of humans than species higher on the marine food chain would.

Additionally, even bones of higher trophic level fish in the Aegean, were not markedly enriched in $\delta^{15}N$ compared to terrestrial animals (Vika and Theodoropoulou 2012). This implied that the consumption of those fish would in turn not have enriched human $\delta^{15}N$ to the extent of fish from other marine regions, such as the Baltic or North Atlantic (see Cramp et al. 2014 for an example). Previous interpretations of Aegean human diets using non-Aegean fish remains as reference may therefore have underestimated the importance of marine foods.

For historical periods, the interpretation of isotopic data has benefitted from the use of contemporary written sources that provide detail and also context for the results (e.g. Vika et al. 2009; Vika 2011; Killgrove and Tykot 2013). This is not possible in our case because much of Aegean prehistory is a period with no written records. Even in its latest part (second millennium BCE), which saw the adoption of hieroglyphic, Linear A and Linear B scripts, the situation remains the same. Hieroglyphic and Linear A have not been deciphered and, perhaps interestingly, Linear B lacks references to fish. The only marine product recorded (and apparently managed by the palaces) are the purple shellfish (here mostly *Hexaplex trunculus*) and the dye that was produced by it (Palaima 2020). Fish and other seafood, much like other categories of commonly consumed foods, such as the pulses (e.g. Sarpaki 1992) seem to have functioned outside the structured, centrally regulated system of the palatial bureaucracy. This does not necessarily reflect their economic or cultural importance (or insignificance) as a food source. It rather implies that fish probably had a different function than basic staples like wheat, barley, wine, or animals which produced economically important by-products (wool, hair, leather) or were involved in cult.

2.6 Synthesis of the Zooarchaeological and Isotopic Data

The task of combining and interpreting zooarchaeological and isotopic data to explore fish consumption in the prehistoric Aegean cannot, at this stage, be direct and straightforward, as there are no archaeological sites where both sets of data are available for the same chronological horizon. To remedy this situation, we adopted a broader approach and focused on two aspects, namely, how geography and social conditions affect the frequency and the type of fish consumption. We discuss these issues by focusing on specific case studies.

2.6.1 Geography and the Availability of High Trophic Level Marine Resources

Because high trophic level fish enrich human bone collagen $\delta^{13}C$ and $\delta^{15}N$ to higher degree than low trophic ones, one might expect that in the northern Aegean, where such fish were consumed more commonly (see discussion above), isotopic analysis would reflect this. Such a hypothesis, however, cannot be verified at the moment, due to lack of isotopic data in that region, with only one case study available at present (Spathes, Fig. 2.1). To explore the applicability of such an approach, however, we examined cases from eutrophic environments, analogous to northern Aegean coastal waters, namely Almyri in Corinthia, and Lerna and Mycenae in the Argolid.

Almyri is a site with very little published archaeological documentation but with 23 skeletons that have been analysed isotopically. Almyri is situated at the western head of the Saronic Gulf, with multiple underwater and coastal freshwater springs (Markantonis and Koumantakis 2011) and, nowadays, a coastal wetland. All these offer favourable conditions for fishing, but there are no geological studies to verify the existence of such features in the Bronze Age. Lerna is a site that, in the Bronze Age, was also situated near eutrophic waters created by a fresh water coastal lake with swampy and marshy patches on the coast (Zangger 1991). Furthermore, bird bones of different dates within the Bronze Age recovered and published from the site include taxa such as cormorants, mallards, ducks that suggest the exploitation of this ecosystem. Mycenae, today a few kilometres north of Lerna, probably also had access to the rich shallow waters head of the Argolic Gulf. Mycenae and its findings, however, will be discussed in detail below.

From the human isotopic data of Lerna and Almyri (Fig. 2.5), we can see some key differences in the dietary strategies practiced at these two sites. Firstly, the average $\delta^{13}C$ and $\delta^{15}N$ at Lerna are both lower than that at Almyri. This difference is statistically distinct for $\delta^{13}C$ ($t_{(71)} = -5.36$, $p < .00001$) and $\delta^{15}N$ ($t_{(71)} = -6.61$, $p < .00001$). The lower mean $\delta^{15}N$ at Lerna suggests that diet was more reliant on plants than at Almyri, where higher $\delta^{15}N$ levels indicate the incorporation of more animal foods. Higher mean $\delta^{13}C$ at Almyri could be indicative of a greater proportion of the diet coming from marine foods. Intra-site variability is greater at Lerna than Almyri, particularly along the nitrogen axis ($F_{(38,\ 33)} = 5.42$, $p < 0.001$). There is no statistically significant difference in variance in $\delta^{13}C$ ($F_{(38,\ 33)} = 1.62$, $p > .05$ between the two sites. This suggests that individuals at Almyri ate more similarly to one another than at Lerna, where dietary composition differed more between individuals.

Overall, it seems that individuals in the Lerna sample consumed a diet that was more reliant on plants and less on animal protein than individuals at Almyri (Fig. 2.5). The greater isotopic variance in the Lerna sample suggests that people consumed more variable diets, with some individuals differing on the nitrogen axis by nearly an entire trophic level. The Almyri individuals are also on average more enriched along the carbon axis, and this combination of higher $\delta^{13}C$ and $\delta^{15}N$ at Almyri is indicative of a greater reliance on upper-trophic marine foods than at Lerna. These observations suggest that geography might be an important factor in fish consumption and should be explored further in future human isotopic analyses.

2.6.2 Social Distinction and Differential Access to Marine Resources

Two case studies from southern Greece (Peloponnese and Crete) best illustrate how intensity and type of fish eating had a strong social dimension. Isotopic analysis has been performed on bones of individuals found in cemeteries belonging to the wealthy powerful palatial centres of Mycenae and Knossos (Fig. 2.6). Both sites have also revealed subsets of individuals who are thought to have consumed marine foods, based on their $\delta^{13}C$ values (Richards and Hedges 2008; Nafplioti 2016), differentiating them from the rest of the Aegean population. Not only is it interesting that these palatial sites have yielded the most uncontested marine signatures in the sample, but it also provides us with an opportunity to compare the specific isotopic patterns of marine consumption between two well-documented sites.

Isotopic analysis of the skeletons from Grave Circles A and B at Mycenae, dating to the sixteenth and seventeenth-sixteenth centuries BCE respectively, produced results for 18 individuals, nine from each Grave Circle (Richards and Hedges 2008). Individuals in these two major burial contexts—Grave Circles A and B— share a common high consumption of animal protein which is expressed in the high

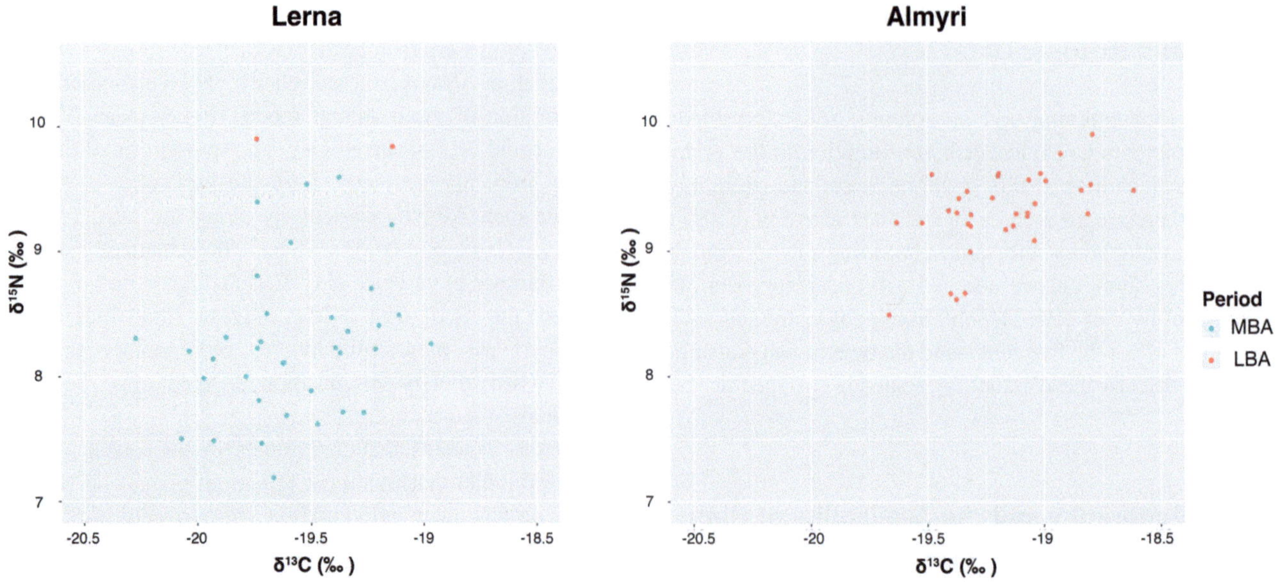

Fig. 2.5 Human isotopic data from the site of Lerna (left) and Almyri (right). (Graphic by A. Allshouse)

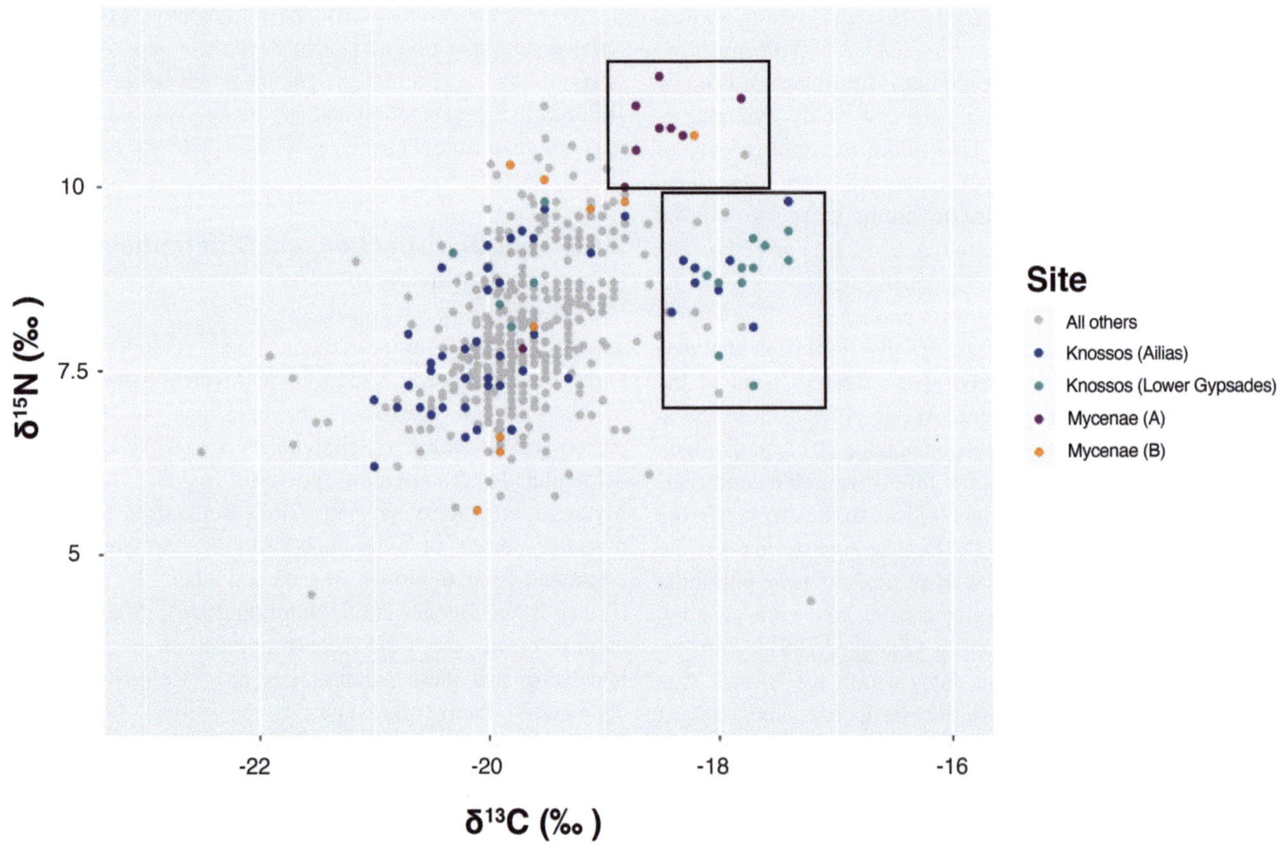

Fig. 2.6 Isotopic values of human bone collagen from the Knossos and Mycenae burial contexts (colour) against other sites (in gray). Clusters of individuals who have been interpreted as marine consumers are outlined with boxes (Richards and Hedges 2008; Nafplioti 2016). (Graphic by A. Allshouse)

(relative to other Bronze Age sites) values of $\delta^{15}N$, but also display some key differences. Of the nine individuals studied from Grave Circle A, eight displayed an isotopic signature well outside the range of the rest of the Aegean sample. These individuals (Fig. 2.6) were found to be enriched in both $\delta^{13}C$ and $\delta^{15}N$ to such an extent that marine contribution was the suggested explanation (Richards and Hedges 2008). Grave Circle B was also slightly enriched above the population mean in $\delta^{15}N$, but only one individual discovered in this burial context had a potential 'marine signature'.

Richards and Hedges (2008) suggested that a good proportion of this isotopic distinction could be attributed to the consumption of higher levels of terrestrial animal protein, while also acknowledging that there was likely a marine component of these individuals' diets. We would argue, based on the data presented above, that these values are most consistent with the regular inclusion of higher trophic level fish in the diet.

The individuals buried in Grave Circles A and B—and especially those of the Grave Circle A—were clearly members of the elite if not actual rulers, judging by their grave goods, which included large amounts of gold and silver in the form of jewellery, weapons, death masks, and other objects of high elaboration and artistry (indicatively Μυλωνάς 1983; Graziadio 1991; Heitz 2008; now displayed in the National Archaeological Museum in Athens). This provides us with a connection, at least at the site of Mycenae, between marine consumption and wealth/status. It should also be noted that the purported marine-consuming individuals at Mycenae are more common in the later (and wealthier: see Mylonas 1966) Grave Circle A than in Grave Circle B.

So far, analysis and publication of fish bone assemblages of this date and region are missing. It is evident that intense consumption of high tropic level fish had been related to high status, as the bones of the interred individuals there are enriched in both $\delta^{13}C$ and $\delta^{15}N$ beyond the level seen for most other sites. However, to date we cannt evaluate this observation in the context of Mycenae due to lack of complementary zooarchaeological data.

More resent research on Crete may offer some further understanding of the social parametres of the consumption of fish. Two Knossian cemeteries, the Ailias and the Lower Gypsades Cemeteries, both of which lie within 3 km of the central palace and exhibit some chronological overlap, offer isotopic evidence for a marine component in human diets. They were both excavated in the 1950s, but the human skeletons have only recently been studied. The Ailias Cemetery, with its chamber tombs, is thought to have been used in Middle Minoan II—Late Minoan I (1800/1700 to 1600/1450 BCE), and Lower Gypsades Cemetery with its tholos (vaulted) tomb and ossuary was used between Middle Minoan III and Late Minoan I (1700/1600 to 1600/1450 BCE) (Hood 1957, 2010; Nafplioti 2016 with further references). There are key differences between them in the burials (e.g. organisation of burial space, management of bones, burial type), but also in the burial offerings. In the case of Ailias, offerings were abundant and, in some cases, elaborate and prestigious in comparison to Lower Gypsades, where the offerings were more moderate (for a synopsis, see Nafplioti 2016: 44). Thirty-nine skeletons from Ailias cemetery and 16 from Lower Gyspades have yielded isotopic data.

The individuals thought to have consumed marine foods at Knossos plot very differently from those at Mycenae (Fig. 2.5), and this is our first point of interest. While the Mycenaean marine consumers were elevated in *both* $\delta^{15}N$ and $\delta^{13}C$, suggesting input from high trophic marine foods, those at Knossos are mainly distinguished by their $\delta^{13}C$, and do not differ significantly in $\delta^{15}N$ from their non-marine consuming peers. The Knossian marine consumers' isotopic values are therefore more consistent with intensive consumption of low-trophic marine foods than with regular inclusion of high trophic marine foods.

If we look at the distribution of these marine consumers across cemeteries, we find that the Lower Gypsades cemetery has a much higher concentration of marine consumers—11 out of 16 individuals compared to Ailias' 10 out of 45 (Nafplioti 2016; Allshouse et al. Forthcoming). If we take this observation one step further, however, and look at individual tombs/burial contexts, we find that the majority of the marine consumers at Lower Gypsades come from the so-called "rectangular structure", a commingled tomb that was added on to the main structure in Late Minoan IA (Hood and Smyth 1981: 7, 11). All eight individuals found in the rectangular structure had this intensive marine signature, suggesting that fish (and shellfish) consumption was a common dietary habit among the members of this group.

This contrasts with what was observed at Mycenae, where marine consumption appeared to be associated with wealth, in terms of grave goods. The exact opposite appears to be the case at Knossos. The wealthier burial contexts of Ailias, with grave goods including Egyptian amethyst, gold and silver adornments (Hood 2010) yielded fewer individuals with a marine signature, and the greatest concentration of marine consumers was found in the Lower Gypsades' "rectangular structure", where grave goods consisted of only ceramic sherds (Hood 1957). Grave goods do not, of course, on their own, confirm a high social status, but this fact can still help us better understand differing patterns of marine exploitation in the prehistoric Aegean. By comparing the human isotopic data to archaeological contexts, we can demonstrate that individuals likely consuming high trophic marine foods, such as those found at Mycenae, appear to be associated with rich burial contexts, while individuals consuming low-trophic marine foods (i.e. those relying primarily on smaller, near-shore species) such as those from the Knossos cemetery, are more commonly associated with poor burial contexts.

No fish bone assemblages from Bronze Age Knossos are available to help us refine this interpretation, but zooarchaeological work on several contemporary sites on the island, and particularly on Neopalatial Mochlos that was mentioned above, is instructive. As noted previously in this chapter, at Mochlos in the Late Minoan IA (1600–1500 BCE) and Late Minoan IB (1500–1450 BCE) periods (Mylona 2022), like everywhere on Crete (Mylona 2020) and most possibly at Knossos as well, fish and shellfish consumption was widespread and common. The majority of the fish and shellfish consumed, however, were small or immature species, of the shallows, all of low trophic level: picarel, bogue, damsel fish, wrasses and combers, as well as a number of immature sparids, parrot fishes and others, along with limpets, topshells, ceriths, purple shellfish and so on (Table 2.1). Remains of these fish and molluscs are found everywhere, in all Neopalatial dwellings on site. Elements of the rudimentary technology required for their capture, e.g. single small bronze hooks, were also found in various houses that appear to have no maritime or fishing character. It has been suggested that the capture of these small low trophic level fish and the shellfish could have been done by anyone living at the site, without any particular specialised knowledge or an elaborate fishing toolkit (Mylona 2022). The archaeo-ichtyological profile of other contemporaneous assemblages on Crete is similar, although in most cases contextual analysis is missing. It is probable that the widespread consumption of this type of marine foods created an isotopic signature similar to what has been observed in many of the individuals in the two Knossian cemeteries, especially the "rectangular structure" at Lower Gypsades, and also on skeletons elsewhere in Greece as discussed above.

At Neopalatial Mochlos, however, there is evidence for fish consumption of a different kind (Mylona 2022). In certain buildings, there were remains of elaborate meals that included a considerable number and variety of large, high trophic level fish. These were found together with the remains of large amounts of sheep and goat and other foods not commonly eaten on site, such as cattle and dog. The fish in these contexts appear to be part of some type of diacritical eating, and they document not only the access of a selected part of the site's population to those fish, but, at the same time, access to the professional sector of fishing. Judging by the distribution of bones of these large fish on site, it is clear that this access was restricted. Such privileged access to large, high trophic level fish, would leave elevated $\delta^{15}N$ and $\delta^{13}C$ values on the consumers' bones, analogous to those detected in the Mycenae Grave Circle A.

2.7 Conclusions

Current research has not yet reached a consensus regarding the importance of aquatic, especially marine, food resources in the diet of prehistoric people in the Aegean. The rich archaeo-ichthyological and archaeo-malacological record supports an optimistic view on the subject, suggesting a systematic, and in some places intensive, exploitation and consumption of aquatic foods by wide sectors of the population of coastal sites. Earlier isotopic studies, by contrast, have failed to identify such a considerable contribution of aquatic resources in the diet.

This review of both approaches highlights the limitations and the possibilities each offers, and pinpoints possible grounds or ways of dealing with the discrepancy in the datasets. Advanced isotopic methodologies which rendered the method more sensitive, along with better and richer reference data, now permit a more nuanced analysis of isotopic finds. At the same time, the zooarchaeological record has also benefitted by advances in its methodologies and by the inception of contextual analysis of data. It appears that a combination of the two, which is based on mutual understanding of research agendas and the nature of the data, might prove very fruitful in understanding the exploitation of marine (and aquatic more generally) resources in the past. In this chapter, such an approach has been attempted, where published isotopic and zooarchaeological data have been examined side by side, each posing questions to the other and attempting in tandem to explore specific issues.

In the zooarchaeological record, a clear geographical patterning on the type of fish consumed can more often be discerned. It appears that in the eutrophic waters of the northern Aegean, high trophic level fish made up the majority of the consumed fish, or at least a very considerable fraction of it. It would be expected that such a consumption pattern would leave a clear isotopic signature. Due to lack of isotopic case studies from this area, we examined two sites from southern Greece, both located in proximity to similar eutrophic seas, and used them as proxies. Isotopic signatures on skeletons from these sites indicate an elevated consumption of marine foods. The geography factor should be explored more systematically in the future.

Isotopic analysis of a few hundred skeletons in various locations around Greece has highlighted some cases as exceptional. The skeletons from Bronze Age Grave Circle A (and to a lesser degree Grave Circle B) at Mycenae and several skeletons from two cemeteries at Knossos, Ailias and Gypsades, show higher markers for consumption of fish and seafood than the average for the Aegean. A correlation of those findings with the excavation data made researchers propose a socially differentiated access to fish and other marine foods. In this study, we added to that a nuanced treatment of isotopic data which is able to detect the consumption of low trophic level fish and we combined that with what is known about the nature of marine foods exploitation in the Bronze Age southern Aegean. This combined approach resulted in a fuller understanding of the miscellaneous data concerning fish consumption, isotopic and zooarchaeological. It appears that there were groups within Aegean populations who embraced a diet rich in marine foods that resulted from the typical and particularly widespread type of fishing, i.e. fishing in the shallow for small, mostly low trophic level fish (and molluscs) (e.g. cases at Gypsades Cemetery at Knossos). At the same time, other groups often consumed larger fish of higher trophic level (e.g. Mycenae Grave Circle A). This appears to be related to the ability of those groups to access certain sectors of the marine economy, namely professional fishermen who appear to also have been in the sphere of the processing of marine products.

The insights gained by the combined approach presented in this chapter will require testing and refinement. So far, isotopic research and zooarchaeology have acted independently, and that is why it is still difficult to find site- or region-specific coincidences of the two. More targeted planning is required to gain optimal data for answering specific research question.

Bibliography

Allshouse, A., Fries-Knoblach, J. Maran, J., Jones, O., Prevedorou, E., Polychronakou-Sgouritsa, N., Stockhammer, P., & Warinner C. Forthcoming. Distinctive Diets: Bronze Age Aegean Food Habits Revealed by a Meta-analysis of Stable Isotope Evidence. *Journal of Archaeological Science*.

Amundson, R.A., Austin, T., Schuur, E.A.G., Yoo, K., Matzek, V., Kendall, C., Uebersax, A., Brenner, D. & Baisden, W.T. 2003. Global Patterns of the Isotopic Composition of Soil and Plant Nitrogen. *Global Biogeochemical Cycles,* 17 (1). https://doi.org/10.1029/2002GB001903.

Bilecenoglu, M. 2009. Growth and Feeding Habits of the Brown Comber, *Serranus hepatus* (Linnaeus, 1758) in Izmir Bay, Aegean Sea, *Acta Adriatica,* 50(1): 105–10.

Bogaard, A., Fraser, R., Heaton, H.E.T., Wallace, M., Vaiglova, P., Charles, M., Jones, G. et al. 2013. Crop Manuring and Intensive Land Management by Europe's First Farmers. *Proceedings of the National Academy of Sciences of the United States of America,* 110(31): 12589–94. https://doi.org/10.1073/pnas.1305918110.

Βεροπουλίδου, P. 2011. 'Οστρεα από τους Οικισμούς του Θερμαϊκού Κόλπου. Ανασυνθέτοντας την κατανάλωση των μαλακίων στη Νεολιθική και την Εποχή του Χαλκού (PhD dissertation, Aristotle Univesity of Thessalonike).

Cheung, C. & Szpak, P. 2021. Interpreting Past Human Diets Using Stable Isotope Mixing Models. *Journal of Archaeological Method and Theory,* 28(4): 1106–42. https://doi.org/10.1007/s10816-020-09492-5.

Chisholm, B.S., Nelson, D.E. & Schwarcz, H.P. 1982. Stable-carbon Isotope Ratios as a Measure of Marine versus Terrestrial Protein in Ancient Diets. *Science,* 216(4550): 1131–32. https://doi.org/10.1126/science.216.4550.113.

Cramp, L.J.E., Jones, J., Sheridan, A., Smyth, J., Whelton, H., Mulville, J., Sharples, N. & Evershed, R.P. 2014. Immediate Replacement of Fishing with Dairying by the Earliest Farmers of the Northeast Atlantic Archipelagos. *Proceedings of the Royal Society B: Biological Sciences,* 281: 20132372–20132372. https://doi.org/10.1098/rspb.2013.2372.

Deniro, M.J. & Epstein, S. 1981. Influence of Diet on the Distribution of Nitrogen Isotopes in Animals. *Geochimica et Cosmochimica Acta,* 45(3): 341–51. https://doi.org/10.1016/0016-7037(81)90244-1.

Dotsika, E., Diamantopoulos, G., Lykoudis, S., Gougoura, S., Kranioti, E., Karalis, P., Michael, D., Samartzidou, E. & Palaigeorgiou, E. 2019. Establishment of a Greek Food Database for Palaeodiet Reconstruction: Case Study of Human and Fauna Remains from Neolithic to Late Bronze Age from Greece. *Geosciences Journal,* 9(4): 165. https://doi.org/10.3390/geosciences9040165.

Dotsika, E., Tassi, M., Karalis, P., Gougoura, S., Kranioti, E.F., Michael, D.E., Kazou, M., Poutouki, A.E. & Diamantopoulos, G. 2022. Reconstruction of Greek Population Diet from Neolithic Period to Modern Times. *American Journal of Biological Anthropology,* 178(2): 328–46. https://doi.org/10.1002/ajpa.24516.

El-Maremie, H. & El-Mor, M. 2015. Feeding Habits of the Bogue, *Boops boops* (Linnaeus, 1758) (Teleostei: Sparidae) in Benghazi Coast, Eastern Libya, *Journal of Life Sciences,* 9(5): 189–96. https://doi.org/10.17265/1934-7391/2015.05.001.

Fraser, R.A., Bogaard, A., Heaton, T., Charles, M., Jones, G., Christensen, B.T., Halstead, P. et al. 2011. Manuring and Stable Nitrogen Isotope Ratios in Cereals and Pulses: Towards a New Archaeobotanical Approach to the Inference of Land Use and Dietary Practices. *Journal of Archaeological Science,* 38(10): 2790–2804. https://doi.org/10.1016/j.jas.2011.06.024.

Froese, R. & Pauly, D. eds. 2022. *FishBase*. World Wide Web electronic publication. www.fishbase.org, version [accessed August 2022].

Graven, H., Keeling, R.F. & Rogelj, J. 2020. Changes to Carbon Isotopes in Atmospheric CO2 over the Industrial Era and into the Future. *Global Biogeochemical Cycles,* 34(11): e2019GB006170. https://doi.org/10.1029/2019GB006170.

Graziadio, G. 1991. The Process of Social Stratification at Mycenae in the Shaft Grave Period: A Comparative Examination of the Evidence. *American Journal of Archaeology,* 95(3): 403–40. https://doi.org/10.2307/505489.

Hedges, R.E.M. 2004. Isotopes and Red Herrings: Comments on Milner et al. and Lidén et al. *Antiquity,* 78(299): 34–37. https://doi.org/10.1017/S0003598X00092905.

Hedges, R.E.M. & Reynard, M.L. 2007. Nitrogen Isotopes and the Trophic Level of Humans in Archaeology. *Journal of Archaeological Science,* 34(8): 1240–51. https://doi.org/10.1016/j.jas.2006.10.015.

Heitz, C. 2008. Burying the Palaces? Ideologies in the Shaft Grave Period. In *Propylaeum-Dok, Digital Repository Classical Studies.* University of Heidelberg: 1–38. Available from https://archiv.ub.uni-heidelberg.de/propylaeumdok/89/1/Heitz.pdf

Hood, M.S.F. 1957. Archaeology in Greece, 1957. *Archaeological Reports,* 4(4): 3–25. https://doi.org/10.2307/581146.

Hood, M.S.F. 2010. The Middle Minoan Cemetery on Ailias at Knossos. In: O. Krzyszkowska, ed. *Cretan Offerings: Studies in Honour of Peter Warren*. BSA Studies 18. London: British School at Athens, pp. 161–68.

Hood, M.S.F. & Smyth, D. 1981. *Archaeological Survey of the Knossos Area*. BSA Supplementary Volume 14. London: Thames & Hudson.

Isaakidou, V., Halstead, P., Stroud, E., Sarpaki, A., Hatzaki, E., Nitsch, E. & Bogaard, A. 2022. Changing Land Use and Political Economy at Neolithic and Bronze Age Knossos, Crete: Stable Carbon ($\delta^{13}C$) and Nitrogen ($\delta^{15}N$) Isotope Analysis of Charred Crop Grains and Faunal Bone Collagen. *Proceedings of the Prehistoric Society,* 88: 155–91. https://doi.org/10.1017/ppr.2022.4.

Jones, A.K. 1986. Fish Bone Survival in the Digestive Systems of the Pig, Dog and Man: Some Experiments. In: D.C. Brinkhuizen & A.T. Classon, eds. *Fish and Archaeology: Studies in Osteometry, Taphonomy, Seasonality and Fishing Methods.* Oxford: British Archaeological Reports, pp. 53–61.

Karachle, K.P. & Stergiou, I.K. 2014. Diet and Feeding Habits of *Spicara maena* and *S. Smaris* (Pisces, Osteichthyes, Centracanthidae) in the North Aegean Sea. *Acta Adriatiica,* 55(1): 75–84.

Killgrove, K. & Tykot, R.H. 2013. Food for Rome: A Stable Isotope Investigation of Diet in the Imperial Period (1st–3rd Centuries AD). *Journal of Anthropological Archaeology,* 32(1): 28–38. https://doi.org/10.1016/j.jaa.2012.08.002.

Kontopoulos, I. & Sampson, A. 2015. Prehistoric Diet on the Island of Euboea, Greece: An Isotopic Investigation. *Mediterranean Archaeology and Archaeometry,* 15(3): 97–111. https://doi.org/10.5281/zenodo.18361.

Lee-Thorp, A.J. 2008. On Isotopes and Old Bones. *Archaeometry,* 50(6): 925–50. https://doi.org/10.1111/j.1475-4754.2008.00441.x.

Mariani, S. 2006. Life-history and Ecosystem-driven Variation in Composition and Residence Pattern of Seabream Species (Perciformes: Sparidae) in Two Mediterranean Coastal Lagoons. *Marine Pollution Bulletin,* 53(1–4): 121–27. https://doi.org/10.1016/j.marpolbul.2005.09.019.

Markantonis, K. & Koumantakis, I. 2011. Contribution on Hydrogeological Investigation of Karstic Systems in Eastern Korinthia. *Advances in the Research of Aquatic Environment,* 1: 511–18. https://doi.org/10.1007/978-3-642-19902-8_60.

Mylona, D. 2003. A Review of the Research on Fish Remains in the Aegean. In: C. Gamble, P. Halstead, Y. Hamilakis & E. Kotjabopoulou, eds. *Zooarchaeology in Greece: Recent Advances.* British London: British School at Athens, pp. 187–93.

Mylona, D. 2008. *Eating Fish in Greece from 500 BC to AD 700. A Story of Impoverished Fishermen of Lavish Fish Banquets?* BAR International Series 1754. Oxford: Archaeopress.

Mylona, D. 2011. Fish Vertebrae. In: A. Sampson, ed. *The Cave of Cyclops. Mesolithic and Neolithic Networks in the Northern Aegean, Greece. Vol. II: Bone Tool Industry, Dietary Resources and the Paleoenvironment and Archaeometrical Studies.* Philadelphia: INSTAP Academic Press, 237–68.

Mylona, D. 2016. Fish and Seafood Consumption in the Aegean: Variations on a Theme. In: T. Bekker-Nielsen & R. Gertwagen, eds. *The Inland Seas: Towards an Ecohistory of the Mediterranean.* Stuttgard: Franz Steiner Publishers, pp. 57–84.

Mylona, D. 2018. Neolithic Fishing at Ftelia, Mykonos. Insight from the Fish Bones. In: A. Sampson & T. Tsourouni, eds. *Ftelia on Mykonos, Greece. Neolithic Networks in the Southern Aegean Basin, Vol. II.* Laboratory of Environmental Archaeology Monograph Series 7. Athens: University of the Aegean, pp. 229–35.

Mylona, D. 2019. The Fish Bones. In: J.A. MacGillivray & L.H. Sackett, eds. *Palaikastro, Building 1.* British School at Athens, Suppl. Vol. 48. Athens: British School at Athens, pp. 373–86.

Mylona, D. 2020. Marine Resources and Coastal Communities in the Late Bronze Age Southern Aegean: A Seascape Approach. *American Journal of Archaeology,* 124(2): 179–213. https://doi.org/10.3764/aja.124.2.0179.

Mylona, D. 2021. Processed Fish, Marine Dyes and the Fishing Domaine of the Bronze Age Aegean. In: R. Laffineur & T.G. Palaima, eds. *Zoia. Animal-Human Interactions in the Aegean Middle and Late Bronze Age.* Aegaeum 45. Liége and Leuven: Peeters, pp. 113–24, Pl. XV.

Mylona, D. 2022. Animal Remains from Neopalatial Mochlos: Exploring Human-Animal Relations. In: J. Soles, ed. *Mochlos IV. The Neopalatial Settlement.* Philadelphia: INSTAP Academic Press, pp. 433–60, Tables 15–44, Plates 115, 117–18.

Mylonas, G. 1966. *Mycenae and the Mycenaean Age.* Princeton (NJ): Princeton University Press.

Μυλωνάς Γ. 1983. *Πολύχρυσοι Μυκήναι.* Αθήνα: Εκδοτική Αθηνών Α.Ε.

Nafplioti, A. 2016. Eating in Prosperity: First Stable Isotope Evidence of Diet from Palatial Knossos. *Journal of Archaeological Science Reports,* 6: 42–52. https://doi.org/10.1016/j.jasrep.2016.01.017.

Nicholson, R.A. 1995. Out of the Frying Pan into the Fire: What Value Are Burnt Fish Bones to Archaeology? *Archaeofauna,* 4: 47–64.

Palaima, T. 2020. *Porphureion* and *Kalheion* and Minoan-Mycenaean Purple Dye Manufacture and Use. In: V. Apostolakou, T.M. Brogan, and P.P. Betancourt, eds. *Alatzomouri Pefka. A Middle Minoan IIB Workshop Making Organic Dyes.* Philadelphia: INSTAP Academic Press, 123–28.

Papathanasiou, A. 2003. Stable Isotope Analysis in Neolithic Greece and Possible Implications on Human Health. *International Journal of Osteoarchaeology,* 13(5): 314–24. https://doi.org/10.1002/oa.705.

Papathanasiou, A. 2015. Stable Isotope Analyses in Neolithic and Bronze age Greece: An Overview. In: A. Papathanasiou, M.P. Richards & C. Fox, eds. *Archaeodiet in the Greek World: Dietary Reconstruction from Stable Isotope Analysis.* Hesperia supplements 49. Princeton: The American School of Classical Studies at Athens, pp. 25–56.

Papathanasiou, A. & Richards, M.P. 2015. Summary: Patterns in the Carbon and Nitrogen Isotope Data through Time. In: A. Papathanasiou, M.P. Richards & S.C. Fox, eds. *Archaeodiet in the Greek World: Dietary Reconstruction from Stable Isotope Analysis.* Hesperia Supplements 49. Princeton: The American School of Classical Studies at Athens, pp. 195–203.

Petroutsa, I.E. & Manolis, K.S. 2010. Reconstructing Late Bronze Age Diet in Mainland Greece Using Stable Isotope Analysis. *Journal of Archaeological Science,* 37(3): 614–20. https://doi.org/10.1016/j.jas.2009.10.026.

Pinnegar, J.K. 2018. Why the Damselfish *Chromis chromis* Is a Key Species in the Mediterranean Rocky Littoral – A Quantitative Perspective. *Journal of Fish Biology,* 92(3): 851–72. https://doi.org/10.1111/jfb.13551.

Powell, J. 2003. The Fish Bone Assemblage from the Cave of Cyclope, Youra: Evidence for Continuity and Change. In: N. Galanidou and C. Perles, eds. *The Greek Mesolithic: Problems and Perspectives.* London: The British School at Athens, pp. 173–81.

Price, G.C., Krigbaum, J. & Shelton, K. 2017. Stable Isotopes and Discriminating Tastes: Faunal Management Practices at the Late Bronze Age Settlement of Mycenae, Greece. *Journal of Archaeological Science Reports,* 14: 116–26.

R Core Team. 2020. R: A Language and Environment for Statistical Computing. *R Foundation for Statistical Computing,* Vienna, Austria. https://www.R-project.org/

Rachedi, M., Derbal, F. & Kara, M.H. 2018. Feeding Habits of the Comber *Serranus cabrilla* (Linnaeus, 1758) (Teleostei, Serranidae) from the Gulf of Annaba (Eastern Coast of Algeria). *Cahiers de Biologie Marine,* 59: 149–58. https://doi.org/10.21411/CBM.A.7B25A05C.

Reese, D.S. 2019. The Invertebrates. In: J.A. MacGillivray & L.H. Sackett, eds. *Palaikastro, Building 1.* BSA Suppl. 48. London: British School at Athens, pp. 387–409.

Richards, M.P. & Hedges, R.E.M. 2008. Stable Isotope Evidence of Past Human Diet at the Sites of the Neolithic Cave of Gerani; the Late Minoan III Cemetery of Armenoi; Grave Circles A and B at the Palace Site of Mycenae; and the Late Helladic Chamber Tombs. In: H. Martlew, Y. Tzedakis & M. Jones, eds. *Archaeology Meets Science: Biomolecular and Site Investigations in Bronze Age Greece.* Oxford: Oxbow, pp. 220–30.

Richards, M.P. & Vika, E. 2008. Stable Isotope Results from New Sites in the Peloponnese: Sykia, Kalamaki and Spaliareika. In: H. Martlew, Y. Tzedakis & M. Jones, eds. *Archaeology Meets Science: Biomolecular and Site Investigations in Bronze Age Greece*. Oxford: Oxbow, pp. 231–35.

Rose, M. 1994. With Line and Glittering Bronze Hook: Fishing in the Aegean Bronze Age (PhD dissertation, Indiana University).

Rose, M. 1995. Fish Remains from Minoan Kommos. In: J. Shaw & M. Shaw, eds. *Kommos I. The Kommos Region and Houses of the Minoan Town 1. The Kommos Region, Ecology, and Minoan Industries*. Princeton: Princeton University Press, pp. 204–39.

RStudio Team. 2022. RStudio: Integrated Development Environment for R. *RStudio*, PBC, Boston, MA. URL http://www.rstudio.com/

Sarpaki, A. 1992. The Palaeoethnobotanical Approach. The Mediterranean Triad or Is It a Quartet. In: B. Wells, ed. *Agriculture in Ancient Greece*. Stockholm: Swedish Institute in Athens, pp. 61–76.

Schoeninger, M.J., DeNiro, M.J. & Tauber, H. 1983. Stable Nitrogen Isotope Ratios of Bone Collagen Reflect Marine and Terrestrial Components of Prehistoric Human Diet. *Science,* 220(4604): 1381–83. https://doi.org/10.1126/science.6344217.

Stergiou, K.I., Christou, E.D., Georgopoulos, D., Zenetos, A. & Souvermezoglou, C. 1997. The Hellenic Seas: Physics, Chemistry, Biology and Fisheries. *Oceanography and Marine Biology,* 35: 415–538.

Stergiou, K.I. & Karpouzi, V.S. 2002. Feeding Habits and Trophic Levels of Mediterranean Fish. *Reviews in Fish Biology and Fisheries,* 11: 217–54.

Szpak, P. 2014. Complexities of Nitrogen Isotope Biogeochemistry in Plant-Soil Systems: Implications for the Study of Ancient Agricultural and Animal Management Practices. *Frontiers in Plant Science,* 5: 288. https://doi.org/10.3389/fpls.2014.00288.

Theodoropoulou, T. 2007. L'Exploitation des Faunes Aquatiques en Égée Septentrionale aux Périodes Pré- et Protohistoriques (PhD dissertation, L'Universitè Paris I. Université de Sorbonne I, Sorbonne).

Theodoropoulou, T. 2018. The Exploitation of Marine Animal Resources at Alepotrypa Cave: Harvesting Strategies, Management and Uses. In: A. Papathanasiou, W.A. Parkinson, D.J. Pullen, M.L. Galaty & P. Karkanas, eds. *Neolithic Alepotrypa Cave in the Mani, Greece*. Oxford: Oxbow, pp. 327–59.

Triantaphyllou, S., Richards, M.P., Zerner, C. & Voutsaki, S. 2008. Isotopic Dietary Reconstruction of Humans from Middle Bronze Age Lerna, Argolid, Greece, *Journal of Archaeological Science,* 35(11): 3028–34. https://doi.org/10.1016/j.jas.2008.06.018.

Vander Zanden, M.J. & Rasmussen, J.B. 2001. Variation in $\delta^{15}N$ and $\delta^{13}C$ Trophic Fractionation: Implications for Aquatic Food Web Studies. *Limnology and Oceanography,* 46(8): 2061–66.

Vika, E. 2011. Diachronic Dietary Reconstructions in Ancient Thebes, Greece: Results from Stable Isotope Analyses. *Journal of Archaeological Science,* 38(5): 1157–63. https://doi.org/10.1016/j.jas.2010.12.019.

Vika, E., Aravantinos, V. & Richards, M.P. 2009. Aristophanes and Stable Isotopes: A Taste for Freshwater Fish in Classical Thebes (Greece)? *Antiquity,* 83(322): 1076–83. https://doi.org/10.1017/S0003598X00099361.

Vika, E. & Theodoropoulou, T. 2012. Re-investigating Fish Consumption in Greek Antiquity: Results from $\delta^{13}C$ and $\delta^{15}N$ Analysis from Fish Bone Collagen. *Journal of Archaeological Science,* 39(5): 1618–27. https://doi.org/10.1016/j.jas.2012.01.016.

Wheeler, A. & Jones, J.A.K. 1989. *Fishes*. Cambridge: Cambridge University Press.

Willis, L.M. & Boehm, A.R. 2014. Fish Bones, Cut Marks, and Burial: Implications for Taphonomy and Faunal Analysis. *Journal of Archaeological Science,* 45: 20–25. https://doi.org/10.1016/j.jas.2014.01.026.

Zangger, E. 1991. Prehistoric Coastal Environments in Greece: The Vanished Landscapes of Dimini Bay and Lake Lerna. *Journal of Field Archaeology,* 18(1): 1–15. https://doi.org/10.1179/009346991791548799.

Fishing Gear and Fish Remains from Vergina: A Case of Inland Fishing in Macedonia

3

Vasiliki G. Stamatopoulou

In memory of my fellow student
Calliope Almatzi,
a researcher of prehistoric fishing

Abstract

In ancient Macedonia, the very land of the renowned royal and elite hunts of Hellenistic times, precious little is known about fishing, especially in non-coastal areas. The fishing gear presented here testifies to fishing activity in the city excavated in Vergina, northern Greece, revealing an elusive aspect of its everyday life. The gear was found in the excavations of the city wall and the acropolis and date to the second half of the second century BCE. The information gained from these finds is enhanced by osteological material of fish offered in the funerary pyre of the famous unplundered 'Tomb II of the Great Tumulus'. Coming from well-dated contexts, these finds contribute to filling a gap in our scarce information about fishing at northern Greek inland sites in the vicinity of rivers and lakes (Mylona 2008: 58 on the paucity of relevant data).

3.1 Introduction

The unearthing of the two unplundered Macedonian tombs of the Great Tumulus, containing burial assemblages of exceptional opulence (Andronikos 1984) considered by many as royal,[1] has rightly overshadowed the wider archaeological research in Vergina and little is widely known about everyday life in this city. However, the excavations carried out by the Aristotle University of Thessaloniki on the acropolis and the city wall of Vergina have brought to light both the fortification and a multitude of portable finds and evidence, providing information about the daily life of its inhabitants. The objects examined here are amongst these.

The ancient city excavated in the area between the modern villages Vergina and Palatitsia, at the northern feet of the Pieria Mountain range, extended on a northward sloping ground, of roughly triangular shape, with its narrowest and highest section to the south, where the fortified acropolis is located and where a number of fishing artefacts was found. Evidence from the excavation has established with certainty that the city wall was erected at the end of the fourth century BCE and that it was abandoned before the middle of the second century BCE (Fig. 3.1) (Faklaris and Stamatopoulou 2010: 113–20; Faklaris and Stamatopoulou 2013; Stamatopoulou 2021: 70).

3.2 Excavation Data

3.2.1 The Net Sinkers Next to the Eastern City Wall

A zone along the inner face of the city wall, after the destruction of the city just before the middle of the second century BCE and the termination of its defensive function, was occupied by slum dwellings. Their occupants made their

[1]Over the past 46 years that followed the first excavation reports on these tombs, ascribing them to the king of Macedonia Philip II and members of the royal family, scholars have criticised nearly every major tenet of this orthodoxy without reaching a consensus except for the fundamental issue of shifting the date of Tomb II to the last 17 years of the fourth century BCE, two decades after Philip II's assassination, excluding him from the candidates. For a summary of this ongoing debate and a review of the literature, see Palagia and Borza 2007 and Faklaris 2011: 345–47.

V. G. Stamatopoulou (✉)
Aristotle University of Thessaloniki, Vergina Excavation, University Campus, Thessaloniki, Greece
e-mail: vstamato@hist.auth.gr

C. Tsouparopoulou, L. Recht (eds.), *Human and Aquatic Beings: Interactions in and beyond the Eastern Mediterranean (3rd–1st Millennia BCE)*, Themes in Contemporary Archaeology, https://doi.org/10.1007/978-3-031-73643-8_3

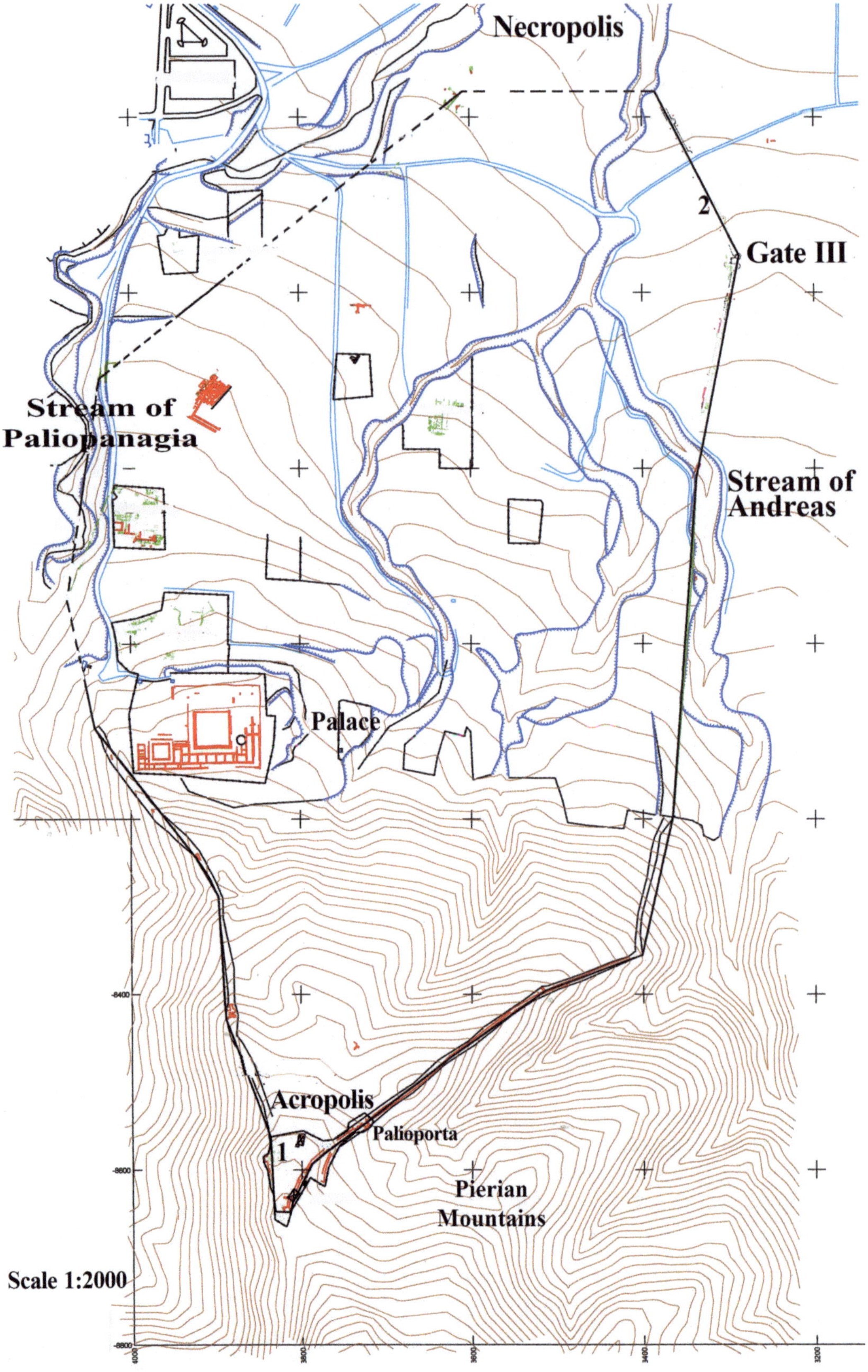

Fig. 3.1 General topographic plan of the ancient city showing the findspots of the fishing gear. 1. The acropolis finds. 2. The city wall sinkers. (Drawn by topographers Giannis Gatzios and Sofia Gatziou)

living mainly by reusing materials from the ruined Hellenistic city and its cemetery (Faklaris and Stamatopoulou 2013). The remnants of the hearth of one such dwelling yielded charred seeds of legumes, cereals and olive stones, together with some lead objects scattered over an area of about 3 m^2, with almost half of them in an area of 1.5 m^2 (Fig. 3.2).

These are 23 lead fishing-net sinkers, made of rectangular-shaped sheets, 0.2–0.3 cm thick, folded lengthwise in half, in a U-shaped cross-section, so that their two long sides do not meet, but have an opening of 0.1 to 0.8 cm, with most of them at 0.2–0.3 cm (Fig. 3.3). Their length ranges between 11 and 10 cm, their width between 1.2 and 1 cm, and their weight between 60.05 and 53.46 grams, with the majority weighing 56–57 grams. Among these, a lead sheet rhomboidal in shape was found, 7.2 cm long, 3.3 cm wide, 0.1 cm thick and weighing 17.24 g, bearing two holes, 0.5 and 0.4 cm in diameter (Fig. 3.4). Many of them were found in a parallel arrangement and at a spacing of ±0.14 cm from each other, indicating that they came from one fishing net that fell folded up on this spot. The debris from this hearth was tested by wet and dry sieving, yielding cereal and legume grains, but no fish remains. Moreover, the soil within the fold of the sinkers was examined for fibre remains from the edge of the net with a negative result (Piercy and Bass 2004: 399, for net fibres preserved in the sinker's fold).

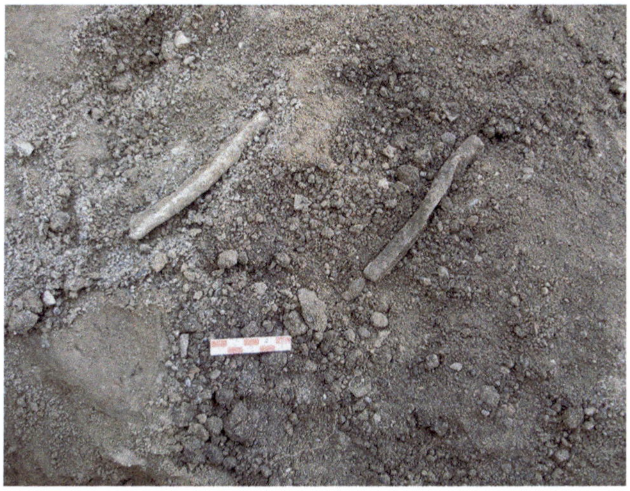

Fig. 3.2 The net sinkers in situ. (©V. Stamatopoulou)

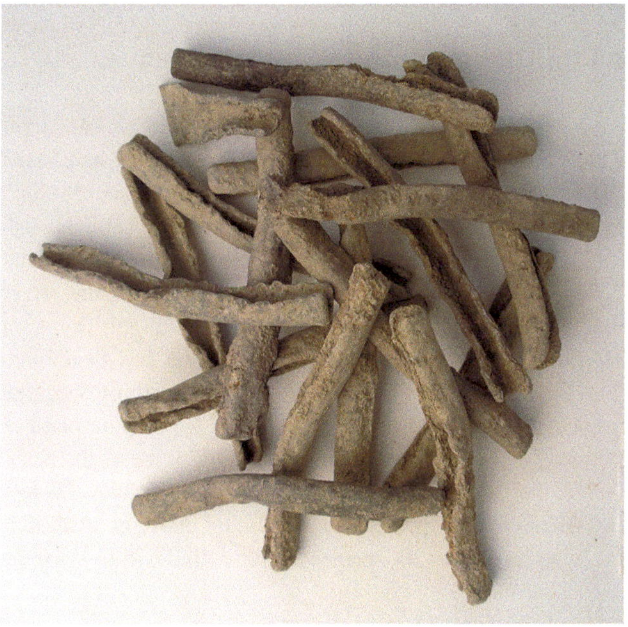

Fig. 3.3 The lead net sinkers from the Vergina city wall. (©V. Stamatopoulou)

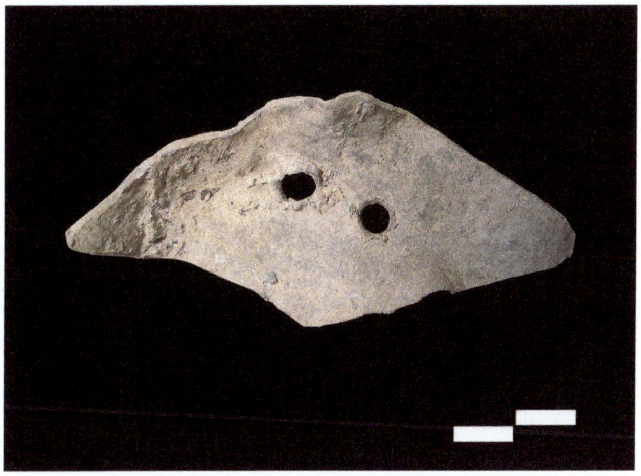

Fig. 3.4 The rhomboidal lead sheet. (©V. Stamatopoulou)

3.2.2 The Fishing Gear on the Acropolis

The acropolis of Vergina was used as a fortress from the end of the fourth century to the middle of the second century BCE. From the mid second century BCE to the first century CE, intensive craft activity is documented on the site, with pottery, textile and metalworking workshops (Faklaris 1997). It has been determined that a significant proportion of the portable finds on the site were gathered from the city as material destined for smelting and repurposing. At the north-eastern end of the acropolis in a second century BCE layer, a bronze fishing hook was found (inv. no BAM347), 6.2 cm in length, 0.6–0.3 cm thick, weighing 5.89 g (Fig. 3.5). It consists of a curved solid copper wire, bent into a curve with an opening of 1.5 cm (Faklaris 1994: 120–21). Its straight part is circular in cross-section, whereas the curved part is ellipsoidal. The barb is 0.7 cm long. Its other end is flattened. Four horizontal grooves are incised immediately before the flattened end, facilitating the tying of the fishing line at this point (Galili et al. 2010: 82).

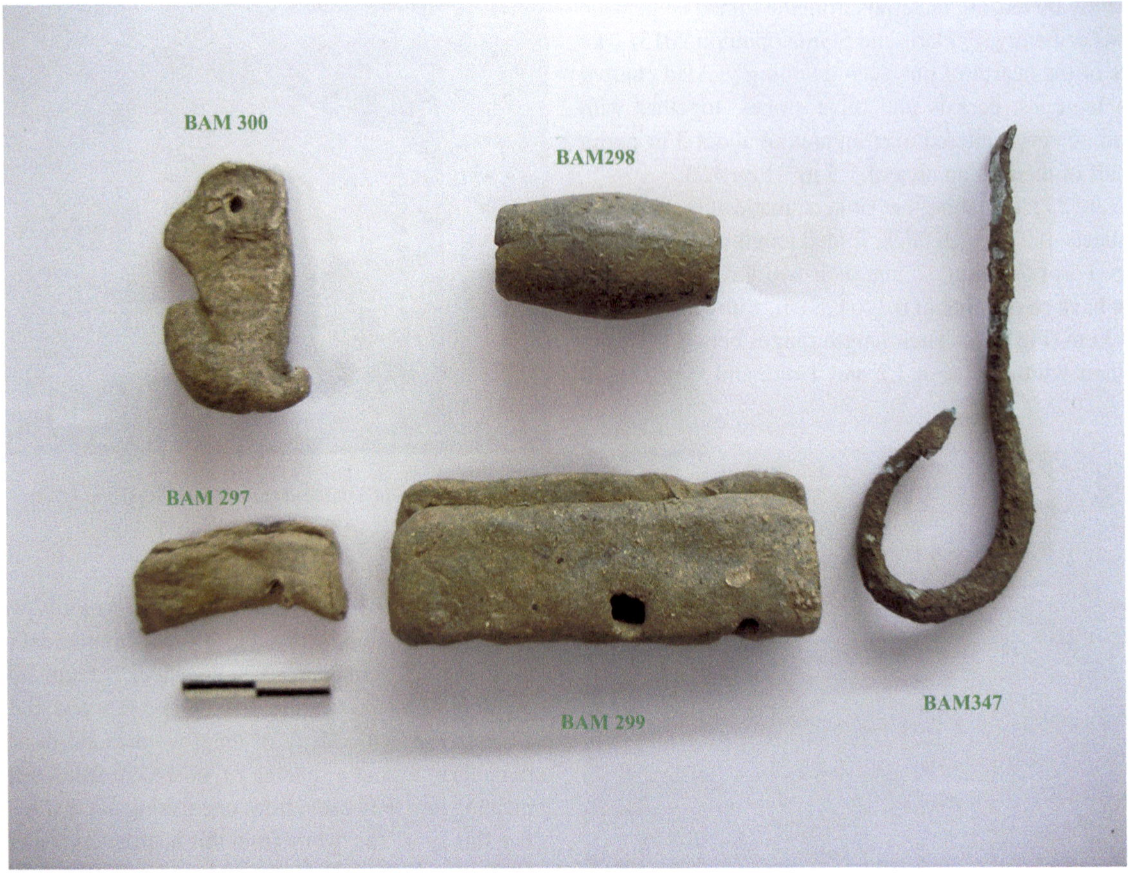

Fig. 3.5 The fishing gear from the acropolis of Vergina. (©V. Stamatopoulou)

Probably associated with this hook is a lead line sinker found near it (inv. no. BAM300), carelessly made and irregularly shaped (Fig. 3.5), 3.2 cm in length and weighing 21.5 g (Faklaris 1994: 121). It is bulged in the lower part, while the upper part is flattened and bears a hole of 0.2 cm in diameter. Moreover, three lead net-sinkers (Fig. 3.5), all different from each other, were found in a layer dated to the second half of the second century BCE (Faklaris 1994: 120):

(a) 2.8 cm in length, 1 cm wide, 0.2 cm thick, weighing 7.31 g (inv. no. BAM297).
(b) 5.3 cm in length, 0.2 cm wide, 0.4 cm, thick, weighing 52.36 g (inv. no. BAM299); found in an undisturbed layer together with a Thessaloniki autonomous mint coin. This is of the same type as the weights found by the city wall.
(c) 2.6 cm in length, 1.6–1.1 cm thick, weighing 28.76 g (inv. no. BAM298). A variation of the previous type, very similar to weights still used today by Greek fishermen.

3.3 Fishing in Vergina

3.3.1 Landscape and Fishing Techniques

Fresh fish have always been a highly desirable food (Davidson 1997: Chap. 1; on the consumption of fish based on written sources and archaeological finds, see Mylona 2008: Chaps. 7–11). In the proximity of fish habitats, fish could be obtained with simple equipment. The fish habitat closest to Vergina is the lower reaches of the river Haliakmon, which in the 39 km before its mouth flows on flat ground, where lowland riverine habitat has developed (Fig. 3.6). Its nearest bank is about 4 km north of Vergina, a distance not shorter in antiquity due to the sloping plain of Vergina, which descends to the north until it meets the river. Today, the fish fauna of the Haliakmon river includes 29 species (Kokkinakis 2007: 110–13; Economidis et al. 1981: 91, who mention 30 species). These finds testify to the

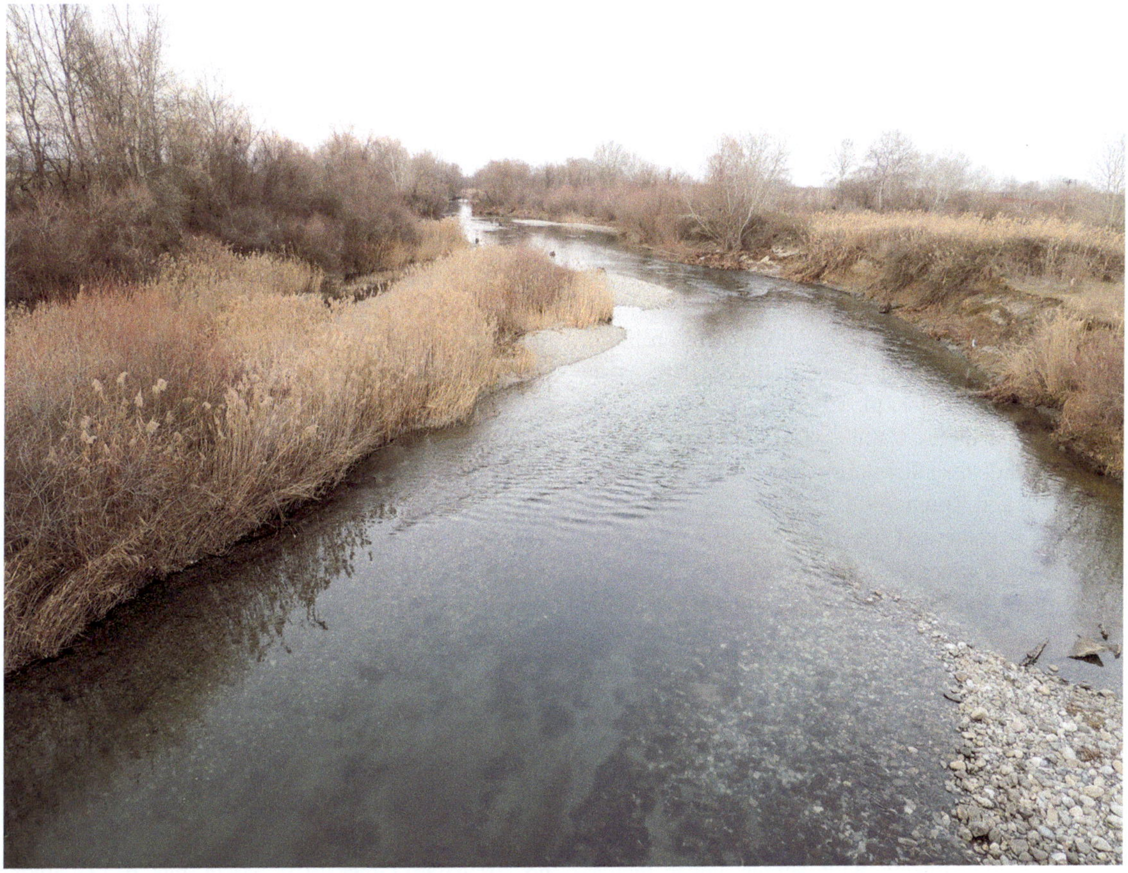

Fig. 3.6 View of the Haliakmon river near Vergina. (©V. Stamatopoulou)

practice of two fishing methods by the inhabitants of Vergina: net fishing and hook fishing (*δικτυεία* and *ἀγκιστρεία*, Ailian, *N.A.* 12, 43).

3.3.2 The Technology of the Vergina Fishing Gear

3.3.2.1 The Net Sinkers

In the hinterland of Macedonia, net-sinkers are not a frequent find. Examples have been found in Aiani, where 46 sinkers have been reported (Karamitrou-Mentesidi and Theodorou 2011: 122; 2013: 187) and in Pieria, one in a Hellenistic farmhouse (Yerofoka 2015: 248, no. M9, Pl. 25) and three in Livithra (Poulaki-Pandermali and Klinaki 2007: 169). In addition to their rarity, especially those from riverside or lakeside areas of the hinterland are not always identified or considered worthy of reference. The sinkers from Olynthus, for instance, were interpreted as lead clamps (Robinson 1941: nos. 1583–1589, 333, Pl. XCIX) and others from Stagira are reported as lead plates (Sismanidis 1994: 283). Difficulty is also observed in the Aiani find (Karamitrou-Mentesidi and Theodorou 2011: 122; 2013: 187) and in Epirus, where four net sinkers, from the Marmara Zervochoriou of Thesprotia, were interpreted as lead *katadesmoi* (inscribed lead curse scrolls, Kanta-Kitsou et al. 2008: 120).

However, eastern Mediterranean studies have dealt comprehensively with net-sinkers and their typological classification. Kuniholm's classification is based on the finds of a seventh century CE Byzantine shipwreck at Yassi Ada, a small island off Halicarnassus, grouping its sinkers into seven types by shape (conical, pear-shaped, spherical, sphenoid, crescent-shaped, triangular, and folded over). The Vergina sinkers fall into the category of folded-over strips of lead (Kuniholm 1982: 303–306). Moreover, a study of 155 weights from Cesarea Maritima, Israel, distinguishes 12 types, the weights from Vergina falling into the twelfth (Oleson 1994: 68–73).

A more comprehensive typology of fishing weights, for nets and lines, is proposed by Galili, Rosen and Sharvit, based on findings from an underwater survey in Haifa, Israel, which yielded 1200 lead and stone sinkers. Material, shape, and manufacture method are the criteria used in their typology (Galili et al. 2002: 182–92; 2010: 88–94). Bernal-Casasola (2010: 96–117) further suggests minor additions to the Haifa typology and summarises observations on lead, ceramic, and stone net-sinkers. According to the Haifa classification, the Vergina net-sinkers are classified as type L2.3, described as folded rectangular lead sinkers (FRLS), the most widespread and long-lasting type throughout the Mediterranean, rendering its independent dating impossible (Szulc-Kajak 2013: 334–37; Alfaro-Giner 2010: 76–79; Sharvit 2013: 242). Indicative of their antiquity and geographic distribution are 73 net-weights from a Late Bronze Age fisherman's grave at Akhziv, Israel (Sharvit 2013), sinkers and hooks from the Prehistoric Aegean (Iakovidis 1989: 249, pl. 36a; Yerontakou 2010: 12; Powell 1996: 118–20, 139–66), 75 Late Helladic sinkers from one fishing net found in a grave at Perati with a bronze hook (Iakovidis 1970: 355–56, pl. 135β), approximately 26 sinkers from a burial at Kamini on Naxos, of Late Helladic IIIC date (Vlachopoulos 2006: 270–71, pl. 73), and sinkers from the Archaic shipwreck at Giglio, Campania (Bound 1991: 26–27, fig. 57) from a mid-seventh-century BCE shipwreck from Dor, Israel (Galili and Rosen 2008: 3–4, fig. 5).

The similarity in shape and size of the Vergina net-sinkers attests to their manufacture by casting. Molds for sinkers of this type have been found in Byzantine layers in Shiqmona and Kastra, Haifa (Galili et al. 2002: 195, figs. 13a,b), whereas the ones found in the Sea of Galilee (Galili et al. 2002: 195, fig. 14; 2013: 151–52), and in Jiyeh, Lebanon (Szulc-Kajak 2013: 336) produced lead sinkers for hook and line. Tools for casting lead weights were found on board the Byzantine shipwreck at Yassi Ada to replenish the supply as needed (Kuniholm 1982: 309). Casting in molds facilitated the decoration of the sinkers with simple linear themes initially and later more elaborate motifs (Galili et al. 2002: 188–90, figs. 9–11; Piercy and Bass 2004: 398–435). A stone mold for casting almond-shaped sinkers was found in Thera (Dumas 2016: 141–42).

The molds produced flat rectangular FRLS-type strips which were subsequently folded around the girth of the net, acquiring a tube-like shape. However, unfolded plates of this type, which had not yet been folded up and remained flat, were found in the same survey in Haifa, indicating that the fishermen carried extra material to supplement their net-sinkers when necessary (Galili et al. 2002: 190–92, fig. 8). Lead is characterised by its plasticity, and the folding of the initially flat plates around the thread of the net is a task that can be accomplished with the use of simple tools, a wooden hammer or pliers. Lead is distinguished by its ease of processing, mainly because of its low melting temperature, which can be achieved in a simple hearth. Pure lead melts at just 327 degrees Celsius, is easily ductile and very malleable, solidifies quickly, and offers considerable resistance to oxidation, thus being suitable for water applications (Pliny, *Natural History* XXXIII 95: 159; Forbes 1964: 226–39; Blümner 1887: 374–76; a small domestic lead workshop was found on Rhodes, dating to the Late Hellenistic Age, with a simple furnace and shapeless fragments of lead, pieces of lead ore, and several lead objects such as loom weights—Kakavoyiannis 1984: 124–40; Treister 1996: 295). The rhomboidal object found by the wall of Vergina is a special accessory of the same net, although no parallels are known. Its two holes would probably function to adjust the tension of the cord around the perimeter of the net.

3.3.2.2 The Net from Vergina

As today, fishermen in antiquity could choose from a wide variety of nets. Depending on the types of fish and waters and the number of fishermen, nets could vary considerably (Gallant 1985: 16–25; Bekker-Nielsen 2002a: 31–32; Galili et al. 2013: 147–48). They were normally knit of linen or spartium junceum thread and referred to by the general terms *diktyon* or *līnon* (*δίκτυον* or *λίνον*)—by their typical material—while net fishing was called *diktyeia* or *diktyia* (*δικτυεία or δικτυΐα*) and diktyothereutikē (*δικτυοθηρευτική*) (*LSJ s.v.* Pollux *Onomasticon Z139*. Ailian, *N.A.*, 12.43. The terms *diktyeia* and *līna* were also used for hunting nets (see also Hull 1964: 10–18 on the knitting of hunting nets).

Ailian (*N.A.* 12, 43) lists the materials required for the manufacture of a net. The knitting of nets created equal-sized square spaces, with a knot at every corner (for variations see Alfaro-Giner 2010: 61–64). These spaces were called *vrohoi* (*βρόχοι*) or *vrohides* (*βροχίδες*) (Pollux *Onomasticon*, E28. *LSJ* s.v.) and varied in size according to the catch aimed for. Fragments of fishing nets have seldom survived (Schlichtherle 1990: Pl. 52; Körber-Grohne and Feldtkeller 1998: Pl. 3c, cat. no. 21; Alfaro-Giner 2010: 64–75, figs. 6–13), among which comes a remarkable find from Akrotiri on Santorini (Michailidis and Aggelidis 2006: 74, fig. 18). However, in Mainland Greece, no organic remains of fishnets have been found, and their sole surviving parts are their sinkers.

The common term for net sinkers was molyvdis (*μολυβδίς*) or *molyvdion* (*μολύβδιον*) (*LSJ* s.v., see also v. *μολύβδαινα-μολυβδαίνη*, lead fishing line sinker) and Ailian (*N.A.*, 1.2, 12.43, 15.5), calls them simply molyvdos (*μόλυβδος*) or molivos (*μόλιβος*) by their material. The three net sinkers from the acropolis of Vergina (inv. nos. BAM297, BAM298, BAM299) come from different contexts, and therefore speculation on the type of nets to which they belonged can only be based on the parameters of the waters and the fish for which they were intended. For the net from the city wall, the number of its weights provides an additional criterion. The minimum number of weights considered indicative of one net is six (Bernal-Casasola 2010: 119). Small nets thrown into the water by hand were equipped with 17–18 sinkers, while the larger ones had several hundred (Szulc-Kajak 2013: 337; Cambitoglou et al. 2001: 732 n. 72), as in the find of more than 300 sinkers in the prow of the Kyrenia shipwreck (306–300 BCE), in an arrangement suggesting that they were the remnants of one folded net (Swiny and Katzev 1973: 45). The number and size of the Vergina city wall sinkers suggest that they all came from one net of relatively moderate dimensions.

Evidently, the Vergina net was not a tool for professional, but for occasional fishing. Apart from its context data, this is also suggested by the distance of about 4 km of the dwelling of our fisher to the nearest point of the river, assuming that a fisher making a living from the river would normally have settled closer to it. Moreover, the shallow waters of the Haliakmon fish habitat lead to the conclusion that we are dealing with a net cast by hand, suitable for shallow waters, used in a manner similar to the current Greek pezovolo (*πεζόβολο*).

The *πεζόβολο* is a relatively small net thrown by one person. The casting begins by folding the net, ensuring that the sinkers move freely without getting tangled up, and the retrieval rope which is fixed in the centre is gathered in large loops. The casting follows, requiring skillful coordination of arms, torso and legs, so that the net is not only launched forward as far as possible, but also that it rotates to unfold, opens in a circle, and falls horizontally on the water (Zervaki 2013: 182 n. 52; Sharvit 2013: figs. 4–5; Bekker-Nielsen 2010: 191–92, fig. 2; illustration of a cast-net on a mosaic from Tunisia). With the equal weight and spacing of the sinkers on the net girth, and with skillful casting, in calm waters, the net should hit the surface horizontally (Bekker-Nielsen 2005: figs. 3–6) and sink with the girth first, like a falling parachute, to trap the fish in the net sack. Cast-net throwing can be complemented by bait casting, prior to casting, to entice fish to the narrow area where the net will drop. Its optimal efficacy is achieved by multiple successive casts (Bekker-Nielsen 2005: 85–86).

This type of net was also used in the ancient Mediterranean, cast in the same manner, as evidenced, among others, by a ring from Rhodes bearing the athletic naked figure of a young fisherman with a net (Fig. 3.7). This work of

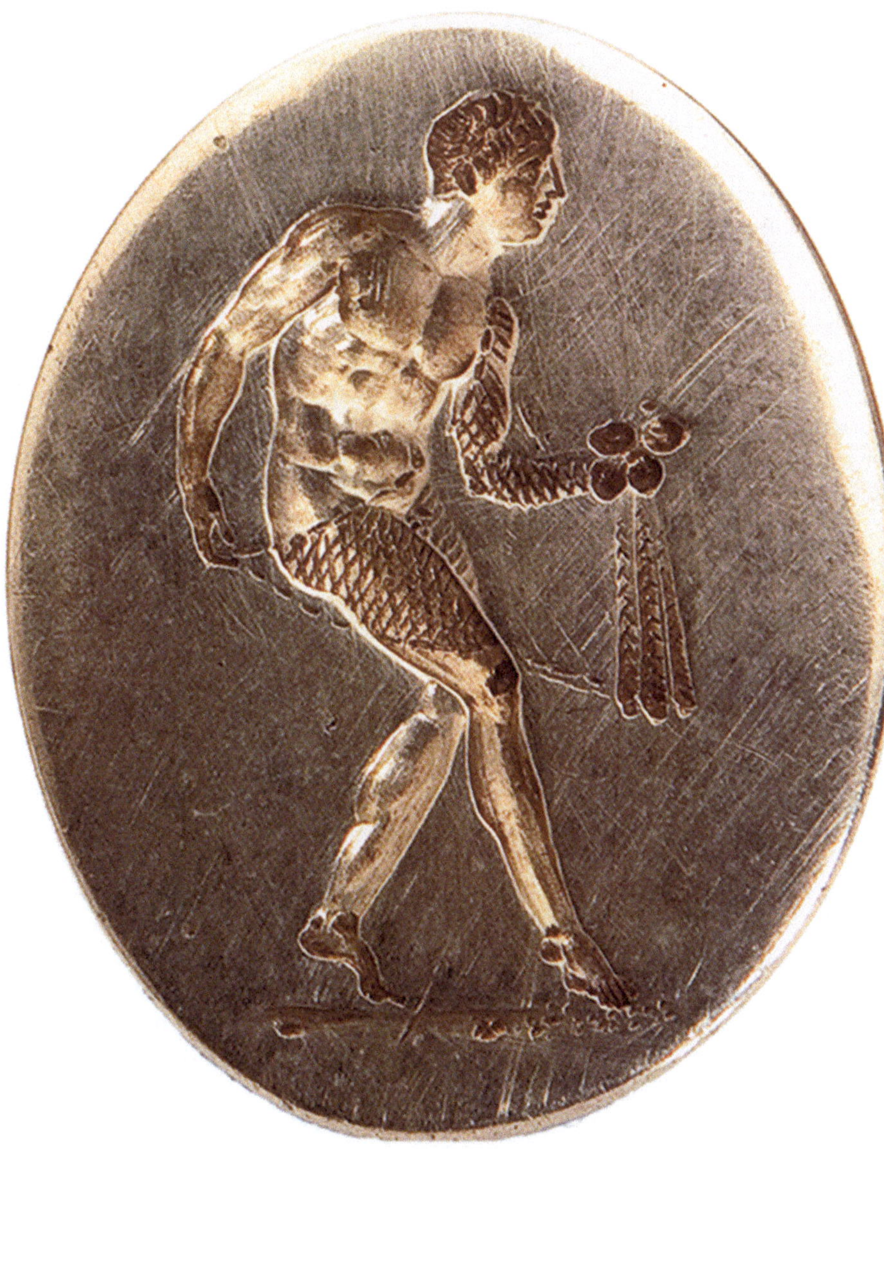

Fig. 3.7 The net-caster ring from Rhodes. (©Ephorate of Antiquities of Dodecanese)

350–300 BCE provides a superb depiction of a cast-net with its sinkers hanging around its folded girth (Zervaki 2013: 176–80). It captures the initial moment of the casting action, just before the throw is released, the fisherman focused and his entire body activated. The fisherman on an Archaic, black-figure kylix from Corinth is depicted at a stage of the casting preceding that shown on the Rhodes ring. He has arranged the net neatly with the large conical sinkers, but the casting motion has not yet started (Alfaro-Giner 2010: 70, fig. 5).

At the end of the second century CE, Oppian lists the main net types of his time (*Hal.* 3. 79–84; Bekker-Nielsen 2005: 83–84). The *ἀμφίβληστρον* (Oppian *Hal.* 3.80, 4.144. *LSJ* s. v.) is a small circular net cast by one person called *δικτυβόλος* (*LSJ* λ. δικτυβόλος), normally standing on the bank or shallows, or more rarely in a boat. Net depictions in vase paintings are uncommon. A notable example can be seen on a red-figure pyxis attributed to the Wedding Painter (470–460 BCE) (Clairmont 1953; Oakley 1982). Scenes of casting an *ἀμφίβληστρον* from a boat are pictured on North African mosaic floors dating from the Roman period (Koutsouflakis 2010: 68, figs. 6, 10, 15). The *ἀμφίβληστρον* wears sinkers all around the girth and, due to its tossing and retrieving technique, needs no floats.

With these in mind, while the attempted matching of names of ancient nets with the present-day nets is not reliable (Bekker-Nielsen 2002b: 215 n. 5), the *ἀμφίβληστρον* is most likely the net which in modern Greek is called πεζόβολο, and in this case, the Vergina city wall net would be an *ἀμφίβληστρον*, cast by hand from the bank or in the shallow. The acropolis net-sinkers coming from different contexts also belong to the FRLS type and were intended for the same waters and fish; they apparently belong to three additional cast nets.

3.3.2.3 The Hook

The line sinker inv. no BAM300 and the hook inv. no BAM347 found on the acropolis of Vergina testify to hook fishing, a method referred to in sources as *ἀγκιστρεία* (Ailian *N.A.* 12. 43. *LSJ* s.v. Gallant 1985: 14–16; Galili et al. 2013: 147, on the techniques of hook-fishing). Fishhooks were in use in Greece since at least the Early Neolithic period (Yerontakou 2010: 13–14; Moundrea-Agrafioti 2003; Powell 1996: 124–28). Fishhooks have been found in Mycenaean tombs at Perati (Iakovidis 1970: 354, fig. 156) and in the lakeside Neolithic settlement of Dispilio in northwest Macedonia (Almatzi 2002; Stratouli 1996, 2002). Their use has continued in historical times (Born 2007: 57–62, no. 172–212, Pl. 16). Their similarity to modern hooks has made identification unproblematic. However, some difficulties do occur, as in bronze finds from Dymokastro in Thesprotia (Kanta-Kitsou et al. 2008: 86), where the artifacts in Kanta-Kitsou et al. 2008: fig. 1 are referred to as probable bronze double fish-hooks or staples do not represent fishing gear. Nevertheless, in Kanta-Kitsou et al. 2008: fig. 2, three bronze fishhooks from the same site are pictured.

A significant number of fishhooks have been found in coastal, lakeside and riverside settlements in the mainland and insular Greece and the entire Mediterranean, of which an example, remarkable for its diachronicity, is a fisherman's basket, with numerous hooks, hooked on its inner side, in the exact way that is still common on the Greek coast for longline fishing called παραγάδι (Pappalardo 1990: 206, figs. 5a, 5b). Particularly noteworthy is also a find from the lower part of the city of Halieis, Argolid, where a hoard of several dozens of fishhooks of unspecified date was discovered (Jameson et al. 1994: 315, fig. 5).

Written sources (e.g. Ailian *N.A.* 13, p. 16) mention iron hooks for the fishing of large fish, such as for example tuna. However, the archaeological finds do not indicate any other material than bronze for the production of hooks in historical times. Iron is archaeologically considered to have appeared in hooks in Medieval times (Bernal-Casasola 2010: 88–89; Thomas 2010: 151, for iron hooks from the Red Sea; Galili et al. 2013: 163, for the first appearance of iron hooks). Hook dimensions and shapes varied depending on the expected catch (Koutsouflakis 2010: 77 n. 66). The length of 6.2 cm of the hook from the Vergina acropolis, in a size classification by Bernal-Casasola (2010: 89) is in the medium range (4–8 cm), while those up to 2.5 cm long are classified as small and those longer than 8 cm as large.

A noteworthy feature of the hooks dating to historical times is the shift in the cross-section of their bent part from circular to ellipsoidal, which is also observed in the Vergina hook. This appears to be a sign of craftsmanship, as the bent part of the hook is a point of weakness, liable to further bending or breakage by a struggling fish or a snagged obstacle on the waterbed. Forging this part of the hook shank to an elliptical cross-section increases the holding ability of the hook (Galili et al. 2010: 83–84).

Their shape remains unchanged to the present day, except for minor variations, mainly in their ends. The evolution of the shape since the Bronze Age has aimed at providing a tighter fastening of the line and a firmer hold of the fish (Cambitoglou et al. 2001: 730). The variations of historical-era hooks have been examined in 2010 by Galili, Rosen and Sharvit, who studied 33 fish hooks from two Roman-era shipwrecks surveyed on the shallow seabed off the beach of Haifa, Israel. They are categorised by the shape of their ends, the presence/absence of a barbed point at one and a flattened or a tapered tying end at the other. No intact hooks with barbless points are known from Greece, and Galili et al. (2010: 80–85) also express reservations about their barbless hooks being made that way or the barbed point being destroyed. Both types can have file grooves for optimising the attachment of the fishing line (Galili et al. 2010: 80–85, figs. 32–34). Under this classification, the Vergina hook is a barbed hook with flattened tying end and grooves (on the fishhook typology see Bernal-Casasola 2010: 90–92; Galili et al. 2013: 150–51; on another type of fishhook with an "eye" at the tying end, see Brewer and Friedman 1989: 26–31; Szulc-Kajak 2013: 338–39).

The hook required attachment to a fishing line, called *ὁρμιά, ὁρμίδιον* ἢ *μήρινθος* (Oppian *Hal.* 2. 123; for fishing with hook and line see Ailian *N.A.* 15.10; Plutarch *De Sollertia Animalium* 26). Multiple lines of evidence suggest that the hook and line fishing technique was already well established in the Aegean as early as the Bronze Age (Rose 1994; Parker 1992: 425 on a hook found attached to a preserved fragment of line found in a Roman shipwreck). Depending on the desired catch, the line was made of plant or animal fibres. Ailian mentions fishing lines of spartium junceum, terebinthus or flax and as suitable animal fibres, horsetail hairs in purple, light blue, or natural in colour (Ailian *N.A.*, 1.2, 1.23, 12.43. Plutarch *Moralia* 976[E]). Male horse tail hairs were preferred because in females the urine compromised the strength of the hair (Ailian *N.A.* 12, 43; Oppian *Hal.* 3.75 and 3.151). The sinking of the fishhook required the use of at least one weight, which also varied depending on the catch aimed for (Ailian *N.A.* 1.2), as was the selection of baits, described by the sources as highly specialised (Koutsouflakis 2010: 72 n. 69).

The indispensable hook-line pair could be supplemented by a rod, which when made of reed was called a *κάλαμος* or *δόναξ* and when of wood was called a *ῥάβδος*. It needed to be thin, flexible, straight, and dry (Homer *Odyssey* 12.251; *Iliad* 16.406–408; Plutarch *Mor.* 976[E]; Oppian *Hal.* 3. 74, 1. 23; Ailian *N.A.* 12. 43; Bernal-Casasola 2010: 126–27). Rod fishing could be practiced from the shore or a boat and was called *ἀπό καλάμου ἁλιεία* (Oppian *Hal.* 7.78, 3.468 ff; Ailian *N.A.* 13.2; Plutarch *Antony* 29.7). The scene on a red-figure kylix of the Ambrose Painter (c. 510 BCE) is regarded as the earliest depiction of rod fishing in Attic vase-painting (c. 510 BCE) (Museum of Fine Arts, Boston, no. 01.8024, Beazley, *ARV*[2], 173, 9; Malagardis 1988: 114, fig. 5a).

Finds of hooks in several sites in Macedonia bear witness to the popularity of this fishing method in marine, riverine and lake habitats. In northern Greece, 94 hooks were found at Olynthus (Robinson 1941: 365–74, nos. 1788–1882, Pls. CXVII-CXIX), and some at Stageira (Sismanidis 1994: 275; 1997: 472, 478, fig. 9) and Pella (Makaronas 1965: 416, Pl. 470[e]). Fishhooks from the now drained Lydia Lake are housed in the Museum of Veroia (Stefani 1993: 378, Pl. 115b). Three hooks are known from Abdera (Lazaridis 1965: 456, Pl. 544e; Triantafyllos et al. 1999: 14), an unspecified number of fishhooks along with net sinkers of Hellenistic times were found in Krania, Pieria (Herakleion) (Poulaki-Pantermali 2001: 337), and two more at Leibithra (Poulaki-Pandermali and Klinaki 2007: 169). Four bronze hooks were found in Torone (Cambitoglou et al. 2001: 730–32, fig. 171) as well as hooks and net-sinkers at the Artemisium of Thasos (Prêtre 2016: 105–107, Pl. XXIX (769, 780, 772, 779, 781, 774) and 124–26, Pl. XXXIII (1015–19)).

The efficiency of hook-and-line fishing is poorer than other techniques and it has been suggested that its practice can only be aimed at supplementing the fisherman's diet (Gallant 1985: 14–16; Bekker-Nielsen 2005). However, to improve its efficiency, multiple hooks or multiple lines with single hooks could also be used. There are very few finds of multiple hooks, mainly double, from the Mediterranean, and the identification of multiple lines with single hooks, which requires the preservation of the line, is extremely rare (Bernal-Casasola 2010: 122; Thomas 2010: 151; Bekker-Nielsen 2005: 89; Oppian *Hal.* 3. 470 and 3.72–78). The fresco on the late sixth century BCE Etruscan tomb known as the "tomb of hunting and fishing" at Monterozzi, Tarquinia is thought to depict this type of fishing (Holloway 1965: 341–47; Chatzidimitriou 2010: 33–34).

3.4 Osteological Remains of Fish from Vergina

Other than the fishing gear found on the acropolis and by the city wall of Vergina, the most renowned monument of the site, the Macedonian 'Tomb II of the Great Tumulus', excavated in 1977 and 1978 (Andronikos 1984), dated to the last years of the fourth century BCE (see Palagia and Borza 2007; Faklaris 2011: 345–47), among its lavish funerary gifts, has yielded a significant piece of evidence not only related to fish consumption in Vergina but also to the use of fish in ceremonial practices. Remains of the funeral pyre were found deposited on the external surface of the vault covering the tomb (Andronikos 1984: 97–98). Bones of various domestic animals, game, stones of fruit, and nutshells were found in this context, all offered in the funeral pyre in honour of the tomb occupant, and perhaps reflecting dishes served at the lavish symposia of the Macedonian elite that came to a peak in this period. During the examination of this debris in the lab years later, several fish vertebrae were found among them by Prof. P. Faklaris (Fig. 3.8) (Faklaris 1994: 121, fig. 12, picturing 41 vertebrae; Antikas 2006: 203–209). Other than their identification as fish remains, no specialised study of them has yet been published or conducted and thus their species is not known. Although data on fish offerings and consumption at shrines are documented (Mylona 2008: 91–99) and offerings of animals and food at funerary pyres are well established (Rohde 1925: 164, 167–70; Garland 1985: 36, 112, 144, 168), this is the first time there is archaeological testimony of fish offered at a funerary pyre.

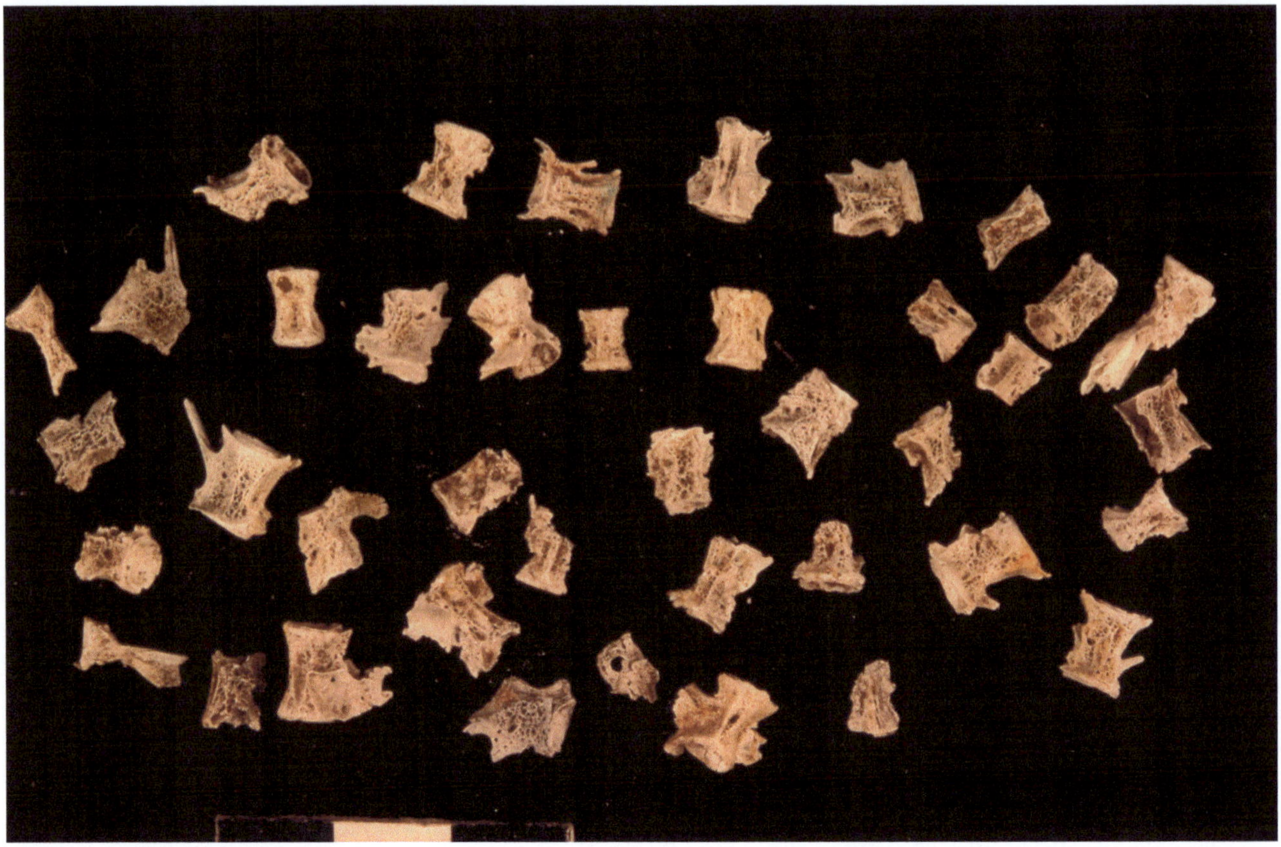

Fig. 3.8 Osteological fish remains from the funerary pyre of the Macedonian Tomb. (© Prof. Pan. Faklaris)

3.5 Conclusion

The bronze hook and its sinker and the lead net-sinkers illustrate for the first time fishing and fish consumption by the inhabitants of Vergina, using two of the oldest fishing methods ever practiced. Vergina fishermen do not seem to have practiced professional or recreational fishing (Koutsouflakis 2010: 60–62), but rather fished occasionally to supplement their diet. Vergina was certainly far from being a fishing village, but utilising the natural resources of the environment was obviously a fundamental factor in the city's economy. The rural population who made a living in the proximity of the inland waters of Haliakmon was able to combine fishing with other means of procuring food, as was the case in many small-scale fishing areas (Mylona 2008: 68).

The fishing finds from Vergina are scanty compared to the excavated city area, and the extent of exploitation of the lower reaches of the Haliakmon as a resource for fish as well as the importance of fishing in the economy and the nutrition of the locals is yet to be assessed. Indeed, even in areas where geographically it is certain that fishing was intensive, the evidence is strikingly scarce (Davidson 1952: 190; Cambitoglou et al. 2001: 729–32). The interaction of the inhabitants of Vergina with the natural environment of their city has not been researched. Fishing finds from the Macedonian hinterland are rare and sometimes puzzling for excavators and the role of the Macedonian river and lake resources awaits more finds and further study.

Acknowledgements I thank Professor Ant. Kokkinankis, a specialist on the fish fauna of Haliakmon River and the archaeologists of the Ephorate of Antiquities of Dodecanese Er. Kaninia, F. Zervaki and F. Seroglou for Fig. 3.6 and permission to present the ring from Rhodes.

Bibliography

Alfaro-Giner, C. 2010. Fishing Nets in the Ancient World: The Historical and Archaeological Evidence. In: T. Bekker-Nielsen & D. Bernal-Casasola, eds. *Ancient Nets and Fishing Gear. Proceedings of the International Workshop on "Nets and Fishing Gear in Classical Antiquity: A First Approach", Cadiz, November 15–17, 2007*. Cadiz: Servicio de Publicaciones de la Universidad de Cádiz/Aarhus University Press, pp. 55–81.

Almatzi, K. 2002. Η αλιεία σ' ένα λιμναίο οικισμό. In: G. Chourmouziadis, ed. *Στο Διοπηλιό, 7500 χρόνια μετά*. Thessaloniki: University Studio Press, pp. 135–43.

Andronikos, M. 1984. *Vergina. The Royal Tombs*. Athens: Ekdotiki Athenon.

Antikas, T. 2006. Symbols of Heroism: Horse Burials in Royal and Common Macedonian Tombs. In: M. Mashkour, ed. *Equids in Time and Space: Papers in Honour of Véra Eisenmann*. Oxford: Oxbow Books, pp. 203–209.

Bekker-Nielsen, T. 2002a. Fish in the Ancient Economy. In: K. Ascani, V. Gabrielsen, K. Kvist & A. Holm Rasmussen, eds. *Ancient History Matters. Studies Presented to Jens Erik Skydsgaard on His Seventieth Birthday*. Rome: L'Erma di Bretschneider, pp. 29–37.

Bekker-Nielsen, T. 2002b. Nets, Boats and Fishing in the Roman world. *Classica et Mediaevalia, Danish Journal of Philology and History*, 53: 215–33.

Bekker-Nielsen, T. 2005. The Technology and Productivity of Ancient Sea Fishing. In: T. Bekker-Nielsen, ed. *Ancient Fishing and Fish Processing in the Black Sea Region*. Black Sea Studies 2. Aarhus: Aarhus University Press, pp. 83–95.

Bekker-Nielsen, T. 2010. Fishing in the Roman World. In: T. Bekker-Nielsen & D. Bernal-Casasola, eds. *Ancient Nets and Fishing Gear. Proceedings of the International Workshop on "Nets and Fishing Gear in Classical Antiquity: A First Approach", Cadiz, November 15–17, 2007*. Cadiz: Servicio de Publicaciones de la Universidad de Cádiz/Aarhus University Press, pp. 187–203.

Bernal-Casasola, D. 2010. Fishing Tackle in Hispania. In: T. Bekker-Nielsen & D. Bernal-Casasola, eds. *Ancient Nets and Fishing Gear. Proceedings of the International Workshop on "Nets and Fishing Gear in Classical Antiquity: A First Approach", Cadiz, November 15–17, 2007*. Cadiz: Servicio de Publicaciones de la Universidad de Cádiz/Aarhus University Press, pp. 83–138.

Blümner, H. 1887. *Technologie und Terminologie der Gewerbe und Künste bei Griechen und Römern*, 4. Leipzig: Teubner.

Born, H. 2007. *Werkzeug und Gerät aus Olympia*. Berlin: De Gruyter.

Bound, M. 1991. The Giglio Wreck: A Wreck of the Archaic Period (c. 600 BC) off the Tuscan Island of Giglio. *Enalia Supplement*, 1: 26–27.

Brewer, D.J. & Friedman, R.F. 1989. *Fish and Fishing in Ancient Egypt*. Warminster: Aris and Philips.

Cambitoglou, A., Papadopoulos, J. & Tudor-Jones, O. 2001. *Torone I, the Excavations of 1975, 1976 and 1978*, Text: Part 2. Athens: Archaeological Society at Athens.

Chatzidimitriou, A. 2010. Η αλιεία και τα προϊόντα της στην εικονογραφία και τις φιλολογικές των αρχαϊκών και κλασικών χρόνων. In: D. Dimitropoulos & E. Olympitou, eds. *Ψαρεύοντας στις ελληνικές θάλασσες*, Τετράδια Εργασίας 33. Athens: Instituto Neoellinikon Ereunon, pp. 29–53.

Clairmont, C. 1953. Danae and Perseus in Seriphos. *American Journal of Archaeology*, 57: 92–94.

Davidson, G.R. 1952. *Corinth XII. The Minor Objects*. Princeton: American School of Classical Studies at Athens.

Davidson, J. 1997. *Courtesans and Fish Cakes: The Consuming Passions of Classical Athens*. London: Fontana Press.

Dumas, C. 2016. *Προϊστορική Θήρα*. Athens: John S. Latsis Foundation.

Economidis, P.S., Kattoulas, M.E. & Stephanidis, A. 1981. Fish Fauna of the Aliakmon River and the Adjacent Waters (Macedonia, Greece). *Cybium*, 5(1): 89–95.

Faklaris, P. 1994. Βεργίνα. Ανασκαφή ακρόπολης 1994. *Archaeologiko Ergo stin Makedonia kai Thraki*, 8: 119–24.

Faklaris, P. 1997. Εργαστήρια στην ακρόπολη της Βεργίνας [Workshops on the Acropolis of Verghina]. In: *Ancient Greek Technology, Proceedings of the 1st International Conference, Thessaloniki, September 1997*. Thessaloniki: Etaireia Makedonikon Spoudon, pp. 193–200.

Faklaris, P. 2011. Κι άλλα προβλήματα του τάφου II της Βεργίνας: το μέγεθος και οι φάσεις κατασκευής του [More Problems of Tomb II at Vergina. Its Size and Construction Phases]. In: P. Valavanis, ed. *Ταξιδεύοντας στην Κλασική Ελλάδα, τόμος προς τιμήν του καθηγητή Π. Θέμελη*. Athens: Etaireia Messiniakon Archaeologikon Spoudon, pp. 345–68.

Faklaris, P. & Stamatopoulou, V. 2010. Βεργίνα. Η αποκάλυψη της οχύρωσης στον αγρό Μπέλα. *Archaeologiko Ergo stin Makedonia kai Thraki*, 24: 113–20.

Faklaris, P. & Stamatopoulou, V. 2013. Η ανασκαφή της οχύρωσης της Βεργίνας κατά το 2013. Η εγκατάλειψη και η καταστροφή του τείχους. Η Βεργίνα των τεχνιτών και των τυμβωρύχων. *Archaeologiko Ergo sti Makedonia kai Thraki*, 27: 173–80.

Forbes, J.R. 1964. *Studies in Ancient Technology* VIII. Leiden: Brill.

Galili, E. & Rosen, B. 2008. Fishing Gear from a 7th-century Shipwreck off Dor, Israel. *The International Journal of Nautical Archaeology*, 37(1): 67–76.

Galili, E., Rosen, B. & Sharvit, J. 2002. Fishing-gear Sinkers Recovered from an Underwater Wreckage Site, off the Carmel Coast Israel. *The International Journal of Nautical Archaeology*, 31(2): 182–201.

Galili, E., Rosen, B. & Sharvit, J. 2010. Artifact Assemblages from Two Roman Shipwrecks off the Carmel Coast. *Atiqot*, 63: 61–110.

Galili, E., Zemer, A. & Rosen, B. 2013. Ancient Fishing Gear and Associated Artifacts from Underwater Explorations in Israel – A Comparative Study. *Archaeofauna*, 22: 145–66.

Gallant, T.W. 1985. A Fisherman's Tale: An Analysis of the Potential Productivity of Fishing in the Ancient World. *Miscellanea Graeca*, 7: 12–25.

Garland, R. 1985. *The Greek Way of Death*. London: Duckworth.

Hull, D.H. 1964. *Hounds and Hunting in Ancient Greece*. Chicago: University of Chicago Press.

Iakovidis, S. 1970. *Περατή. Το νεκροταφείον*. Athens: Archaeological Society at Athens.

Iakovidis, S. 1989. *Γλας I. Η ανασκαφή 1955–1961*. Athens: Archaeological Society at Athens.

Jameson, M.N., Runnels, C.N. and van Andel, T.H. 1994. *A Greek Countryside. The Southern Argolid from Prehistory to the Present Day*. Stanford: Stanford University Press.

Kakavoyiannis, E. 1984. Παραγωγή μολύβδου από λιθαργύρους στην ελληνιστική Ρόδο. *Athens Annals of Archaeology*, 17(1–2): 124–40.

Kanta-Kitsou, A., Palli, O. & Anagnostou, I. 2008. *Igoumenitsa Archaeological Museum*. Igoumenitsa: Ministry of Culture.

Karamitrou-Mentesidi, G. & Theodorou, D. 2011. Από την έρευνα στο φράγμα Ιλαρίωνα (Αλιάκμων): Η ανασκαφή στο Μέγα Άη Γιώργη και άγιο Κωνσταντίνο Δήμητρας Γρεβενών. *Το αρχαιολογικό έργο στην Άνω Μακεδονία*, 1 (2009): 109–29.

Karamitrou-Mentesidi, G. & Theodorou, D. 2013. From the Research at the Ilarion Dam (Aliakmon): The Excavation at Megas Ai Giorgis and Aghios Konstantinos near Dimitra, Grevena, the Archaeological Work in Upper Macedonia. *The Archaeological Work in Upper Macedonia*, 2 (2011): 173–203.

Kokkinakis, A.K. 2007. Environmental Risks for the Fish Fauna in the Main Rivers of Central Macedonia (Greece). *Journal of Environmental Protection and Ecology*, 8(1):107–18.

Körber-Grohne, U. & Feldtkeller, A., 1998. Pflanzliche Rohmaterialien und Herstellungstechniken der Gewebe, Netze, Geflechte sowie anderer Produkte aus den neolithischen Siedlungen Hornstaad, Wangen, Allensbach und Sipplingen am Bodensee. *Forschungen und Berichte zur Vor-und Frühgeschichte in Baden-Württemberg*, 68: 131–242.

Koutsouflakis, G. 2010. Η Αλιεία στον Ελληνορωμαϊκό Κόσμο: Σημειολογία Κειμένων και Αλιευτικής Εικονογραφίας. In: D. Dimitropoulos & E. Olympitou, eds. *Ψαρεύοντας στις ελληνικές θάλασσες*, Τετράδια Εργασίας 33. Athens: Instituto Neoellinikon Ereunon, pp. 55–83.

Kuniholm, P.I. 1982. The Fishing Gear. In: G.F. Bass, H. Frederick & F.H. Van Doorninck Jr., eds. *Yassi Ada I. A Seventh-century Byzantine Shipwreck*. Nautical Archaeology Series 1. Texas: Texas A&M University Press, pp. 296–310.

Lazaridis, D. 1965. Ανασκαφή Αβδήρων. *Archaeologiko Deltion, Chronika*, 20: 465, pl. 544e.

Makaronas, C. 1965. Ανασκαφαί Πέλλης. *Archaeologiko Deltion, Chronika,* 20: 412–21.
Malagardis, N. 1988. Images du monde rural attique à l' époque archaique. *Archaeologiki Ephemeris,* 127: 95–134.
Michailidis, I. & Aggelidis, P. 2006. Οι συνθήκες διατήρησης των οργανικών υλικών φυτικής προέλευσης στον προϊστορικό οικισμό του Ακρωτηρίου Θήρας. *Περιοδικό Άλς (Εταιρεία Στήριξης Σπουδών Προϊστορικής Θήρας),* 4: 61–81.
Moundrea-Agrafioti, A. 2003. Mesolithic Fish Hooks from the Cave of Cyclope, Youra. In: N. Galanidou & C. Perles, eds. *The Greek Mesolithic: Problems and Perspectives.* BSA Studies vol. 10. London: British School at Athens, pp. 131–41.
Mylona, D. 2008. *Fish-eating in Greece from the Fifth Century B.C. to the Seventh A.D.: A Story of Impoverished Fishermen or Luxurious Fish Banquets?* BAR International Series 1754. Oxford: BAR.
Oakley, J.H. 1982. Danae and Perseus on Serifos. American Journal of Archaeology, 86: 111–15.
Oleson, J.P. ed. 1994. *The Harbours of Caesarea Maritima. Results of the Caesarea Ancient Harbour Excavation Project 1980–85*, II. *The Finds and the Ship*. BAR International Series 594. Oxford: BAR.
Palagia O. & Borza, E.N. 2007. The Chronology of the Royal Macedonian Tombs at Vergina. *Jahrbuch des Deutschen Archaologischen Instituts,* 122: 81–125.
Pappalardo, U. 1990. L'eruzione pliniana del Vesubio nel 79 d.C.: Ercolano. In: C. Albore Libadie & F. Widemann, eds. *Volcanology and Archaeology: Proceedings of the European Workshops of Ravello, November 19–27, 1987, and March 30–31, Strasbourg 1989.* Strasbourg: Council of Europe, pp. 197–215.
Parker, A.J. 1992. *Ancient Shipwrecks of the Mediterranean and Roman Provinces.* BAR International Series 594. Oxford: BAR.
Piercy, G.V. & Bass, G. 2004. Fishing Gear. In: G. Bass, S. Matthews, J.R. Steffy & F. Van Doorninck, Jr., eds. *Serçe Limani. An Eleventh-century Shipwreck, Vol. I. The Ship and Its Anchorage, Crew and Passengers.* Texas: Texas A&M University Press, pp. 398–435.
Poulaki-Pandermali, E. & Klinaki, E. 2007. Λείβηθρα. *Archaeologiko Ergo sti Makedonia kai Thraki,* 21: 161–69.
Poulaki-Pantermali, E. 2001. 1997–2001: Όργα εθνικά και άλλα στην περιοχή του μακεδονικού Ολύμπου. *Archaeologiko Ergo sti Makedonia kai Thraki,* 15: 331–46.
Powell, J. 1996. *Fishing in the Prehistoric Aegean.* Studies in Mediterranean archaeology and literature, 137. Jonsered: Astrom Editions.
Prêtre, C. 2016. *La fibule et le clou, ex-voto et instrumentum de l'Aremision.* Études Thasiennes 23. Athens: École française d'Athènes.
Robinson, D.M. 1941. *Excavations at Olynthus, Part X: Metal and Minor Miscellaneous Finds.* Baltimore: Johns Hopkins Press.
Rohde, E. 1925. *Psyche. The Cult of Souls and the Belief in the Immortality among the Greeks* (Translated by W.B. Hillis). London: Kegan Paul, Trench, Trubner & Co. Ltd & New York: Harcourt, Brace & Company Inc.
Rose, M. 1994. With Line and Glittering Bronze Hook: Fishing in the Aegean Bronze Age (PhD dissertation, Indiana University).
Holloway, R. 1965. Conventions of Etruscan Painting in the Tomb of Hunting and Fishing at Tarquinii. *American Journal of Archaeology,* 69: 341–47.
Schlichtherle, H. 1990. *Siedlungsarchäologie im Alpenvorland I: Die Sondagen 1973-1978 in den Ufersiedlung Homstaad-Homle I. Befunde und Funde zum frühen Jungneolithikum am westlichen Bodensee.* Stuttgart: Konrad Theiss.
Sharvit, J. 2013. Lead Fishing-net Weights from the 'Fisherman's Grave', Akhziv. *Atiqot,* 74: 9–17 & 242–243.
Sismanidis, K. 1994. Αρχαία Στάγειρα. *Archaeologiko Ergo sti Makedonia kai ti Thraki,* 8: 275–87.
Sismanidis, K. 1997. Ανασκαφικά και αναστηλωτικά αρχαίων Σταγείρων 1997. *Archaeologiko Ergo sti Makedonia kai ti Thraki,* 11: 469–79.
Stamatopoulou V. 2021. Excavating the Hellenistic Fortification of Vergina, Northern Greece. In: M. Eisenberg & R. Khamisy, eds. *The Art of Siege Warfare and Military Architecture from the Classical World to the Middle Ages.* Oxford: Oxbow Books, pp. 64–72.
Stefani, E. 1993. Νομός Ημαθίας. *Archaeologiko Deltio,* 48, Chron.: 378–79.
Stratouli, G. 1996. Die Fischerei in der Agais wahrend des Neolithikums: Zur Technik und zum potentiellen Ertrag. *Prähistorische Zeitschrift,* 71(1): 1–27.
Stratouli, G. 2002. Τα εργαλεία από λειασμένο λίθο, οστό και κέρατο. In: G.C. Chourmouziades, ed. *Στο Δισπηλιό, 7500 χρόνια μετά.* Thessaloniki: University Studio Press, pp. 155–74.
Swiny, H.W. & Katzev, M.L. 1973. The Kyrenia Shipwreck: A Fourth-century B.C. Greek Merchant Ship. In: D.J. Blackman, ed. *Marine Archaeology: Proceedings of the Twenty-third Symposium of the Colston Research Society Held in the University of Bristol, April 4th to 8th, 1971.* London: Butterworths, pp. 339–60.
Szulc-Kajak, A. 2013. Fishing Gear from Jiyeh (Porphyreon). Preliminary Report. *Polish Archaeology in the Mediterranean,* 22: 334–40.
Thomas, R. 2010. Fishing Equipment from Myos Hormos and Fishing Techniques on the Red Sea in the Roman Period. In: T. Bekker-Nielsen & D. Bernal-Casasola, eds. *Ancient Nets and Fishing Gear. Proceedings of the International Workshop on "Nets and Fishing Gear in Classical Antiquity: A First Approach", Cadiz, November 15–17,* 2007. Cadiz: Servicio de Publicaciones de la Universidad de Cádiz/Aarhus University Press, pp. 139–59.
Treister, M.Y. 1996. *The Role of Metals in Ancient Greek History.* Leiden: Brill.
Triantafyllos, D., Kallintzi, N. & Soulakis, N. 1999. Αρχαιολογικό Μουσείο Αβδήρων. *Archaeologiko Ergo sti Makedonia kai ti Thraki,* 13: 1–14.
Vlachopoulos, A. 2006. *Υστεροελλαδική ΙΙΙΓ περίοδος στη Νάξο. Τα ταφικά σύνολα και οι συσχετισμοί τους με το Αιγαίο,* vol. A. Athens: Kardamitsa.
Yerofoka, E. 2015. *Η αγροικία στη θέση Τρία Πλατάνια του Νομού Πιερίας. Συμβολή στη μελέτη της αρχαίας αγροικίας.* Thessaloniki: University Studio Press.
Yerontakou, E. 2010. Ο ψαράς και η ψαριά του στο προϊστορικό Αιγαίο. In: D. Dimitropoulos & E. Olympitou, eds. *Ψαρεύοντας στις ελληνικές θάλασσες,* Τετράδια Εργασίας 33. Athens: Instituto Neoellinikon Ereunon, pp. 11–27.
Zervaki, F. 2013. Ψαρεύοντας στα ρηχά: Δαχτυλίδι με παράσταση πεζόβολου από τη νεκρόπολη της αρχαίας Ρόδου και η εικονογραφία της αρπαγής του Παλλαδίου. In: A. Giannikouri, ed. *Όλβιος Άνερ, Τόμος προς τιμήν του Γρηγόρη Κωνσταντινόπουλου.* Athens: Hellenic Ministry of Culture, pp. 171–92.

Abbreviations

Beazley, ARV2 Beazley, J. D. (1963). *Attic Red-figure Vase-Painters.* Clarendon Press.

List of Primary Sources

Ailian, *De Natura Animalium.,* 1.2, 1.23, 12.43, 13, 13.2, 15.5, 15.10
Homer, *Odyssey* 12.251. *Iliad* 16.406–408
Oppian, *Halieutica* 1.23, 2.123, 3.151, 3.468, 3. 470, 3.72–78, 3.74, 3.75, 3.79–84, 3.80, 4.144, 7.78
Pliny, *Natural History* XXXIII 95
Plutarch, *Antony* 29.7. *De Sollertia Animalium* 26. *Moralia* 976^{E}
Pollux, *Onomasticon Z139,* E28

4 The Marine Aspect of the Temple Repositories in the Palace of Knossos

Anastasios Eleftheriou, Marina Panagiotaki, and David S. Reese

Abstract

The so-called Temple Repositories in the West Wing of the palace of Knossos are famous for their faience 'snake goddesses'. However, along with hundreds of other objects, there were also remains of molluscs, with about 6392 marine shells of more than 35 species and subspecies. These are unique in the sense that even if the molluscs were originally collected to be consumed, their shells were not discarded as food debris by the Minoans. On the contrary, they were used in the shrine to which the 'snake goddesses' and all the other objects belonged as something important, since over 600 of them were painted. The painted ones may have been displayed on the altar or arranged on panels on the walls of the shrine. Others bear deliberate perforations at the umbo (the 'beak' of a bivalve) to be suspended perhaps in garlands with flowers and beads, while others display similar perforations caused by biogenic processes. Their importance is also evident in the fact that after the destruction of the shrine they belonged to, they were collected along with all the other important artefacts (made of faience, gold, ivory) and were carefully arranged inside the Temple Repositories and covered in soil—a ceremonial burial at the heart of the palace. Along with an analysis of these important remains in their context, the marine shells will be compared with the shells living today in the Aegean Sea. They will also be discussed according to their natural setting and habitat through an overview of the development and changes of the shores of the Knossos region.

Supplementary Information: The online version contains supplementary material available at https://doi.org/10.1007/978-3-031-73643-8_4.

A. Eleftheriou (✉)
Institute of Marine Biology, Biotechnology and Aquaculture, Hellenic Centre for Marine Research, Crete, Greece
e-mail: telef@hcmr.gr

M. Panagiotaki
Department of Mediterranean Studies, University of the Aegean, Rhodes, Greece

D. S. Reese
Division of Anthropology, Peabody Museum of Natural History, Yale University, New Haven, CT, USA

4.1 Introduction

In 1903, two large, stone-built, underground cists were discovered under the later gypsum floor of a room at the centre of the West Wing of the Knossos Palace by Sir Arthur Evans and his archaeologist assistant Duncan Mackenzie (Evans 1902–1903, 1921). Inside the cists were hundreds of objects of different materials, arranged in layers according to type, together with the faience "snake goddesses" that prompted Evans to call the cists the Temple Repositories (TR). They are dated to the Middle Minoan (MM) III-Late Minoan (LM) IA or early Late Minoan IA (part of the Neopalatial period, ca. 1650–1500 BCE) (Evans 1921: 550; Panagiotaki 1998: 198; 1999: 151).

The central "figure" of the deposit was the faience so-called "snake goddesses" that were accompanied by many faience artefacts in the shape of land animals, fruits and flowers, as well as marine forms such as fish, argonauts, and cockles. However, the most numerous objects in the TR were the remains of molluscs, 6392 marine shells of more than 35 different species (Fig. 4.1).

The TR shells are unique in the sense that even if the molluscs were originally collected to be consumed, their shells were not discarded as debris by the Minoans. On the contrary, they were used in the shrine to which the "snake goddesses" and all the other objects are thought to have belonged as something important, since over 600 of them were painted. The painted ones may have been displayed on the altar or arranged on panels on the walls of the shrine. Others bear deliberate perforations at the umbo (the "beak" of a bivalve), to be suspended perhaps in garlands with flowers and beads, while others display similar perforations caused by biogenic processes. Their importance is also evident in the

C. Tsouparopoulou, L. Recht (eds.), *Human and Aquatic Beings: Interactions in and beyond the Eastern Mediterranean (3rd–1st Millennia BCE)*, Themes in Contemporary Archaeology, https://doi.org/10.1007/978-3-031-73643-8_4

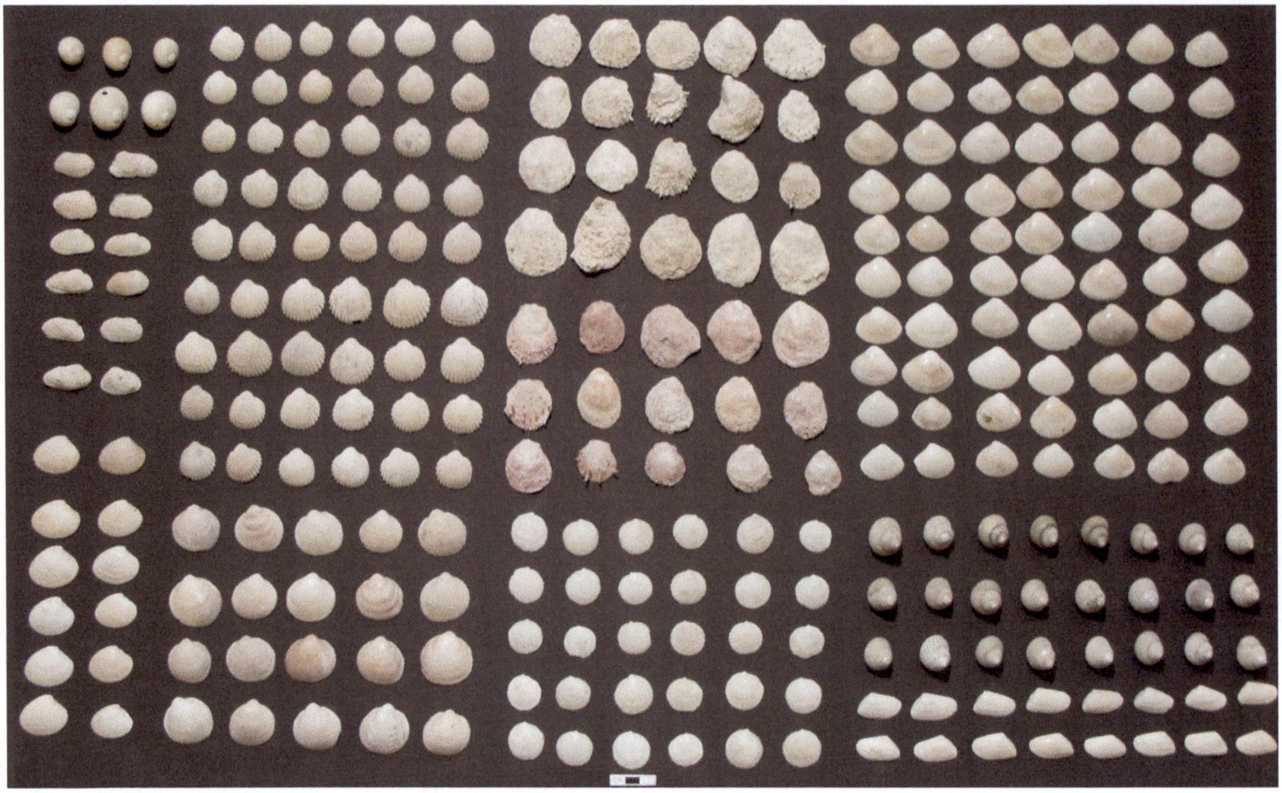

Fig. 4.1 Sample of shells found in the Temple Repositories of Knossos. (Photograph by M. Kallergis, IMBC)

fact that, after the destruction of the shrine they belonged to, they were collected, together with all the other important artefacts (made of faience, gold, ivory) and were carefully arranged inside the TR and covered in soil as part of a ceremonial burial at the heart of the palace.

"In the deposit fragments of porcelain began to come out along with various kinds of seashells predominant among which was the variety called in modern Greek 'petalida'". Thus wrote Mackenzie in his daybook on the discovery of the seashells in the East Repository (Mackenzie DM/DB, 1903: vol. ii, p. 87, 28 May). Evans, although more interested in finds, did not mention the shells in his notes. In the original publication, however, he wrote "The shells from the Repository, bushels of which were taken out, were the ordinary seashells of the neighbouring coast, many varieties being included, though cockles were the most abundant" (Evans 1902–1903: 43). In his illustrations, about 100 shells are visible (Evans 1902–1903: 42, figs. 21, 63; 1921: figs. 377–378), while 183 were put on display in the Heraklion Museum (HM), together with the faience figures. There are many more shells on display in the HM today.

However, neither the shells illustrated nor those on display can account for the "bushels" alluded to by Evans. This made Panagiotaki search for the missing shells in the HM storerooms. Two wooden boxes containing shells were located but were classified as "uncatalogued"; the label on the boxes stated that they came from Knossos in 1904 and were originally stored in Case 64 in HM. Panagiotaki treated these shells as the missing material from the TR for the following reasons:

(a) Many of the shells in the boxes are painted in the same way as the known ones (through the published photographs of Evans) from the TR.
(b) Not all species recorded by Evans (1902–1903: 43) appear in his published photographs, although they are included among the uncatalogued shells together with many more.
(c) Together with the shells were a small number of faience beads of the same shape, texture and colour as those from the TR.
(d) Case 64, referred to as the previous storage place of the shells, contained almost exclusively objects recovered from the TR; the year 1904 noted on the label of the wooden boxes may refer to the year of arrival at the HM and not the year of excavation (note that this was the case in other instances as well; moreover, in the excavations of 1904 at Knossos, no large numbers of shells were recorded by the excavators).

The Latin names of the shells published by Evans were made by Walter Frank Raphael Weldon (1860–1906), who in

1903/04 was a professor of evolutionary biology at University College, London. Weldon examined the shells "although not all specimens could be identified because of the 'worn state of the valves'" (Evans 1902–1903: 43). In his paper, Evans mentioned that "The following is a list of the varieties found: *Dolium galea, Trochus lineatus, Cardium edule, Pectunculus glycimeris, Spondylus gaederopus, Venus verrucosa, Venus multilamellata, Mactra stultorum, Tellina* (worn), tube of Serpulid worm and piece of an Echinid". The list of Evans, compared to the present study, is far from complete; this can be explained by the fact that it is not possible for a non-specialist to distinguish many of the species, which suggests that Evans must have submitted to the specialist only those specimens which could be easily distinguished by a non-specialist. The study of the whole collection of the seashells recovered by Evans from the TR was undertaken by the zoologist Eleftheriou; a list of the shells and a brief discussion was provided in the publication of the TR material by Panagiotaki (1999: 128–31, 152–55), who recognised the two-fold importance of the shells: biological and archaeological. She concentrated on their archaeological importance and fully discussed their possible functions. The present study concentrates on the biological importance of the shells. It presents the whole collection in which even more species are revealed, and a full discussion of the shells and their habitat.

4.2 The Shell Material

The material kept in the HM (on display and in wooden boxes) consists of a large mixture of shells and also one sea urchin or echinoid test. The initial sorting separated entire or slightly damaged specimens and those which were in a relatively good state of preservation from the broken shells which were still identifiable and the small fragments which generally could not be identified. The shell material examined is representative of the typical eastern Mediterranean fauna of the coastal areas. It includes several benthic species which belong to distinct assemblages in which bivalves and gastropods are important components. They occupy the medio-littoral and sub-littoral/infra-littoral zones of the coast, inhabiting fine sediments as well as the splash zone of hard substrata. Molluscs could also have been collected from deeper water using specific tools, fishing techniques and facilities (Karali 1999). However, dead shells are thrown up by wave action and litter the strand zone of all beaches where they can be collected as empty shells. Severe storms (as in the winter of 1992) cause mass mortalities of molluscs, dislodging them from their natural environment and washing them up alive on the shoreline, thus increasing the numbers of available dead shells.

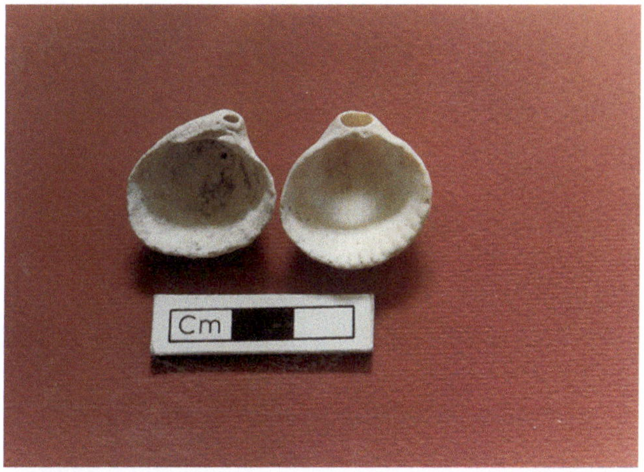

Fig. 4.2 Two water-worn valves naturally holed at the umbo (*Cardites* on left; *Cerastoderma* on right. (Photograph by N. Panagiotakis)

The preservation of shells also varies. A large number look worn, but they exhibit no traces of any obvious use (such as rubbing or scraping); it is more likely that they were abraded on the shore. Furthermore, some shells display distinct holes at the umbo area, caused by long-term abrasion against the sand and pebbles of the seashore (Fig. 4.2). Quite a number of the shells are pierced on different parts of their valves (Fig. 4.3). The perforations were created by a predatory gastropod (*Naticarius*), seven of which were found among the rest of the shell material. A large range of species which share the same habitat as *Naticarius* were affected in a similar way.

Some of the shells (228) were burnt, with 94 badly burnt (Fig. 4.4), thus excluding the possibility that they were roasted in order to be consumed. The burning is more likely to have been caused by fire in the place where the shells were "stored". It is therefore very possible that they were caught in the fire which destroyed the shrine in which the TR objects were kept. This burning is also obvious in other objects recovered from the TR. Summarised results of the shells examined are given below and provide information on the number of each species, the intact and also the damaged valves, the valves with holes, the burnt valves, as well as the number of painted specimens. The total of the identifiable material—excluding fragments less than 5 mm—amounted to 6392 (including the shells from the museum display), with 6249 bivalves and 143 gastropods.

4.2.1 The Bivalves

The bivalves include 25 species in 15 families. No attempt was made to use the valve pairing technique for estimating the number of individuals as the state of preservation of colour, hinges and hinge teeth was rather poor.

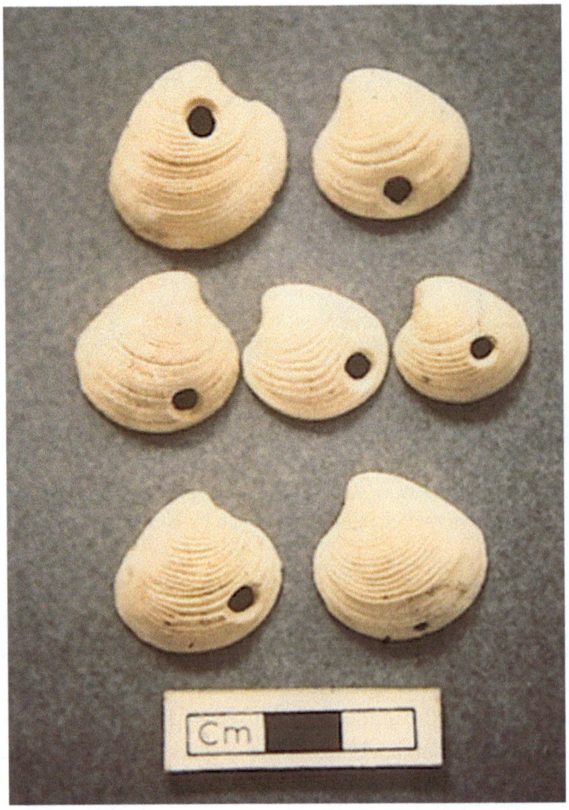

Fig. 4.3 *Chamelea gallina valves* with perforations made by predatory gastropod *(Naticarius)* activity. (Photograph by N. Panagiotakis)

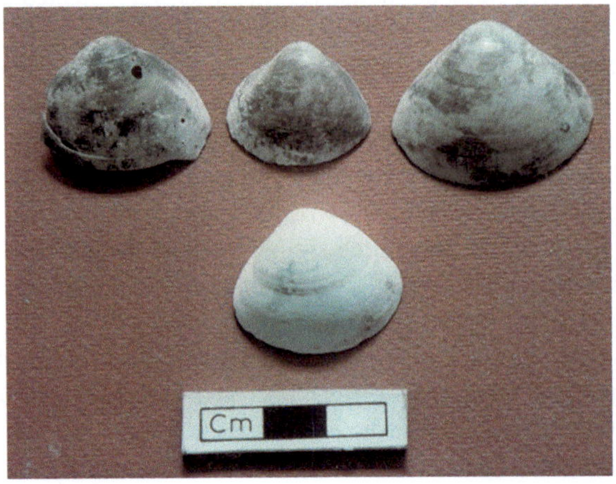

Fig. 4.4 *Mactra* valves with traces of burning. (Photograph by N. Panagiotakis)

The bivalves are:

1523 *Mactra stultorum* (24.37%; 751 valves [32 painted {Figs. 4.4, 4.5, and 4.6}], 745 broken, 27 holed [Fig. 4.5 shows man-made holes])—trough shell.

1297 *Acanthocardia tuberculata* (20.75%; 993 valves [210 painted {Figs. 4.9 and 4.10}, 289 broken, 15 holed)—(red-nosed) cockle.

991 *Glycymeris nummaria* (15.85%; 774 valves [33 painted], 139 broken, 78 holed) (WFRW/AE as *Petunculus glycimeris*; Panagiotaki 1999: 128 as *G. insubricus*)—dog cockle.

740 *Chamelea gallina* (11.84%; 703 valves [255 painted {Fig. 4.7}], 13 broken, 24 holed (Panagiotaki 1999: 128 as *C. gallina gallina*)—venus shell.

616 *Venus verrucosa* (9.85%; 538 valves [1 painted], 41 broken, 37 holed) (WFRW/AE as *V. multilamellata*)—venus shell.

508 *Loripes orbiculatus* (431 valves [22 painted], 23 broken, 54 holed) (Panagiotaki 1999: 128 as *L. lacteus*)—lucinid clam.

143 *Spondylus gaederopus* (120 valves, 23 broken)—spiny or thorny oyster.

107 *Glycymeris glycymeris* (85 valves [22 painted], 7 broken, 15 holed) (WFRW/AE as *Petunculus glycimeris*)—dog cockle.

84 *Donax trunculus* (82 valves, 2 broken)—wedge shell (this is possibly the *Tellina* of WFRW/ AE).

70 *Cardites antiquatus* (66 valves, 4 holed) (Panagiotaki 1999: 128 as *Venericardia antiquata*)—cardite.

63 *Callista chione* (46 valves [6 painted], 17 broken)—smooth clam.

23 *Barbatia barbata* (22 valves, 1 broken)—ark shell.

20 *Cerastoderma glaucum* (20 valves) (WFRW/AE as *Cardium edule*)—(common) cockle.

15 *Dosinia lupinus* (15 valves) (Panagiotaki 1999: 128 as *Dosinia exoletus*)—smooth artemis.

10 *Ostrea edulis* (10 valves) (Panagiotaki 1999: 128 as *Ostrea* sp.)—oyster.

7 *Spisula subtruncata* (6 valves, 1 broken) (not in Panagiotaki 1999: 128)—trough shell.

6 *Arca noae* (5 valves, 1 broken)—ark shell.

5 *Cardita calyculata* (5 valves)—cardite.

4 *Ruditapes decussatus* (3 valves [3 painted], 1 broken) (Panagiotaki 1999: 128 as *Venerupis decussatus*)—carpet shell.

4 *Anomia ephippium* (4 valves, 4 holed) (Panagiotaki 1999: 128 as Anomiidae)—jingle shell or saddle oyster.

4 *Lima lima* (4 valves)—file clam.

3 *Chama gryphoides* (3 valves)—jewel box.

3 *Mimachlamys varia* (2 valves, 1 broken) (Panagiotaki 1999: 128 as *Chlamys varia*)—(variegated) scallop.

2 *Abra alba* (1 valve, 1 broken)—white furrow shell.

1 *Chamelea striatula* (valve) (Panagiotaki 1999: 128 as *C. gallina striatula*)—venus shell.

As mentioned above, there are a number of bivalves which were pierced at different parts of the valve, some on the umbo area, which was due to predation by the predatory gastropod *Naticarius*. The holes, perfectly symmetrical with bevelled

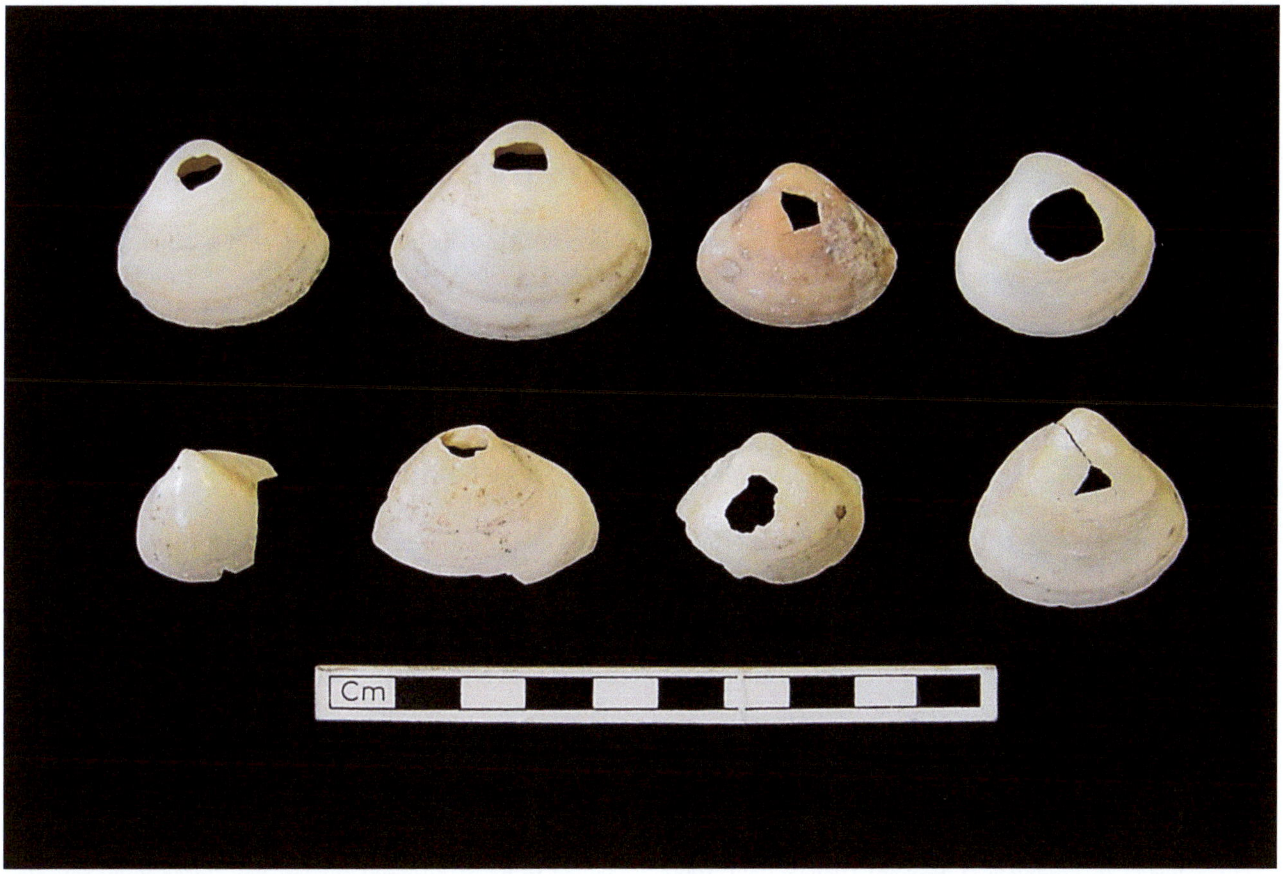

Fig. 4.5 *Mactra* valves with man-made perforations. (Photograph by M. Kallergis)

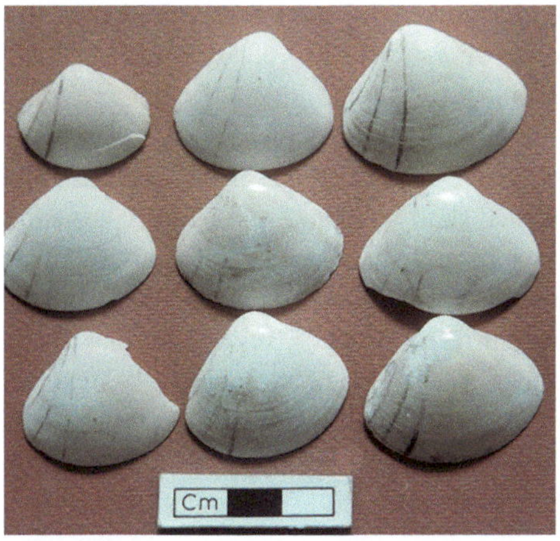

Fig. 4.6 *Mactra* valves with painted vertical lines. (Photograph by N. Panagiotakis)

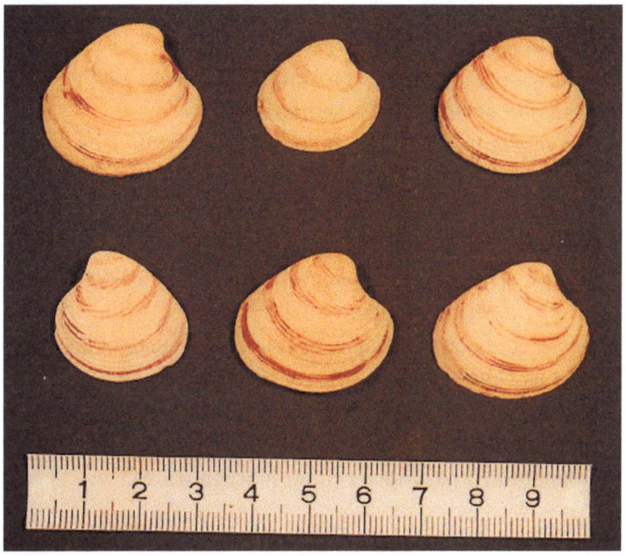

Fig. 4.7 *Chamelea gallina* valves with red concentric bands. (Photograph by M. Kallergis, IMBC)

sides, conform with the radula action of the gastropod. Several valves of *Mactra*, *Acanthocardia*, *Glycymeris*, *Chamelea* (Fig. 4.3) and *Loripes* bear similar holes, some of which were in various stages of completion. However, it should be kept in mind that predation by octopus both on bivalves and gastropods produces similar holes (Hanlon and Messenger 1996). Besides these naturally occurring holes, on a small number of species, mainly *Mactra* (Fig. 4.5) are holes of imperfect contours, cracks and scratches with clear evidence of unsuccessful human attempts to drill regular holes in the valves.

4.2.2 The Gastropods

Gastropods are present in relatively small numbers, with 13 species belonging to 11 families. The total number of shells amounted to 124, to which nineteen unidentified specimens have been added, amounting to a total of 143 specimens. Ten species occurred with less than ten specimens each. The related top shells *Phorcus turbinatus* and *P. articulatus* were numerically dominant followed by smaller numbers of *Patella.*

The gastropods are:

46 *Phorcus turbinatus* (46 complete, 44 painted black) (Panagiotaki 1999: 128 as *Monodonta turbinata*; WFRW/AE as *Trochus lineatus* [now *Phorcus lineatus*, and probably misidentified])—top shell.
18 *Phorcus articulatus* (18 complete, 18 painted black) (Panagiotaki 1999: 128 as *Monodonta articulata*)—top shell.
15 *Thylacodes arenarius* (14 broken, 1 calcified) (Panagiotaki 1999: 128 as *Vermetus arenarius*)—vermetid, worm shell (could also be *Vermetus triqetrus*) (this is the "tube of Serpulid worm" of WFRW/AE).
12 *Patella caerulea* (11 complete, 1 broken)—limpet.
9 *Zonaria pyrum* (8 complete, 1 broken)—cowrie (Panagiotaki 1999: 128 as Cypraeidae)—(pear) cowrie.
7 *Luria lurida* (6 complete, 1 broken) (Panagiotaki 1999: 128 as Cypraeidae)—(lurid) cowrie.
7 *Naticarius stercusmuscarum* (6 complete [1 with black staining, 1 holed on body], 1 broken) (Panagiotaki 1999: 128 as *N. millepunctatus*)—moon shell.
3 *Semicassis granulata* (3 broken) (Panagiotaki 1999: 128 as *Phalium granulatum*)—helmet shell.
2 *Columbella rustica* (2 complete)—dove shell.
2 *Tritia gibbosula* (2 complete [1 holed on ventral]) (Panagiotaki 1999: 128 as *Nassarius gibbosulus*)—nassa or basket shell.
1 *Conus ventricosus* (broken) (Panagiotaki 1999: 128 as *C. mediterraneus*)—cone shell.
1 *Tonna galea* (apical fragment calcified) (Panagiotaki 1999: 128 as *Dolium galea*)—tun or cask shell.
1 *Hexaplex trunculus* (open body and broken mouth) (Panagiotaki 1999: 128 as *Murex trunculus*)—murex.
19 unidentified gastropods (from museum case).

The number of gastropods indicates that they were not a favoured form. Most specimens were in a good state of preservation with only a small number of broken shells. Fragments were tentatively identified to genera or species, while the *Thylacodes*/vermetids cannot be counted with any confidence due to their tubes being broken up. Some specimens were discoloured or calcified but, with the exception of *Naticarius* shells with evidence of manmade holes, the rest of the material lacked holes. On the whole, paint was not applied to the gastropods. The exceptions were on *Phorcus,* where black paint was applied to the whorl of the shell.

4.2.3 The Echinoderm

The deposit also included 13 sea urchin test (body) fragments (Fig. 4.12):

1 *Paracentrotus lividus* (Panagiotaki 1999: 128 as *Paracentrotus* sp.)—sea urchin or echinoid.

4.3 The Painted Shells

A total of 647 painted shells (of which 584 were bivalves and 63 gastropods) was recorded in the TR collection, including shells in the museum exhibit. Many of the species are painted in the manner described by Evans (1921: 519) "streaked and banded with brilliant artificial tints – crimson, venetian red, orange, brown, green and black – tastefully applied in unison with the natural lines and hues". The paint is applied only on the dorsal surface, in red, green, and black only, which agrees with Mackenzie's notes (DM/DB, 1903: vol. ii, p. 87, 28 May "... red, green, black"). At first glance, the paint seemed to be of different shades, but closer study revealed that the exact shade depends on its state of preservation, being darker where it is better preserved. The orange colour mentioned by Evans may have been mistaken with the natural horizontal orange striations of a few *Acanthocardia* valves which are in a very good state of preservation, and thus vividly coloured. All of the *Phorcus* were also painted black. One *Naticarius* exhibits some black staining.

The painted shells present a variety of types: a few such as *Mactra* have been given just three black slightly curved lines from the umbo to the periphery of the shell only on half of their dorsal surface (Fig. 4.6), suggesting that the decoration of the shells may not be complete. A few others are completely black. Usually, the paint follows the striations of the shell—obviously an easier way to apply it. On *Chamelea,* for instance, horizontal concentric bands (of almost the same width) appear from the umbo to the widest part of the shell in red or in green (5–6 bands), following the natural concentric growth lines of the shell (Fig. 4.7). The same arrangement can be observed in *Dosinia* (Fig. 4.8). *Acanthocardia*, on the other hand, exhibits vertical lines and bands, once again following the natural vertical rib lines of the shell. This type of shell presents a variety of designs which have all been very skilfully executed; either the shell is entirely painted in green, or only the depressions are painted alternately, again in green (Fig. 4.9); alternatively, red paint is applied in the depressions

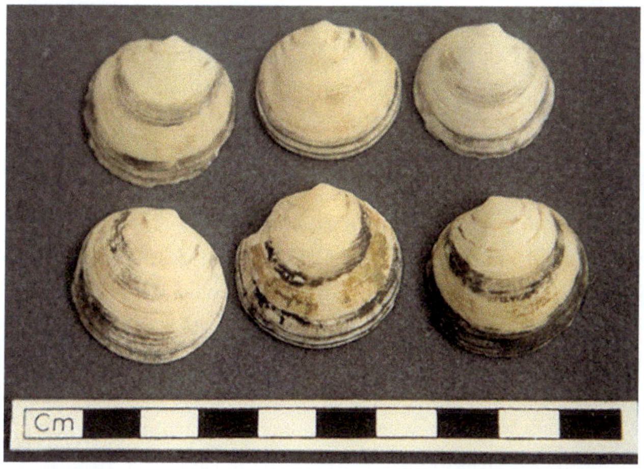

Fig. 4.8 *Dosinia* with black concentric bands. (Photograph by N. Panagiotakis)

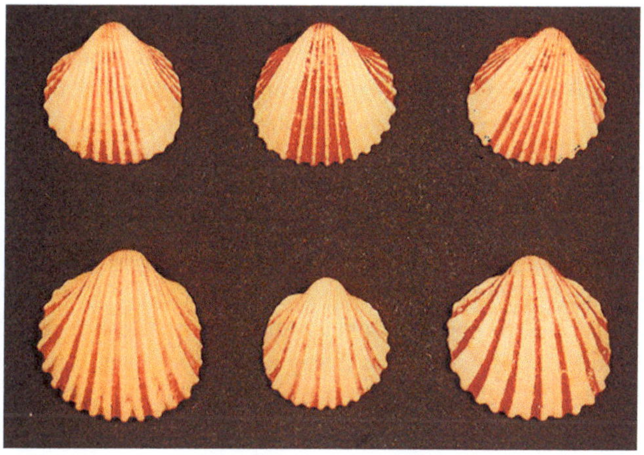

Fig. 4.10 *Acanthocardia* valves with different painted designs. (Photograph by M. Kallergis, IMBC)

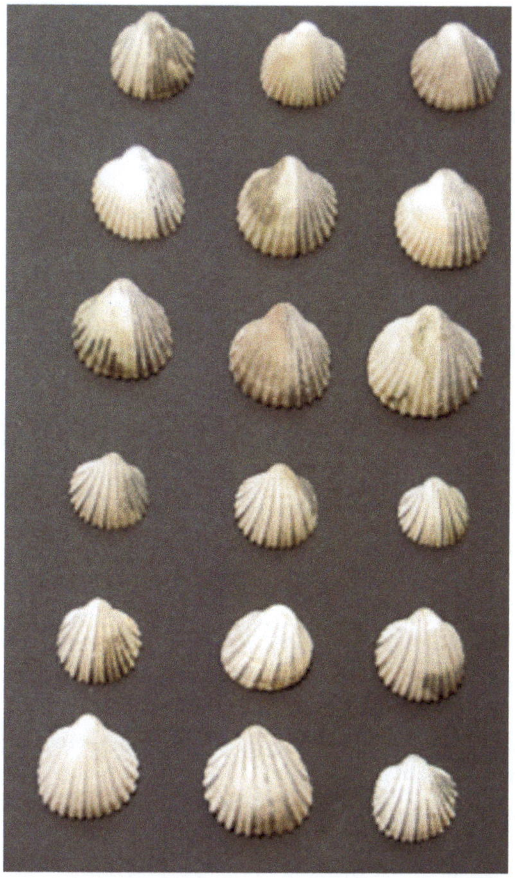

Fig. 4.9 *Acanthocardia* valves with green vertical lines. (Photograph by M. Kallergis, IMBC)

alternately, or a tripartite arrangement is created by a rather wide central band (following the natural vertical depressions of the shell) flanked by two slighter ones (Fig. 4.10). The uniformity of style and paint of the shells points to the work of a single person; it is, therefore, highly possible that they were all painted by one person over a relatively short period of time (Panagiotaki 1999: 129). It is impossible to answer the question why they painted shells that have a natural colour of their own. Were the colours chosen, red, green and black, more symbolic than the natural hues of the shells? Or did they suit the taste and the general aesthetics of the Minoans? The designs created on some of the shells certainly show the 'delicate' and 'fresh' Minoan aesthetics (on the colours used in Minoan paintings, Peters 2008; on the colours on Minoan faience, Tite et al. 2009).

The inside of the valves is never painted, with one exception: one *Glycymeris* valve has traces of red paint at the distal part of its ventral surface, as if it was used as a small vessel to carry the paint—a kind of palette (see below). Painted marine invertebrates are rarely found. Neolithic Phaestos produced a *Charonia nodifera* (trumpet shell) with red ochre staining on the exterior and with an open apex: it is thought to be a trumpet (Levi 1957–1958: fig. 201d [no. 1872]; Vagnetti 1975: 95, fig. 31:3). Also of note is a fossil bivalve from the MM III–LM IA antechamber deposit of the Poros tomb which had some red paint on it (Reese 1992: 180–81).

4.3.1 Analytical Work on the Pigments

Permit for analytical work on the pigments applied to the shells was granted by the Greek Ministry of Culture and the HM to Panagiotaki, who arranged in 1990 for their analysis at the Laboratory of Archaeometry of the National Centre for Scientific Research "Demokritos", conducted by Drs Yannis Maniatis and Maria Seremetaki (see Appendix 4.1, online supplementary material). Six painted samples were submitted: one valve of *Chamelea gallina* decorated with four red horizontal bands on its dorsal surface (Sample 1); an *Acanthocardia* fragment decorated with red vertical bands (Sample 2); about half a valve of an *Acanthocardia* decorated with vertical bands in green and black (Sample 3); half of a valve of *C. gallina* (slightly burnt) and decorated with four black horizontal bands (Sample 5); and two fragments (*C. gallina* and *Mactra*), both decorated with black (Sample 6). A *Glycymeris* valve was also submitted in order to analyse its content: traces of red pigment inside its ventral surface (almost up to the rim) (Sample 4). The sampled shells were returned to the HM collection (by Panagiotaki) after the specks of paint needed for the analytical work were obtained.

It should be pointed out that some of the shells, especially Samples 1 and 2, looked as if they were covered with a white powder (over the paint). The analysis showed that the white powder was calcium coming from the shell itself. The analytical work on the red, green and black pigments showed that they were all natural pigments: the red is hematite, the green is malachite and the black is charcoal (carbon) black. Hematite is native to Crete, malachite, on the other hand, must have been imported, possibly from Laurion on the Greek mainland, Egypt, or the Timna Valley of Israel.

Apart from the above shells, three more complete shells without any colour were also submitted to Demokritos to be dated using the conventional Carbon-14 method. Two of the samples, after being treated mechanically and chemically to remove surface depositions and mineral alterations, did not produce enough material for analysis. The third sample did produce a date; however, the calibration with the marine curve produces a high degree of uncertainty and this together with the lack of knowledge of the local reservoir constant and the original location of the shells make the result not useful for a precise dating of the specific context of Knossos. Hence, it is not reported.

4.4 Comparative Assemblages

4.4.1 Contemporary Shells from Elsewhere at Knossos

It may be useful to compare the TR shells with the other MM III–LM I shells found at Knossos to indicate just how unique the numbers and species of shells are in the TR.

The Royal Road North at Knossos (excavated 1958–1961) produced one MM III shell (*Spondylus* upper valve) and 17 LM I shells: 8 *Spondylus* valves (2 articulating from one deposit, 1 collected dead), 4 *Hexaplex* (from three deposits), 2 *Antalis dentalis* (tusk shell; 2 worn), 1 *Pinna* fragment, 1 *Tritia* (water-worn, holed), and 1 *Tonna* (small). The Knossos Royal Road South produced three MM III shells from two deposits: 1 *Spondylus* (collected dead), 1 *Bolinus brandaris* (murex), and 1 *Charonia variegata* (trumpet shell; body fragment). The MM IIA–LM IA produced 1 *Spondylus* (collected dead) and the LM IA 1 *Spondylus* (fresh) (Reese analysis).

At the Stratigraphic Museum Site at Knossos (1979–1982 excavations) there are 10 MM III shells from ten deposits: 3 *Patella* (2 gigantic), 2 *Glycymeris* (1 water-worn), 2 *Ostrea*, 2 *Phorcus*, 1 *Spondylus* (broken) (Reese analysis).

From the MM IIIB–LM IA at the SM site there are 27 shells from 18 deposits: 5 *Patella* (four deposits), 2 *Spondylus* (two deposits: water-worn, broken; upper valve), 2 *Cerastoderma* (two deposits: right valve, small; broken), 2 *Ostrea* (2 deposits: 1 collected dead, broken), 2 *Hexaplex* (1 worn; 1 broken), 2 *Charonia* (two deposits: lip fragment; body fragment, very small individual), 2 *Steromphala varia* (two deposits), 1 *Mactra* (right valve), 1 *Glycymeris* (water-worn), 1 *Donax*, 1 *Dosinia* (left valve), 1 *Ruditapes* (fragment), 1 *Columbella* (lip/body fragment), 1 *Pisania striata* (apex fragment), 1 *Tarantinaea lignaria*, and 2 *Antalis* (two deposits, 1 broken) (Reese analysis).

At the Unexplored Mansion at Knossos (excavated 1971–1972), there is 1 MM III–LM IA *Conus* (ground-down labial side and holed) from Room N and 1 MM III–LM II *Conus* (ground-down labial side and holed) from Room H (Reese analysis).

4.4.2 Large Collections from Other Minoan Sites

It may also be useful to compare the size of the TR shell collection (6392) with the other MM III–LM I large samples from other contemporary Greek sites. It should be noted that these are all collections of edible marine shells.

The five largest Kommos MM III–LM I shell deposits are:

- MM III–LM I, Space 42, floor and fill, South Area: 1130 *Patella* (Reese 1995: 253).
- LM I, Room 7b, Central Hillside: 855 *Patella* (Reese 1995: 253).
- LM IA Early; Building T Room 19, floor, SA (Trench 53A/ pails 44–45,50 + 62D/94): 496 *Patella* + 28 others (Reese 1995: 253; Ruscillo 2006: 825–26).
- LM I, Space 21, Building T, fill, SA: 437 *Patella* (Reese 1995: 253).
- LM IA Final; Building T Room 22 west end, final use, SA (52A/51–52 + 56A1/102–103): 180 *Patella* + 3 others (Reese analysis; Ruscillo 2006: 827).

For Pseira, the Block AF, Room AF 3A/B, produced three relevant samples:

- Mixed MM–LM IA debris from collapse of building (Unit AF 3B-1 + 2 + 2 SE): 109 *Patella*, 8 *Phorcus* + 24 others (Reese 2009: 136).
- LM IA floor level (Unit AF 3B-3): 96 *Patella*, 13 *Columbella*, 9 *Phorcus* + 21 others (Reese 2009: 136).

- LM IA floor level (Unit AF 3B-4): 133 *Patella*, 6 *Phorcus* + 7 others (Reese 2009: 136).

On the Greek mainland, the largest relevant collection is a MH II-III/LH I floor deposit from Ayios Stephanos (Laconia) of 155 *Donax* individuals (not valves) (Reese 2008: table 25.3 [Nu 2 {1974} pails 25,28–29, 31–37]).

4.5 Environment and Shell Collection

4.5.1 The Physical and Biotic Environment of Crete

The physiography of the Cretan coastline has been shaped over the millennia by important changes in sea level as well as sudden tectonic events. Nevertheless, the island acquired its present outline since the Pleistocene. Since Minoan times, the coastline has been affected only by small eustatic changes which resulted in the subsidence of the coastline up to 1.5–2 m, as evidenced in the submerged Minoan port facilities (Manolioudis 2013) in the Anisara area. Erosion processes caused by ancient rivers washed down alluvial material into the sea, material which was deposited in the shallows of the fringing shore. These soft sediments in the littoral and sub-littoral have a rich and varied marine fauna of which molluscs form communities of considerable extent and complexity (Eleftheriou et al. 1993). Many species of bivalves burrow in the well-sorted sediments, where they extract nourishment from organic particles and plankton. Some species are epibenthic, attached to hard substrata (*Spondylus*, *Barbatia*, *Lima*) apart from the free-living *Mimachlamys* which can be found in different substrata. Gastropods occupy a variety of habitats, some on hard substrata, feeding on algae and detritus, while some are carnivorous or carrion feeders (*Naticarius*, *Semicassis*, *Conus*, *Hexaplex*, *Tritia*) which prey on the bivalve communities.

From the climatic evidence which shows temperature uniformity over the last thousands of years in the eastern Mediterranean, it can be confidently deduced that there have been no significant changes in the overall composition of the marine fauna in Cretan waters, and although there is some evidence of lower abundance of certain species, there is no evidence of species replacement. Yet the present-day sparsity of pectinids and the absence of oysters from Cretan waters suggest possible recent changes in the ecosystem and productivity of the Cretan Sea. It is nonetheless relevant to mention that the size of the species of molluscs in Minoan times, both bivalves and gastropods, is similar to the size found in present day fauna.

4.5.2 'Fishing' Zones

The molluscs from the TR are common throughout the northern shores of Crete, occurring principally in shallow water. The majority of bivalves are found in the shallow fine sediments of the medio-littoral and infra-littoral zones, while the gastropods are confined to the hard substrata and are found mainly in the supra-littoral/medio-littoral zones of the rocky outcrops. Both groups are accessible and can be collected by hand or by means of a simple implement for digging them out or by scraping hard surfaces. However, a number of molluscs occur in deeper water and these require the use of appropriate fishing tools, or they can be accessed by free diving.

4.5.3 Seasons of Collection

Molluscs can be harvested all year round, in order to satisfy consumer demands. This can be verified by their presence in markets which specialise in shellfish sales. However, since molluscs are basically micro-feeders, feeding on plankton and organic particles, they reach the peak of their condition in growth and gonad development with the onset of the seasonal increase of plankton production, from early spring to the summer months. Thus, many groups of molluscs (such as pectinids, mytilids, ostreids, venerids, etc.) are primarily harvested and marketed when molluscs are in their best condition.

4.5.4 Tools Used to Collect the Shells

Crete is an island with substantial marine resources. Considering the very close contact of its modern inhabitants with the sea, it would be unthinkable that the Minoan people did not equally have such a close relationship. In the HM there is ample evidence in the rich iconography on vessels, vases and receptacles as well as a collection of metal hooks which testify to the close bond of the Minoans with the sea. These include scenes with fishermen, fishing by line and hook, or using spears as well as catching tuna with shore nets, or catching fish and gastropods by traps or pots.

Unfortunately, we do not have any information about the equivalent of contemporary modern techniques and tools such as dredges or grabs for collecting shellfish. However, as most of the molluscs occur in shallow water, we surmise that there would have been no need of specialised equipment as they could be collected by hand or shallow diving.

4.6 Reflections on the Shells from the Temple Repositories

The question is how the shells found their way to the palace. Most of the identified species are edible (apart from 16 cowries, 15 vermetids, 7 *Naticarius*, 2 *Columbella*, 2 *Tritia*, 1 *Conus*). Consequently, some may have been collected alive and were then consumed, while others seemed to have been dead in the sea long enough to allow marine invertebrates to live on the inside of the valves (Fig. 4.11) or to become water and beach worn. Many may have been collected for their significance as seashells and perhaps their symbolic value.

The next question to be addressed concerns the identity of those who brought them to the palace. Regardless of how the shells were taken to the palace, can it be surmised that they were offerings to the shrine by fishermen to ensure a good catch? Were the shells painted by the same people who collected them? Were they subsequently painted at the shrine for decorative or religious purposes? The faience beads present exactly the same problem as, like the shells, they seem to have been made by the same hand or school. Was (a) the shrine/palace workshop responsible for their making or (b) a private town workshop? Were they made to order, or made and sold to pious people to offer in the shrine? Only one thing is clear: the destination for both the faience beads and the shells was the palace shrine (Panagiotaki 1999: 130).

What was the meaning of the shells in the shrine? To answer this question, one has to consider similar finds in the Minoan/Aegean world. Shells are found in every site from the Neolithic onwards in Crete, on other Aegean islands as well as on the mainland. Whether natural or man-made, in various materials such as faience or clay, they were used as ornaments, objects of adornment (necklaces, pendants, etc.)

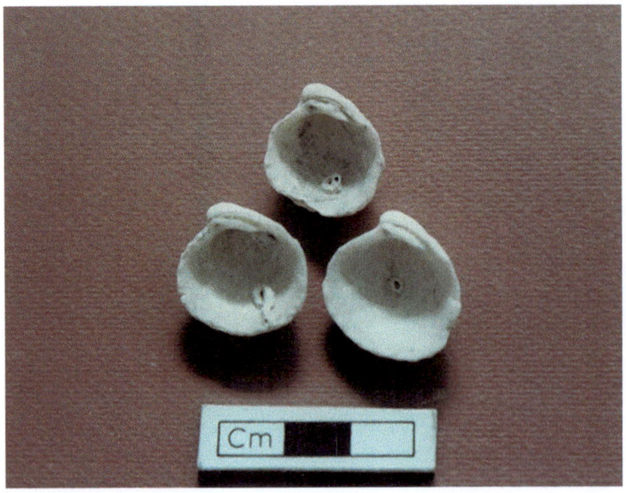

Fig. 4.11 *Chamelea* with marine incrustations inside the valves. (Photograph by N. Panagiotakis)

(Reese 1984; Karali 1999). However, with the exception of the painted shells, the remaining shell material found in the sacred areas connected with objects with religious connotations, assumes religious significance and could have been used as votive offerings. Similarly, the shells deposited in the Knossos TR were also found in votive contexts (Juktas peak sanctuary, Kato Syme sanctuary and in tombs at Knossos, Poros, Archanes; Reese analyses).

Evans suggested that the shells from the TR had decorated the floor or the altar of the shrine, in much the same way that the Double Axe shrine altar was decorated with water-worn pebbles (Evans 1902–1903: 43). However, since the majority of the shells could not withstand the weight of footsteps (*Glycymeris* and *Chamelea* may have), they would have been quite unsuitable as a floor cover. In order to form part of the floor decorations, they would also have to have been embedded in plaster or cement and there is no trace of such materials on the shells. They could on the other hand have decorated the altar of the shrine or they could have been incorporated into panels, probably attached to them by means of glue which has not survived. The perforated ones may have been used in garlands in combination with the faience beads (Panagiotaki 1999: 130); a similar situation is evident in a much later shrine (Archaic-Hellenistic) at Kythnos, where cockles still kept part of their suspension wires in gold, silver or bronze (Theodoropoulou 2013: 202).

The *Glycymeris* valve with traces of red paint in its interior considered by Panagiotaki "as a small vessel to carry the paint: a kind of palette" (1999: 129) should be given more thought. This particular valve was not the largest in the collection.

4.7 Other Marine Objects in the Temple Repositories

4.7.1 Fish Vertebrae

In a photograph not published by Evans (published in Panagiotaki 1993: 54i fig. B; 1999: 119, 172, pl. 17) almost 50 fish vertebrae are displayed at the most central spot, in front of the marble cross and between the two faience "snake goddesses". They are combined with the skull of a weasel (*Mustela nivalis*) and arranged in a way that recalls a curled snake.

Seven fish vertebrae were identified in the HM storerooms among the TR objects (published by Panagiotaki 1999: 119, 172). The largest one, also visible in Evans's photograph, belongs to a porbeagle shark, *Lamma nassa* (D 29.7 mm [*Lamna cornubica* is written on the bone, the old name of this species]). Thirty-nine smaller fish vertebrae (and some fragments) were found among the marine shells in storage (Fig. 4.12). The presence of fish vertebrae in the TR is

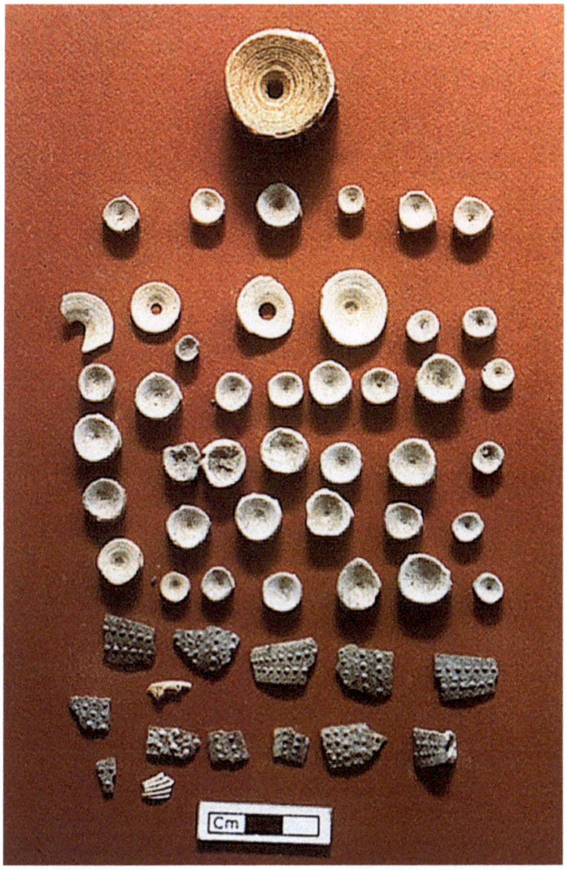

Fig. 4.12 Fish vertebrae and *Paracentrotus* fragments. (Photograph by N. Panogiotakis)

not surprising, considering that the marine items recovered from them far exceed in numbers all the other objects.

4.7.2 Faience Marine Objects

Among the many different kinds of objects recovered from the TR were marine ones made of faience. They include fish and, most importantly, imitations of shells (Panagiotaki 1993: 66–68; 1999: 78–81, 104). They can be identified as (a) flyingfish (*Hirundichthys rondeletii*, the blackwing flyingfish or *Exocoetus volitans*, the blue flyingfish or tropical two-wing flyingfish); (b) argonauts made in the round; (c) cockle shells also in the round, and (d) rocks.

Evans presented the marine faience items on a panel (Evans 1902–1903: figs. 46, 63; 1921: figs. 377, 379), because one of the rock fragments was in a shape that could fit in the corner of a panel. However, the argonauts and cockles, being three-dimensional, were meant to be free-standing. They could thus either decorate the surface of a piece of furniture or they were part of a game, especially the cockles (Panagiotaki 1999: 80, 104). Note that Evans had identified among the various faience plaques in the TR some of geometric shape that may have been part of a game board (Evans 1921: 481, fig. 344a; Panagiotaki 1999: 88–90, 104), analogous to that identified at Mycenae (Evans 1921: 483, fig. 346).

One aspect that should be stressed is that the fine shell of an argonaut is created to house the eggs of the cephalopod. The presence of seven argonauts in the TR certainly adds to the fertility aspect, a theme obvious in the TR in other media. If the different size and portrayal of the argonauts suggest difference in age, then again the TR items point to difference in age in other media as well—the snake figurines themselves can be identified as older and younger ones (Panagiotaki 1999: 98, 104).

The presence of these marine objects in the TR certainly matches the themes evident in the majority of the materials recovered: (a) display, evident not only in the marine faience items, but also in the rock crystal, the ivory and most importantly the faience "snake goddesses" themselves; (b) the natural world, evident in the animal plaques with mother and young goats and cows, in flowers and fruits; the gentle and nurturing side of nature is present. In fact, all the cosmos is present in the TR: the land, the sea and the air (the air is evident in the marble cross, if correctly interpreted as an astral symbol and the "sun" symbol made of rock crystal and precious metals) (Panagiotaki 1999: 149). In all this, the sea is more emphatically represented in the faience marine objects, the sealings depicting fish and shells (Panagiotaki 1998, 1999; Shapland 2022) and the thousands of actual shells. This emphasis on the sea, on the marine life, may point to the respect, if not the veneration, of the sea by the Minoans (Panagiotaki 1999: 104). One of the five or six faience "goddesses" may have been responsible and cared for the sea. And as already pointed out, the shells may have been offerings by pious people (even children/youngsters) as a non-blood sacrifice, a "smaller-scale offering" (Panagiotaki 1999: 149). The "myriad painted seashells and beads could be interpreted" as small-scale offerings brought to the shrine in a ceremony along the central court with court ladies participating and religious emblems carried in display, with the "astral disc . . ." "to attract the light and dazzle the people present and thus announce perhaps the Goddess's epiphany" (Panagiotaki 1999: 150–51).

4.8 Conclusions

In conclusion, we can reflect upon the importance of these shells which could be two-fold: (a) archaeological, and (b) biological/ecological.

(a) Their archaeological importance lies in the fact that they were not retrieved as food debris scattered all over the site; on the contrary, they were recovered from a closed deposit where they were deliberately placed. They accompanied the faience figures and objects of art, arranged at the bottom of the East Repository. They should thus be seen as directly connected with the shrine of the palace to which these objects belonged. After the destruction of the shrine, the material (broken or well preserved) was carefully collected, obviously out of respect, and deliberately arranged inside the TR (Panagiotaki 1999: 150). The fact that among the carefully collected items there were so many natural shells (forming in fact the largest collection), shows the value of these items to the Minoans and the importance of the sea to an island people, perhaps even veneration of the sea. The fact that they painted some of the shells shows not only the love of the Minoans for colour, but also the deeper implications and religious or magical connotations of colour.

(b) Regarding the biological/ecological importance of the shells, it is worth commenting on the resilience of this fauna over the millennia. Evans observed that the seashells were of types that one can still see on the island shores. Recent studies of the marine fauna of the Cretan seas (Eleftheriou and Smith 1993; Karakassis and Eleftheriou 1997) point to the same conclusion. Furthermore, excavations at many sites on Crete and elsewhere (Reese 1987; Wilkens 1996) have underlined the universality of this fauna, with the skeletal remains of molluscs being identical with the shells from the TR.

Despite the lack of information about conditions in the Minoan seas, it is contemplated that some changes might be discernible. Having accepted that the shell fauna has remained more or less the same, it would be important to examine the additional biotic indices of the molluscs from the TR and the contemporary populations. However, it is important to know whether and to what extent the eastern Mediterranean was more productive in ancient times than it is today. In those days, the eastern Mediterranean was fed by important riverine systems (the Nile, and many small rivers in the Aegean area and Asia Minor) whose inputs would be responsible for a higher production of the marine ecosystem of the whole area. However, it should be added that no available data exist that could support such an assertion.

Furthermore, assessment of the state of richness of the seas in Minoan times, by measuring the shell size of the specimens of the two populations, was not appropriate due to the insufficient number of specimens. Nevertheless, an overall comparison of the sizes of a small number of species (*Mactra*, *Acanthocardia*, *Loripes*, *Phorcus*) did not reveal any significant differences over time between the two populations.

Acknowledgements Anastasios Eleftheriou would like to thank Margaret Eleftheriou, his wife, for her support and patience during the writing of this paper, to Manolis Kallergis (Institute of Marine Biology of Crete) for taking many of the TR photographs used here, and to Dr. Vivian Mara of the University of Crete for the sorting of a very large amount of shell material. Marina Panagiotaki is grateful to the British School at Athens and the Greek Ministry of Culture for permitting the study of the material recovered by Sir Arthur Evans in the Central Palace Sanctuary area at Knossos and to the Heraklion Museum for providing the facilities for the study of the shells and the archaeologist Nikos Panagiotakis for the photographs of the shells.

Bibliography

Eleftheriou, A. & Smith, C. 1993. Preliminary Investigations of the Benthic Ecosystem from the Aegean Shelf (Eastern Mediterranean). In: N. Della Croce, ed. *Symposium Mediterranean Seas 2000*. Istitito Scienze Ambentali Marina. Genova: University of Genova, pp. 105–17.

Eleftheriou, A., Smith, C. & Tselepidis, A. 1993. *Food Chains in the Aegean Sea*. NATO, SFS Programme, Final Report. Heraklion: IMBC.

Evans, A. 1902–1903. The Palace of Knossos. *The Annual of the British School at Athens*, 9: 1–153.

Evans, A. 1921. *The Palace of Minos at Knossos* 1 *The Neolithic, Early and Middle Minoan Ages*. London: McMillan.

Hanlon, R.T. & Messenger, J.B. 1996. *Cephalopod Behaviour*. Cambridge: Cambridge University Press.

Karakassis, I. & Eleftheriou, A. 1997. The Continental Shelf of Crete: The Structure of Macrobenthic Communities. *Marine Ecology Progress Series*, 160: 185–96.

Karali, L. 1999. *Shells in Aegean Prehistory*. BAR International Series 761. Oxford: Archaeopress.

Levi, D. 1957–1958. Gli scavi a Festòs nel 1956 e 1957. *Annuario della Scuola Archeologica*, XXXV–XXXVI: 193–361.

Manolioudis, S. 2013. *From the Ancient Quarries to the Monuments of Civilisation*. Heraklion (privately published).

Panagiotaki, M. 1993. The Temple Repositories of Knossos: New Information from the Unpublished Notes of Sir Arthur Evans. *The Annual of the British School at Athens*, 88: 49–91.

Panagiotaki, M. 1998. Dating the Temple Repositories Vases. *The Annual of the British School at Athens*, 93: 188–98.

Panagiotaki, M. 1999. *The Central Palace Sanctuary at Knossos*. BSA Suppl. Vol. 31. London: The British School at Athens.

Peters, M. 2008. Colour Use and Symbolism in Bronze Age Crete: Exploring Social and Technological Relationships. In: C.M. Jackson & E.C. Wager, eds. *Vitreous Materials in the Late Bronze Age Aegean*. Sheffield Studies in Aegean Archaeology Vol. 9. Oxbow: Oxford, pp. 187–208.

Reese, D.S. 1984. Topshell Rings in the Aegean Bronze Age. *The Annual of the British School at Athens*, 79: 237–38.

Reese, D.S. 1987. The EM IIA Shells from Knossos with Comments on Neolithic to EM IIIA Shell Utilisation. *The Annual of the British School at Athens*, 82: 207–11.

Reese, D.S. 1992. Fauna from the Poros Tomb. In: P. Muhly, ed. *Minoikos laxeftos tafos ston Poro Herakleio (Anaskafis 1967)*. Vivliothiki tis Arkeologikis Eterias, no. 129. Athens: Archaeological Society, pp. 180–81.

Reese, D.S. 1995. The Marine Invertebrates. In J.W. Shaw & M.C. Shaw, eds. *Kommos* I. *The Kommos Region, Ecology, and Minoan Industries* 1. Princeton: Princeton University Press, pp. 240–73.

Reese, D.S. 2008. The Marine and Fresh-water Invertebrates, 1973–77. In: W.D. Taylour & R. Janko, eds. *Ayios Stephanos: Excavations at a Bronze Age and Medieval Settlement in Southern Laconia*. CD 79–84. BSA Suppl. Vol. 44. London: British School at Athens, pp. 519–24.

Reese, D.S. 2009. Faunal Remains from Block AF. In: Philip P. Betancourt et al., eds. *Pseira X. The Excavation of Block AF*. Prehistory Monographs 28. Philadelphia: INSTAP Academic Press, pp. 131–42.

Ruscillo, D. 2006. Faunal Remains and *Murex* Dye Production. In: J.W. Shaw & M.C. Shaw, eds. *Kommos* V. *The Monumental Minoan Buildings at Kommos*. Princeton: Princeton University Press, pp. 776–840.

Shapland, A. 2022. *Human-Animal Relations in Bronze Age Crete: A History through Objects*. Cambridge: Cambridge University Press.

Theodoropoulou, T. 2013. The Sea in the Temple? Shells, Fish and Corals from the Sanctuary of the Ancient Town of Kythnos and Other Marine Stories of Cult. In: G. Ekroth & J. Wallensten, eds. *Bones, Behaviour and Belief. The Zooarchaeological Evidence as a Source for Ritual Practice in Ancient Greece and Beyond*. ActaAth-4°, 55. Stockholm: Skrifter utgivna av Svenska Institutet i Athen, pp. 197–222.

Tite, M.S., Maniatis, Y., Kavoussanaki, D., Panagiotaki, M., Shortland, A.J. & Kirk, S.F. 2009. Colour in Minoan Faience. *Journal of Archaeological Science,* 36: 370–78.

Vagnetti, L. 1975. L'Insediamento Neolitico di Festos. *Annuario della Scuola Archeologica,* L-LI (1972–73): 7–138.

Wilkens, B. 1996. The Fauna from Italian Excavations on Crete. In: D.S. Reese, ed. *The Pleistocene and Holocene Fauna of Crete and Its First Settlers*. Madison (WI): Prehistory Press, pp. 241–62.

"ἐς πόντον ἰχθυόεντα": Marine Creatures in Aegean and Cypriot Pictorial Pottery at the End of the Late Bronze Age

5

Anna Lekka

Abstract

This paper examines iconography of marine creatures on Late Bronze Age pictorial pottery from the Aegean and Cyprus. The figurative motifs of Aegean and Cypriot pictorial pottery at the end of the Late Bronze Age derive mainly from the animal kingdom and include marine creatures such as fish, octopuses and seahorses. Fish represent one of the most common themes of Aegean Late Helladic IIIC pottery and Cypriot Pictorial White Painted Wheelmade III ware. Mycenaean pictorial motifs and local practices contributed to the establishment of a fish iconography on Cyprus.

Abbreviations

AA	Archälogischer Anzeiger
AnatSt	Anatolian Studies
ASAtene	Annuario della Scuola Archaeologica di Atene e delle Missioni Italiane In Oriente
AURA	The Athens University Review of Archaeology
BSA	Annual of the British School of Athens
BCH	Bulletin de Correspondance Hellénique
LH	Late Helladic
LM	Late Minoan
SIMA	Studies in Mediterranean Archaeology

A. Lekka (✉)
Hellenic Ministry of Culture, Directorate of Documentation and Protection of Cultural Goods, Department of Private Archaeological Collections and Antique Shops, Athens, Greece
e-mail: alekka@culture.gr

5.1 Introduction

There is a great variety of fish representations in Aegean art already from the Early Bronze Age. During the Late Bronze Age, decoration consisting of marine creatures is frequently found in wall paintings and is the dominant decorative theme on Minoan pottery constituting a distinct style, the so-called Marine Style (Mountjoy 1974, 1977, 1985; Hatzaki 2011). In addition to pottery, portrayal of marine creatures is also found in other categories such as seals and even floor decoration (Gill 1985).

Fish representations become more popular during the Late Helladic III period (LH III, c. 1400–1090 BCE) in Mainland Greece.[1] At the end of the Late Bronze Age, common cultural elements and different social and economic conditions resulted in regional styles in the Aegean, such as a late marine style that combines the Octopus and the Close styles, with the former being dominant in the central Aegean and the latter in the Argolid. On Cyprus, the Late Cypriot IIIA (LC IIIA, c. 1200–1100 BCE) pictorial pottery,[2] which succeeds the Pastoral style of the thirteenth century BCE, where representations of bovids are pre-dominant, includes a variety of motifs and is richer in subject matter and more complex and ambiguous in terms of presenting ideas. Workshops, like the one at Hala Sultan Tekke, combine motifs, images, and symbols to produce compositions with an underlying mythological or ritual significance. During this period, in the twelfth century BCE, scenes with marine creatures are more common.

[1] For the discussion on the absolute chronology of the Late Helladic period, see Mountjoy, 1999: 16. For a comparison of the chronology of the Late Bronze Age and corresponding periods in the eastern Mediterranean, see Lekka 2022: tabl. 1.

[2] For the pictorial pottery of twelfth century BCE from Cyprus, see Lekka 2015.

C. Tsouparopoulou, L. Recht (eds.), *Human and Aquatic Beings: Interactions in and beyond the Eastern Mediterranean (3rd–1st Millennia BCE)*, Themes in Contemporary Archaeology, https://doi.org/10.1007/978-3-031-73643-8_5

5.2 The Motifs

The motifs usually appear isolated, while there are also attempts at narration, as can be seen in examples of fishing scenes. Both in the Cypriot examples and in the Aegean ones, the bodies of the animals are decorated in a similar manner with various types of linear patterns.

5.2.1 Octopus

On Cyprus, there are no octopus representations. In contrast, in the Aegean area, octopus decoration on pottery, which appears in the early phases of the Late Bronze Age, continues to be used and becomes popular during the LH IIIC period (c. 1200–1090 BCE), giving its name to the so-called Octopus style (Michimasa 2006). The stirrup jars with octopus decoration found in the cemeteries of Naxos, Aplomata (Kardara 1977) and Kamini (Vlachopoulos 2008), constitute part of a category of ceramics the development and dispersion of which in the mature phase of LH IIIC in the Aegean indicates the creation of a new cultural *koine* with Attica, the Cyclades and the Dodecanese playing a leading role, while there are also ties with Cyprus and the eastern Mediterranean. The octopus motif matches well with the globular shape of the body of the stirrup jar, with its tentacles spreading over the surface of the vessel, while other marine creatures decorate the space between the tentacles. Octopus decoration can also be found on other types of vases such as kylikes and kraters (Güntner 2000: pl. 72–82; Michimasa 2006: 263–79).

5.2.2 Seahorse

On Cyprus, schematic seahorses decorate the homonymous 'Seahorse Krater' from Hala Sultan Tekke (Åström 1988). The lozenge-shaped body of the seahorse has a hooked upper end for the head and a curved lower part for the tail. The famous seahorses on the interior of the LH IIIC kalathos from Tomb 14 at Langada in Kos are clearly naturalistic, rendered with a completely different combination of shapes (Morricone 1965–1966: 118, fig. 100–101). In contrast, the Hala Sultan Tekke examples are so highly stylised that they could also be complementary linear motifs.

5.2.3 Crab

An example of a representation of crabs (FM 8:10)[3] comes from Kalymnos (Mountjoy 1999: 1134, no. 19, fig. 465). It is a stirrup jar where solidly painted crabs are depicted along with other marine and terrestrial creatures between octopus tentacles.

5.2.4 Other Marine Creatures

A unique, to my knowledge, fish representation occurs on a Vapheio-type cup from Maa-*Palaeokastro* (Karageorghis and Demas 1988: 329–330, Pls. CLXXV, XLIX, LXXVII) (Fig. 5.1). The solidly painted fish is positioned vertically and features large, crescent-shaped fins on either side of the head. It appears to be attacking the bird pictured directly above it. The scene is unusual, unnatural, and generally problematic. The motif itself looks more like a gastropod—cuttlefish or squid—rather than a fish. In fact, it may be attacking the fish depicted directly below rather than the bird, which may be represented flying overhead in a 'bird's eye' perspective. The fins/tentacles, which point up instead of down, appear to support this view. In my opinion, it looks very much like *clarias* and *heterobranchus*, a species of fish with mixed characteristics of cuttlefish and fish with slender body, flat head and a broad mouth with pairs of barbels.[4]

5.2.5 Fish

This motif (FM 20), which probably originated in Nilotic scenes, is one of the most common themes of LH IIIC on pottery, more common than birds in the Aegean, though not on Cyprus. Fish decorate a great number of vases from the Greek mainland and the Aegean, which mainly belong to the Close style. They are more frequently found on the interior surface of kalathoi, the shoulder of stirrup jars, and between the tentacles of octopuses in the Octopus Style. They also occur as supplementary motifs on kraters and other shapes of the Close style.

Most of the body of LH IIIC fish is covered with parallel, horizontal, straight and wavy lines. Alternating filled-in sections and bands occur rarely. They often appear combined with other animals. A popular combination is the fish and bird, either as separate motifs or in scenes of waterfowl chasing fish.[5] For example, on vases from Aplomata in Naxos dating to the LH IIIC period, fish are rendered in three ways: by a single or double outline, in silhouette, and as a combination of both techniques (Kardara 1977: 64). Gills are rendered by two or four concentric arcs and fins either in outline or silhouette. The body is decorated with horizontal

[3] FM is the standard abbreviation for Furumark Motif, see Furumark 1972.

[4] For *clariidae* see Brewer and Friedman 1989: 60–63.

[5] Depictions of the fish-bird pattern occur in Mesopotamia as early as the fourth millennium BCE (Mackay 1931: 259, 263). It is also a very popular decorative combination represented on Bichrome ware of the Syro-Palestinian coast during the Bronze Age (Amiran 1969: fig. 140), while it is one of the most characteristic decorative patterns in Nilotic landscapes in the frescoes of Amarna (Mekhitarian 1989: Pl. 6).

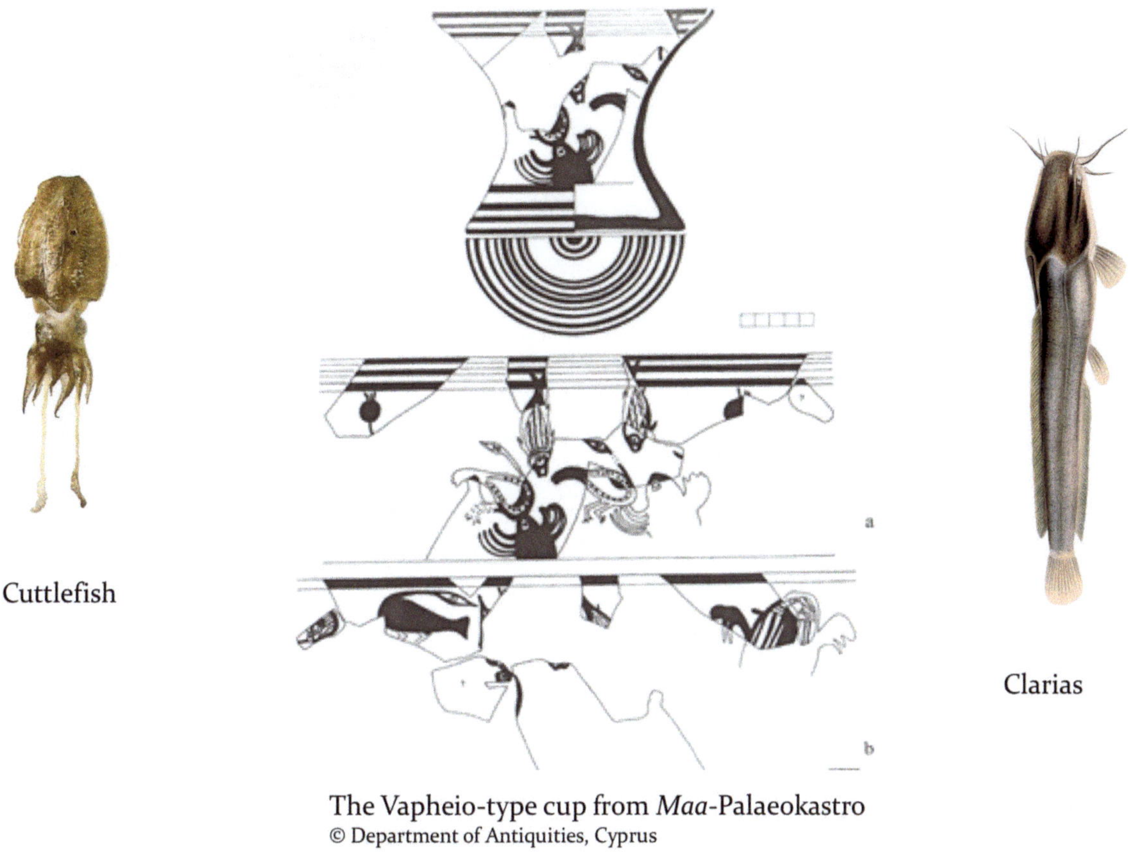

Fig. 5.1 The Vapheio-type cup from Maa-*Palaeokastro*

wavy lines or with vertical superimposed arched lines. The tail is either triangular or split.

In the Pictorial style of the twelfth century on Cyprus, the majority of fish motifs come from Maa-*Palaeokastro* (Fig. 5.2). The rendering of the scales varies, usually with wavy lines or short strokes and dots. Concentric arcs denote the gills, dots indicate the eyes, and the snout is pointed. The fins are either triangular or rendered with oblique strokes. Because of the stylised, conventional rendering, it is not possible to identify the species.

The body of the fish is decorated in a variety of ways: wavy lines, dashed lines, vertical or oblique lines, random short strokes, dots, cross-hatching, while some have solid-painted body. There are many examples of wavy line (FM 20: 12) decoration from various sites on Cyprus: Enkomi (Dikaios 1969–1970: 845, 849, 309:280, 309, 312, 313, pl. 81:26; Coche de la Ferté 1951: 24, pl. III.5), Hala Sultan Tekke (Åström 1998: 63, fig. 106), Potamia (Vermeule and Karageorghis 1982: 68, 208–9, VI.62), Alassa (Hadjisavvas 1991: 178), Maa-*Palaeokastro* (Karageorghis and Demas 1988: 329–30, pls. CLXXV, XLIX, LXXVII), and Palaepaphos (Maier and von Wartburg 1986: 184, fig. 67a-b).

In Mainland Greece, wavy lines are used to decorate the body of fish from LH IIIA onwards. LH IIIC examples come from Mycenae (Sakellarakis 1992: 105, no. 232, 233), Tiryns (Güntner 2000: 295, pl. 65:3), Midea (Demakopoulou 2006: 42, fig. 29left) and Asine (Mountjoy 1999: 188, no. 448, fig. 59) in the Argolid, Perati in Attica (Iakovidis 1969–1970: B, 140, nos. 229, 232, 248, 569, 679, 746, 909, 1210, fig. 21: α-γ), Kos (Mountjoy 1999: 1113, 1115, no. 135, fig. 456), Aplomata (Kardara 1977: 64, fig. 25:2, 4, 7, nos. 943, 912, 934), Kamini (Mountjoy 1999: 949, no. 37, fig. 387) and Grotta in Naxos (Vlachopoulos 2003: 227, fig. 10; Vlachopoulos 2008: fig. 43.13:b), from the Mycenaean well on the Athenian Acropolis (Broneer 1939: fig. 37:c), and several fragments from Lefkandi (Crouwel 2006: pl. 66: F1, F9, F10).

Equally characteristic are the specimens depicted on Late Minoan III (LM III) larnakes (Kanta 1980: pls. 65–66). Very similar fish decorate pottery from the Syro-Palestinian coast, the shoulder of a Philistine jug with birds from Tell ʻAitun (Dothan 1982: 109, 198, fig. 3:9, 10, 12:2, pl. I:6), and pottery from Tarsus in Cilicia (Goldman 1956, pl. 335:1330).

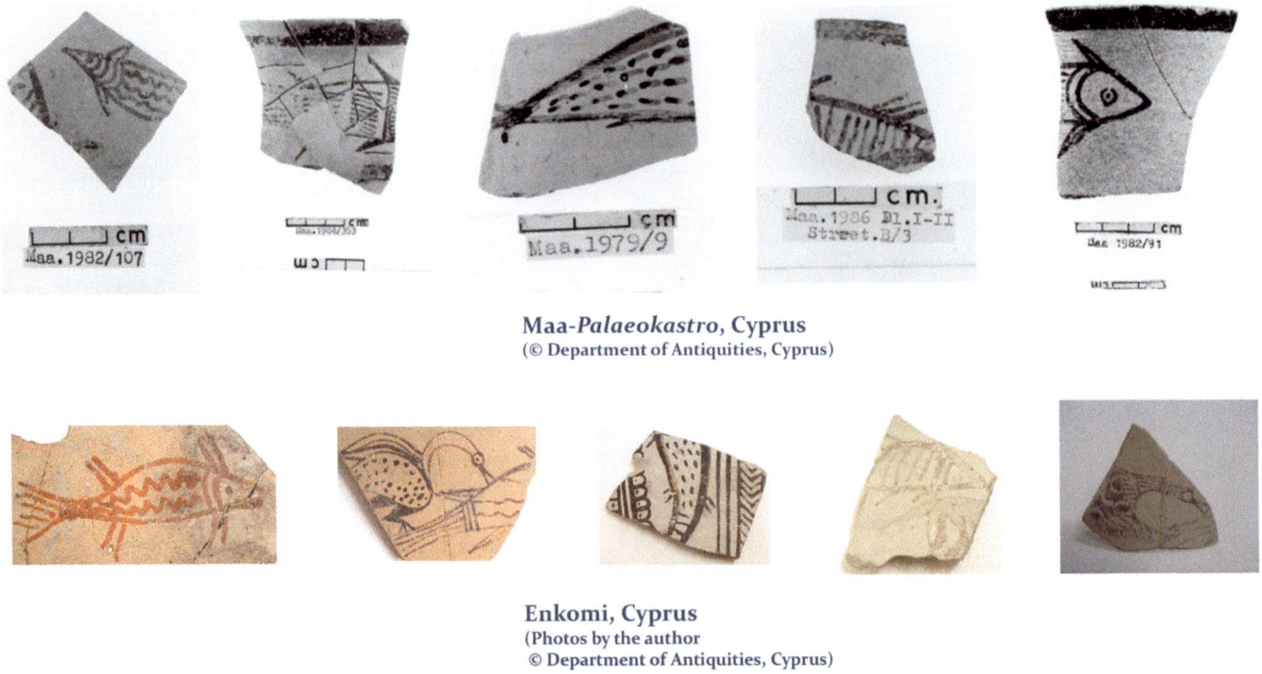

Fig. 5.2 Representations of fish on Cypriot pictorial pottery of the twelfth century BCE

Dashed lines adorn fish on two bowl fragments from Maa-*Palaeokastro* (Karageorghis and Demas 1988: 104, 129, pl. XLIII, LXXVII, CLXX, CXCII). A similar fish decorates a krater fragment from Mycenae's acropolis (Vermeule and Karageorghis 1982: X.88). Two examples from Maa-*Palaeokastro* (Karageorghis and Demas 1988: 138, 147, pls. XCVI, CVIII), the krater B1003 from Kourion-Bamboula (Benson 1960: 64, pls. I, III) and two examples from Kition (Karageorghis 1981: 4, nos. 20, 26, pls. II.20, III.26) are decorated with vertical or oblique lines. A similar fish adorns a kalathos from the Athenian Acropolis (Sakellarakis 1992: 107, no. 240), a stirrup jar from Aplomata Naxos (Kardara 1977: 11–12, pl. 2) and a bowl from Kos (Morricone, 1972–1973: 382, fig. 380a). Güntner groups them together in his Rough Drawing Style (Güntner 2006: 60–61). Contrary to the previous examples, where the lines are carelessly rendered, the horizontal lines decorating the body and the split tail on the fragment of a closed vessel from Hala Sultan Tekke are carefully executed by a master vase painter (Åström 1998: fig. 106), as are also the horizontal lines on the rear of a fish's body from Enkomi (Dikaios 1969–1970: 845, pl. 303: 303, 313).

Dots are a development of a more naturalistic rendering of fish skin (FM 20:1) from LH IIIA2 onwards. They may have been used to render the skin of fish such as *scarus* or *pagrus* (bluespotted seabream, *stiktofangri* in Greek). The scales are rendered with small and large dots.

On Cyprus, a fish of this type is depicted on the 'Horned God Krater' from Hala Sultan Tekke (Fischer 2011: 80–82, fig. 15). Similar fish appear on the reverse side of the 'Creature Krater' (Fischer 2013: 56, fig. 8b), on another krater (Åström 1998: 63, fig. 106) and on a shallow bowl (Fischer 2011: 93, fig. 3:7) from Hala Sultan Tekke. On both sides of the ring-based krater from Kition, similar fish are arranged vertically (Karageorghis 1977). This arrangement recalls the stirrup jar from the cemetery at Palaiokastro in Arcadia (Mountjoy 1999: pl. 1: a). Other examples are the fragments of a closed vase from Kition (Karageorghis 1981: 5, no. 35, pls. III.35, X.28), the Vapheio-type cup from Maa-*Palaeokastro* (Karageorghis and Demas 1988: 329–30, pls. CLXXV, XLIX) and the large fish on the shoulder of a strainer jug from *Pano Mandilaris* at Alassa (Hadjisavvas 1991: 178, fig. 17.5). A similar fish, only larger, with thick outline and small, pointed snout decorates the fragment of an open vase from Tarsus (Goldman 1956: 226, no. 1332, fig. 335).

Fish with cross-hatched bodies are uncommon. Cross-hatched decoration is also an attempt to render the scales naturalistically. Examples include the fish on a bowl from Maa-*Palaeokastro* (Karageorghis and Demas 1988: 199, pls. CXLIX, CXVIII) and an almost identical example is one of the three fish that decorate the fragments of a ring-based krater from Kition (Karageorghis 1981: 8, no. 25, pls. V.25, XII.17).

Cross-hatching decorates the fish on a LH IIIC krater fragment from the acropolis at Mycenae (Sakellarakis 1992: 109, no. 243) and fragments from Kynos-Livanates (Dakoronia 2006: fig. 2), and Phocis (Mountjoy 1999: 783, fig. 309: 256). A more naturalistic example is depicted on the interior of a LH IIIC Middle shallow bowl also from the Mycenae acropolis (Mountjoy 1999: 173–74, no. 369, fig. 49). A solidly painted, yet naturalistically rendered fish decorates the shoulder of an early LH IIIC stirrup jar from Rhodes (Mountjoy 1999: 1053, no. 192), examples from Kos (Mountjoy 1999: 1119, fig. 458, 461). Another example is found on the four-handled krater from the late phase at Enkomi (Courtois 1971: 268–72, fig. 105–106).

5.3 Fishing

Since this chapter examines images of marine creatures, it would also be useful to discuss marine habits. The observation of the landscape and dietary habits are the sources of inspiration for fishing scenes.[6] The Egyptians used various fishing methods, illustrated in tomb scenes, drawings, and papyrus documents (Brewer and Friedman 1989). Representations in frescoes and miniatures[7] reflect a developing fishing technology. In the tomb of two royal servants, Niankhkhnum and Khumhotep, at Beni Hasan of the Fifth Dynasty (2494–2345 BCE), Khumhotep, portrayed on a heroic scale on a small papyrus boat, is spearing two fish with his harpoon (Feucht 1992). Other scenes in Egyptian frescoes represent fishing nets, like the ones in Mereruka's[8] Tomb in Saqqara of the Sixth Dynasty (2345–2181 BCE) and in the Tomb of Urarna of the Second Intermediate Period (1782–1570 BCE) (Brewer and Friedman 1989: 24–26, 42, fig. 2.32). The rich variety of marine motifs on Minoan frescoes and pottery implies that the Minoans also had a deep familiarity with marine life.

In the Aegean, depictions of group fishing are found in the iconography of the LH IIIC period. A very characteristic example is found on a strainer jug from the LH IIIC cemetery at Aplomata in Naxos (Hadjianastasiou 1996). On the so called 'Fishermen's Jug', four fish are depicted moving in pairs towards a net. On the other side of the vase there are six fishermen behind two groups of fish, surrounded by a fishing net rendered by a line. Another example of fishing with a net is depicted on fragments of a krater from Kynos-Livanates (Dakoronia and Kounouklas 2015: 38–39, fig. 9) and a scene of a fisherman on a krater from Kalapodi in Phthiotis (Mountjoy 1999: 815, no. 28, fig. 325).

On a Vapheio-type cup from Miletus, fish are depicted in combination with tongue-shaped and vertical motifs which probably represent nets or other fishing tools (Mountjoy 1998: fig. 3:5; Mountjoy 2006: 112–13, fig. 4), a scene reminiscent of the fish and nets depicted on a Middle Minoan II Kamares ware pithos from Phaistos (Betancourt 1985: pl. 11F).

Fish, as an important source of protein, is attested to have been part of the Cypriot diet in the Late Bronze Age and fish from the Nile were also imported at this time (Karageorghis 2007: 258).[9] Many types of fish[10] have been identified among the excavated material from Enkomi, Kition Athienou and Hala Sultan Tekke, where a significant number of fishing tools was also discovered (Reese 1984, 2003, 2008; Åström 1989: 204; Lernau 2018).[11] The types of fish found in excavations give us an idea of the fishing methods (Fig. 5.3). The vertical heraldic arrangement of a pair of fish depicted on a krater from Potamia is of great interest (Vermeule and Karageorghis 1982: 68, 208–209, no.VI62). The fish are depicted on either side of a vertical stemmed chevron which probably renders a fishing mullein (*euphorbia characias*). The inhabitants of the coasts of Asia Minor knew the properties of *euphorbia characias* and used it as a kind of chemical fishing method (Φανουρίου 2013). Noteworthy is the prevalence of fish decoration on pottery from the west coast of Cyprus (Maa-*Palaeokastro*), whereas depictions of birds prevail on the east coast (Enkomi, Kition). Large-scale fishing of species like mackerel and tuna[12] took advantage of seasonal migration and these preferences are probably related to fisheries in the west and crossings of migratory birds in the east. Each spring, millions of tunafish entered the Mediterranean in search of warm waters (Fagan 2017: 163). Lagoons, such as the one in Larnaca, were valuable fishing grounds, especially on stormy days, when many fish would try to find shelter.

References to sea life and fishing are found in Homer even though Homeric heroes are described as eating primarily meat and not fish (Berdowski 2008) (Table 5.1).

[6]For fishing methods see e.g., Buchholz et al. 1973; Powell 1992, 1996; Fagan 2017; Shapland 2022. For fish and fishing in the Near East, see Potts 2012.

[7]Cl. Two boats pulling a seine. Tomb of Meketre, Sixth Dynasty, Egyptian Museum, Cairo (Brewer and Friedman 1989: 46, fig. 2.42)

[8]Vizier of the king Teti I (2345–2333 BCE), see Clayton 2006: 64.

[9]For fish exploitation and preservation in the eastern Mediterranean, see Zohar and Artzy 2019.

[10]For instance: sea bass, drum fish, meagre, red mullet, seabream, parrotfish, grey mullet, Nile perch, shark, barracuda, mackerel.

[11]Personal communication with David Reese 27.01.2020, 05.06.2020, 29.09.2021, 14,12,2021.

[12]For fishing of tuna in the Aegean, see Mylona 2021.

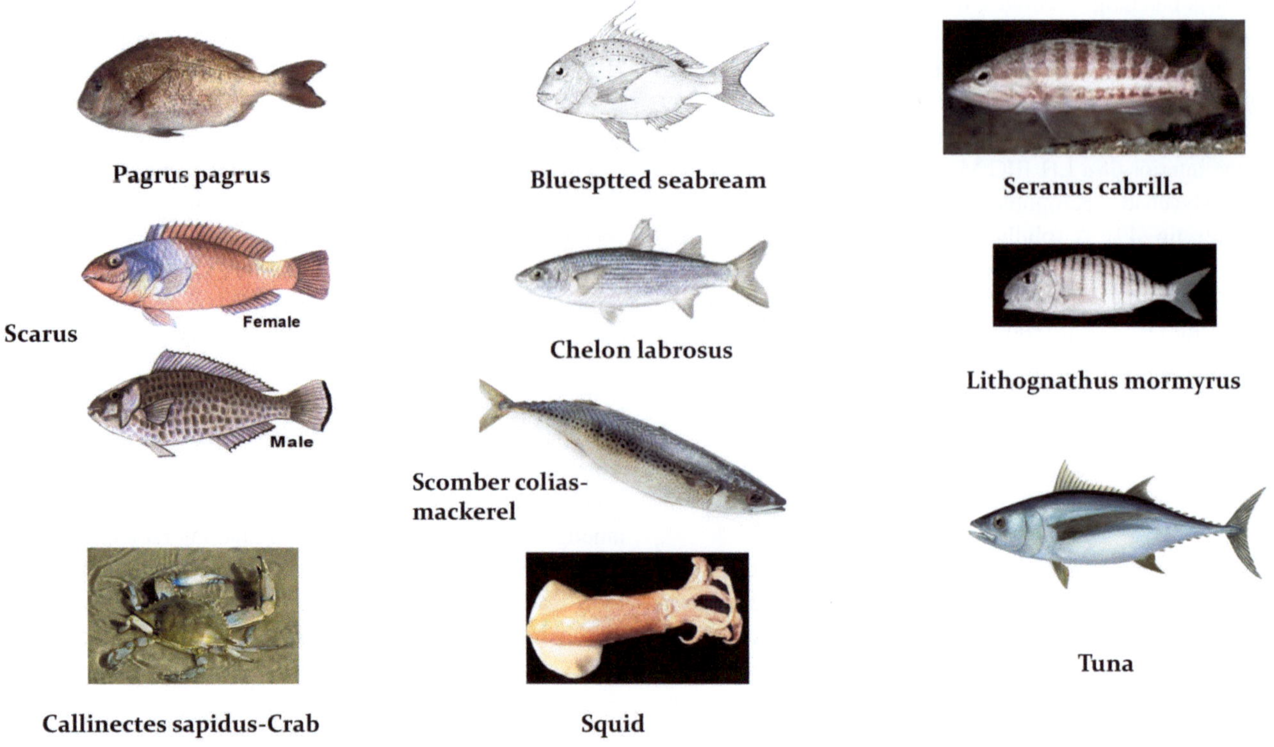

From the official page of the European Commission https://fish-commercial-names.ec.europa.eu/fish-names/home_en. Accessed 22 Nov 2022

Fig. 5.3 Fish and molluscs of the Eastern Mediterranean

Table 5.1 Selected references to sea life and fishing in Homer

"..On to Pieria he stepped from the upper air, and swooped down upon the sea, and then sped over the wave like a bird, the cormorant, which in quest of fish over the dread gulfs of the unresting sea wets its thick plumage in the brine.." (Odyssey, Rhapsody 5, 50–53) "..But if I swim on yet further in hope to find shelving beaches[a] and harbors of the sea, I fear me lest the storm-wind may catch me up again, and bear me, groaning heavily, over the teeming deep; or lest some god may even send forth upon me some great monster from out the sea—and many such does glorious Amphitrite breed." (Odyssey, Rhapsody 5, 417–22)
"..and just as, when a cuttlefish is dragged from its hole, many pebbles cling to its suckers, even so from his strong hands [435] were bits of skin stripped off against the rocks.." (Odyssey, Rhapsody 5, 431–33)
"..and as a fisher on a jutting rock, when he casts in his baits as a snare to the little fishes, with his long pole lets down into the sea the horn of an ox of the steading, and then as he catches a fish flings it writhing ashore.." (Odyssey, Rhapsody 12, 251–54)
"..and spearing them like fishes they bore them home, a loathly meal..." (Odyssey, Rhapsody 10, 123–24)
"..Now so long as my men had grain and red wine they kept their hands from the kine, for they were eager to save their lives.[b] But when all the stores had been consumed from out the ship, and now they must needs roam about in search of game, fishes, and fowl, and whatever might come to their hands—fishing with bent hooks, for hunger pinched their bellies—then I went apart up the island that I might pray to the gods in the hope that one of them might show me a way to go..." (Odyssey, Rhapsody 12, 327–34)
"..and the sea yields fish.." (Odyssey, Rhapsody 19, 113)
(Homer. The Odyssey with an English Translation by A.T. Murray, PH.D. in two volumes. Cambridge, MA., Harvard University Press; London, William Heinemann, Ltd. 1919)

[a]https://www.perseus.tufts.edu/hopper/text?doc=Perseus:text:1999.01.0136:book=5:card=408
[b]https://www.perseus.tufts.edu/hopper/text?doc=Perseus:text:1999.01.0136:book=12:card=327

5.4 Symbolism

Although the interpretation of the images starts from the environment and its relation to human activity and dietary habits, images visualise ideas. A symbol of life and rebirth, with chthonic, apotropaic, and purifying properties, the fish has a long tradition in the Near East, dating back to ancient rituals in Mesopotamia (Sahrhage 1999). Fish appear in religious scenes in the art of the Fertile Crescent, either to underline the life-giving waters of the god Enki-Ea (Leick 1998: 40–44) or together with other, female deities

(Tsukimoto 2014: 23, 30, figs. 21, 22). Fish sacrifices to chthonic deities obviously involved the invocation of rebirth after death (Burkert 1972: 204–12; Leick 1998: 55–57, 128–29). They are linked to female deities of the Aegean (Savoldi 1996: 61–90; Panagiotaki 1999: 81; Cichon 2022: 126),[13] forerunners of Aphrodite,[14] the *Potnia Delphinia*[15] or *Potnia Ichthyon* (Picard 1922: 60), who is also depicted on seals, and of the East (Astarte, Ishtar, Inanna) (Black and Green 1998: 108–109; Leick 1998: 86–93, 96–99; Wright 1990: 40; Natan-Yulzary 2020). Tablets from Ugarit mention rituals with fish offerings in honour of Astarte as Goddess of the Sea (Pritchard 1969: 132–33; Oden Jr 1977; Langdon 1989: 196–97; Pardee 2002: 53–55).[16] The son of the Syrian goddess Atargatis, who was transformed into a fish, was called Ichthys (Roscher 1890–1894: col. 94; Rostovtseff 1933; Oden Jr 1977: 47–107). Egyptian wall paintings depict male figures offering fish in honour of the dead (Porter and Moss 1960: 289, no. 182). Certain fish are associated with Egyptian deities, such as Isis and Hathor (Andrews 1994: 91–92). Sacred trees, connected with the cult of the goddess, are depicted in combination with fish or crabs (Keel 1998: fig. 54, 55a-b; Keel and Uehlinger 1998: 210–48). A krater fragment from Ugarit shows a warrior with sword attached to his waist placing a fish on a tall structure, possibly an altar, crowned by horns of consecration (Vermeule and Karageorghis 1982: XIII29). In the Philistine pantheon, Dagon is associated with the worship of fish (Singer 1992), although representations of fish are relatively rare on Philistine pottery.[17] They were also associated with religious beliefs and funerary rituals, as suggested by representations on Minoan larnakes (Mavriyannaki 1972: 55–57; Buccholz et al. 1973: 145, no. 131, 146, nos 144–147, p. 151, no. 58; Kanta 1980: 177, pls 65:3–4, 66:1–2, 73:9–10). These scenes provide glimpses into the beliefs and myths surrounding the deceased's voyage to the underworld. On Cyprus, fish bones, probably the remains of funerary banquets, have been found in tombs at Kalavassos (South 2000: 352, 361). Smaller quantities of fish bones have been identified in Cypriot sanctuaries (Webb 1999: 250–52).

The religious syncretism of the twelfth century BCE is reflected in cult, with the interchange of deities with common characteristics, symbols, and veneration.[18] Among other creatures, the octopus had a strong symbolic value. Like fish, it also belongs to the realm of the Mother Goddess. As "monster" of the sea, like other archetypal creatures, it must be considered to be one of the terrors of the sea. It could represent the mystic centre, the unfolding of creation and the rebirth, since if one of its tentacles breaks off, another will grow in its place. The crab is connected with the moon and Mother Goddess.[19] It moves forward and backward, oriented by lunar tides, balancing the regression.

5.5 Conclusion

At the end of the Late Bronze Age, the sea connected the peoples of the Mediterranean. The islanders' experience of navigation put them in an advantageous position with regard to the turbulent developments of this transitional period. The lives of people living near the sea are always decisively determined by it. Iconography provides us with indirect clues about the relationship between humans and the marine world, while at the same time giving us information about the various sea creatures, fishing methods, and so on. The presence of marine motifs in such a wide range of depictions proves that both the artists and the members of the community had great knowledge of the different marine species, which in turn indicates that the sea was an integral part of their life and diet.

The use of isolated motifs in pottery, such as sea creatures, is often merely decorative. The similarity in the depiction of these motifs between the Late Helladic IIIC pottery and the twelfth century BCE pictorial pottery of Cyprus shows that, despite the upheavals, trade and exchanges continued, while at the same time there is a strong Mycenaean tradition that combines with Cypriot elements to create local styles. Representations inspired by nature were occasionally combined with religious elements and narratives. The relationship between ideology and iconography was important in Late Bronze Age societies. The combination of motifs leads to narrative scenes that demonstrate the syncretism of this era. In conclusion, it is worth reiterating Nicholas Coldstream's belief that animals were the *lingua franca* of sacred iconography in the eastern Mediterranean (Coldstream 1995: 41).

[13] For Greek deities, the sea and sea-creatures, see Lesky 1947; Bevan 1985: 131–49.

[14] Aphrodite is sometimes represented sitting on a dolphin, as a goddess born of the sea.

[15] Seals often depict a marine goddess at the centre of the composition holding or surrounded by dolphins, her sacred companions.

[16] Lucian refers to a lake with fish near the sanctuary of the goddess in Hieropolis (Oden Jr 1977: 1–46).

[17] For fish decoration on Philistine pottery and related bibliography, see Bunimovitz and Lederman 2010; Meiberg 2013.

[18] For the connection between the Middle Eastern deities and the Mother Goddess of Cyprus and in particular between Aphrodite Kypris-Ashtar/Atargatis-Hera, see Budin 2004, 2014.

[19] According to Greek mythology, a crab attacked Hercules following the orders of Hera during his battle against Lernaia Hydra. Hercules killed the crab which Hera transformed into a constellation. On the connection between the mythological cycles of Hercules, Marduk and Ninurta and the fresh water, see Seguda 2023.

Bibliography

Amiran, R. 1969. *Ancient Pottery of the Holy Land. From Its Beginnings in the Neolithic Period to the End of the Iron Age*. Jerusalem: Massada Press.

Andrews, C. 1994. *Amulets of Ancient Egypt*. London: Trustees of British Museum.

Åström, P. 1988. The Hippocampus Krater. *Report of the Department of Antiquities, Cyprus (RDAC)*, 1988 (Part 1): 173–76.

Åström, P. 1989. Trade in the Late Cypriot Bronze Age. In: E. Peltenburg, ed. *Early Society in Cyprus*. Edinburgh: Edinburgh University Press, pp. 202–208.

Åström, P. 1998. *Hala Sultan Tekke 10. The Wells*. SIMA 45:9. Göteborg: Astrom Editions.

Benson, J.L. 1960. A Syrian Krater from Bamboula at Kourion. *Palestine Exploration Quarterly*, 92: 64–69.

Berdowski, P. 2008. Heroes and Fish in Homer. *Palamedes*, 3: 75–91.

Betancourt, P.P. 1985. *The History of Minoan Pottery*. Princeton, New Jersey: Princeton University Press.

Bevan, E. 1985. Representations of Animals in Sanctuaries of Artemis and Other Olympian Deities (PhD dissertation, University of Edinburgh).

Black, J. & Green, A. 1998. *Gods, Demons and Symbols of Ancient Mesopotamia*. London: The British Museum Press.

Brewer, J.D. & Friedman, R. 1989. *Fish and Fishing in Ancient Egypt*. The Natural History of Egypt: Volume II. Warminster: Aris and Phillips.

Broneer, O. 1939. A Mycenaean Fountain on the Athenian Acropolis. *Hesperia*, 4: 317–433.

Buchholz, H.G., Jöhrens, G. & Maull, I. 1973. *Jagd und Fischfang*. Archaeologica Homerica II:J. Göttingen: Vandenhoeck and Ruprecht.

Budin, S. 2004. A Reconsideration of the Aphrodite-Ashtart Syncretism. *Numen*, 51: 95–145.

Budin, S. 2014. Before Kypris was Aphrodite. In: David Y. Sugimoto, ed. *Transformation of a Goddess. Ishtar – Astarte – Aphrodite*. Fribourg: Academic Press Fribourg & Göttingen: Vandenhoeck & Ruprecht, pp. 195–215.

Bunimovitz, S. & Lederman, Z. 2010. A Unique Philistine Fish Motif from Tel Beth-Shemesh. *Israel Exploration Journal*, 60(1): 58–71.

Burkert, W. 1972. *Homo Necans. Interpretationen altgriechischer Opferitten und Mythen*. Berlin: De Gruyter.

Burkert, W. 1985. *Greek Religion, Archaic and Classical*. Cambridge (MA): Harvard University Press.

Cichon, J.M. 2022. *Matriarchy in Bronze Age Crete. A Perspective from Archaeomythology and Modern Matriarchal Studies*. Oxford: Archaeopress.

Clayton, P. 2006. *The Complete Pharaohs. The Reign-by-reign Record of the Rulers and Dynasties of Ancient Egypt*. Cairo: The American University Press.

Coche de la Ferté, E. 1951. *Essai de classification de la céramique mycénienne d' Enkomi (Campagnes 1946 et 1947)*. Paris: Librairie orientaliste Paul Geuthner.

Coldstream, N.J. 1995. The Originality of Ancient Cypriot Art. In: *Kypros: Apo ten proestorea stous neoterous chronous*. Nicosia: Cultural Foundation of the Bank of Cyprus, pp. 37–55.

Courtois, J.C. 1971. Le sanctuaire du dieu au lingot d' Enkomi-Alasia. In: C.F.A. Schaeffer, ed. *Alasia I*. Mission archéologique d'Alasia, tome IV. Paris: Klincksieck, pp. 151–362.

Crouwel, H.J. 2006. Late Mycenaean Pictorial Pottery. In: R.D.G. Evely, J.H. Crouwel & M.R. Popham, eds. *Lefkandi IV. The Bronze Age. The Late Helladic IIIC Settlement at Xeropolis*. BSA supplementary volume no. 39. London: British School at Athens, pp. 233–55.

Dakoronia, F. 2006. Mycenaean Pictorial Style at Kynos, East Lokris. In: E. Rystedt & B. Wells, eds. *Pictorial Pursuits. Figurative Painting on Mycenaean and Geometric Pottery. Papers from Two Seminars at the Swedish Institute at Athens in 1999 and 2001*. Acta Instituti Atheniensis Regni Sueciae. Stockholm: Svenska Institutet i Athen, pp. 23–29.

Dakoronia, F. & Kounouklas, P. 2015. Fishing Technology: The Kynos Contribution. In: D. Panagiotoloulos & I. Kaiser, eds. *Ein Minoer in Exil. Festschrift für Wolf-Dietrich Niemeier*. Universitäts Forschungen zur prähistorischen Archäologie. Band 270. Bonn: Verlag Dr. Rudolph Habelt, pp. 35–50.

Demakopoulou, K. 2006. Mycenaean Pictorial Pottery from Midea. In: E. Rystedt & B. Wells, eds. *Pictorial Pursuits. Figurative Painting on Mycenaean and Geometric Pottery. Papers from Two Seminars at the Swedish Institute at Athens in 1999 and 2001*. Acta Instituti Atheniensis Regni Sueciae. Stockholm: Svenska Institutet i Athen, pp. 30–43.

Dikaios, P. 1969–1970. *Enkomi Excavations 1948–1958*, vol. I–III. Mainz am Rhein: Philipp von Zabern.

Dothan, T. 1982. *The Philistines and Their Material Culture*. Jerusalem and New Haven: Israel Exploration Society.

Fagan, B. 2017. *Fishing. How the Sea Fed Civilization*. New Haven and London: Yale University Press.

Fanouriou, X. 2013. Όνας αρχαίος τρόπος χημικού ψαρέματος. *vimatisko.gr. Vema tes Ko*. https://psarema.net/enas-archeos-tropos-chimikou-psarematos/ [accessed 18 November 2022].

Feucht, E. 1992. Fishing and Fowling with the Spear and Throw-stick Reconsidered. In: U. Luft, ed. *The Intellectual Heritage of Egypt*. Studia Aegyptiaca XIV. Budapest: Eötvös Lorànd University, pp. 157–69.

Fischer, P.M. 2011. The New Swedish Cyprus Expedition 2010. Excavations at Dromolaxia-Vizatzia/Hala Sultan Tekke. Preliminary Results. *Opuscula*, 6: 69–98.

Fischer, P.M. 2013. The New Swedish Cyprus Expedition 2012. Excavations at Hala Sultan Tekke. Preliminary Results. *Opuscula*, 8: 49–79.

Furumark, A. 1972. *Mycenaean Pottery. I. Analysis and Classification*. Acta Instituti Atheniensis Regni Sueciae, Series in 4o, XX:1. Stockholm: Svenska Institutet i Athen.

Gill, M.A.V. 1985. Some Observations on Representations of Marine Animals in Minoan Art and Their Identification. In: P. Darcque & J.C. Poursat, eds. *L'iconographie minoenne*. BCH Supplement 11. Paris: École Française d'Athènes, pp. 63–81.

Goldman, H. 1956. *Excavations at Gözlü Kule, Tarsus, Vol.2. Text and Plates. From the Neolithic through the Bronze Age*. Princeton, NJ: Princeton University Press.

Güntner, W. 2000. *Figürlich bemalte mykenische Keramik aus Tiryns*. Tiryns Forschungen und Berichte 12. Mainz am Rhein: Philipp von Zabern.

Güntner, W. 2006. Mycenaean Pictorial Vase Painters. In: E. Rystedt & B. Wells, eds. *Pictorial Pursuits. Figurative Painting on Mycenaean and Geometric Pottery. Papers from Two Seminars at the Swedish Institute at Athens in 1999 and 2001*. Acta Instituti Atheniensis Regni Sueciae. Stockholm: Svenska Institutet i Athen, pp. 51–61.

Hadjianastasiou, O. 1996. A Mycenaean Pictorial Vase from Naxos. In: E. De Miro, L. Godart & A. Sacconi, eds. *Atti e memorie del secondo congresso internazionale di micenologia. Roma – Napoli, 14–20 Ottobre 1991*. Archaeologia, Vol. 3. Rome: Gruppo Editoriale Internazionale, pp. 1433–41.

Hadjisavvas, S. 1991. LCIIC to LCIIIA without Intruders: The Case of Alassa-*Pano Mandilaris*. In: J.A. Barlow, D.L. Bolger & B. Kling, eds. *Cypriot Ceramics. Reading the Prehistoric Record*. Philadelphia: University of Pennsylvania Museum of Archaeology, pp. 173–180.

Hatzaki, E. 2011. From LM IB Marine Style to LM II Marine Motifs. Stratigraphy, Chronology and the Social Context of a Ceramic Transformation: A Response to Maria Andreadaki-Vlazaki. In: T.M. Brogan & E. Hallager, eds. *LM IB Pottery: Relative Chronology and Regional Differences.* Monographs of the Danish Institute at Athens 11. Athens: Danish Institute at Athens, pp. 75–87.

Iakovidis, S. 1969–1970. *Perati*, vol. I–III. Vivliotheke tes en Athenais Archaiologikes Hetaireias (BAE) (= The Library of the Archaeological Society at Athens), n. 67. Athens: Archaeological Society at Athens.

Kanta, A. 1980. *The LMIII Period in Crete. A Survey of Sites, Pottery and Their Distribution.* SIMA 58. Göteborg: Astrom Editions.

Karageorghis, V. 1977. A Cypro-Mycenaean IIIC:1 Amphora from Kition, In: K.H. Kinzl, ed. *Greece and the Eastern Mediterranean in Ancient History and Prehistory. Studies Presented to Fr. Schachermeyr on the Occasion of His 80. Birthday*. Berlin: De Gruyter, pp. 192–98.

Karageorghis, V. 1981. *Excavations at Kition IV. The Non-Cypriote Pottery*. Nicosia: Department of Antiquities, Cyprus.

Karageorghis, V. 2007. *A Lifetime in the Archaeology of Cyprus.* Stockholm: Medelhavsmuseet.

Karageorghis, V. & Demas, M. 1988. *Excavations at Maa-Palaeokastro 1979–1986*, Nicosia: Department of Antiquities, Cyprus.

Kardara, C. 1977. *Naxos I. Aplomata Naxou. Kinita evrimata taphon A and B*. Vivliotheke tes en Athenais Archaiologikes Hetaireias (BAE) (=The Library of the Archaeological Society at Athens), n. 88. Athens: Archaeological Society at Athens.

Keel, O. 1998. *Goddesses and Trees, New Moon and Yahveh: Ancient Near Eastern Art and the Hebrew Bible*. Journal of the Study of Old Testament, Supplement 261. Sheffield: Sheffield Academic Press.

Keel, O. & Uehlinger, C. 1998. *Gods, Goddesses and Images of God in Ancient Israel*. Trans. Th. H. Trapp. Philadelphia: Fortress Press.

Langdon, S. 1989. The Return of the Horse Leader. *American Journal of Aarchaeology,* 93: 185–201.

Leick, G. 1998. *A Dictionary of Ancient Near Eastern Mythology*. London: Routledge.

Lekka, A. 2015. Η εικονογραφική κεραμική του 12ου αιώνα π.Χ. στην Κύπρο (PhD dissertation, National and Kapodistrian University of Athens).

Lekka, A. 2022. Pottery Fashion in the Late Bronze Age. An Overview of the Stylistic Similarities between Cypriot and Other Local Wares in the Mediterranean. In: G. Bourogiannis, ed. *Beyond Cyprus: Investigating Cypriot Connectivity in the Mediterranean from the Late Bronze Age to the End of the Classical Period.* Aura supplement 9. Athens: Faculty of History and Archaeology, National and Kapodistrian University of Athens, pp. 153–77.

Lernau, Omri. 2018. Fish Remains. In: P.M. Fischer & T. Bürge, *Two Late Cypriot City Quarters at Hala Sultan Tekke. The Söderberg Expedition* 2010–2017. *SIMA* CXLVII. Uppsala: Astrom Editions, pp. 563–66.

Lesky, A. 1947. *Thalatta. Der Weg der Griechen zum Meer.* Wien: Rudolf M. Rohrer Verlag.

Mackay, J.H.E. 1931. Report on Excavations at Jemdet Nasr, Iraq. *Field Museum of Natural History, Anthropology Memoirs,* I(3): 217–303.

Maier, F.G. & von Wartburg, M.L. 1986. Ausgrabungen in Alt-Paphos 13. Vorläufiger Bericht: Grabungskampagne 1983 und 1984. *Archäologischer Anzeiger,* 1986: 145–93.

Mavriyannaki, C. 1972. *Recherches sur les larnakes minoennes de la Crète occidentale*. Incunabula Graeca LIV. Roma: Edizioni dell' Ateneo.

Meiberg, L. 2013. Casting a Wide Net: Notes on the Inspiration for the Fish Motif on Philistine Pottery. *Tel Aviv,* 40: 55–71.

Mekhitarian, A. 1989. *Ägyptische Malerei*. Geneva: Verlag Wasmuth and Zohlen.

Michimasa, D. 2006. The Octopus Style. A Study of Octopus-painted Aegean Pottery of 12th-11th Centuries BCE, Its Regional Styles, Development and Social Significance (PhD dissertation, University College London).

Morricone, L. 1965–1966. Eleona e Langada. Sepolcreti della tarda éta del Bronzo a Coo. *Annuario della Scuola Archeologica di Atene e delle Missioni italiane in Oriente (ASAtene),* 43–44: 5–311.

Morricone, L. 1972–1973. Coo. scavi e scoperte nel Serraglio e in località minori (1935–1943). *Annuario della Scuola Archeologica di Atene e delle Missioni italiane in Oriente (ASAtene),* 50–51 (N.S. 34–35): 139–396.

Mountjoy, P.A. 1974. A Later Development in the Late Minoan IB Marine Style. *Annual of the British School at Athens,* 69: 177–80.

Mountjoy, P.A. 1977. Attributions in the LM IB Marine Style. *American Journal of Archaeology,* 81: 557–60.

Mountjoy, P.A. 1984. The Marine Style Pottery of LM IB/LH IIA: Towards a Corpus. *Annual of the British School at Athens*, 79: 161–219.

Mountjoy, P.A. 1985. Ritual Associations for LM IB Marine Style Vases. In: P. Darcque & J.C. Poursat, eds. *L'iconographie minoenne*. BCH Supplement 11. Paris: École Française d'Athènes, pp. 231–42.

Mountjoy, P.A. 1998. The East Aegean – West Anatolian Interface in the Late Bronze Age: Mycenaeans and the Kingdom of Ahhiyawa. *Anatolian Studies,* 48: 33–67.

Mountjoy, P.A. 1999. *Regional Mycenaean Decorated Pottery*. Berlin: Verlag Marie Leidorf.

Mountjoy, P.A. 2006. Mycenaean Pictorial Pottery from Anatolia in the Transitional LH IIIB2-LHIIIC Early and the LHIIIC Phases. In: E. Rystedt & B. Wells, eds. *Pictorial Pursuits. Figurative Painting on Mycenaean and Geometric Pottery. Papers from Two Seminars at the Swedish Institute at Athens in 1999 and 2001.* Acta Instituti Atheniensis Regni Sueciae. Stockholm: Svenska Institutet i Athen, pp. 107–21.

Mylona, D. 2021. Catching Tuna in the Aegean: Biological Background of Tuna Fisheries and the Archaeological Implications. *Anthropozoologica,* 56(2): 23–37.

Natan-Yulzary, S. 2020. Lady Athirat of the Sea. A New Look at KTU 1.4. ii 3-11. *Aula Orientalis,* 38(1): 131–46.

Oden, R.A. Jr. 1977. *Studies in Lucian's De Syria Dea*. Harvard Semitic Museum Monograph 15, Missoula, Mont: Scholars Press.

Panagiotaki, M. 1999. *The Central Palace Sanctuary at Knossos.* London: The British School at Athens.

Pardee, D. 2002. *Ritual and Cult at Ugarit.* Atlanta (GA): Society of Biblical Literature.

Picard, C. 1922. *Ephèse et Claros. Recherches sur les sanctuaires et les cultes de l' Ionie du Nord.* BÈFAR 123. Paris: Éditions de Boccard.

Porter, B. & Moss, R.L.B. 1960. *The Theban Necropolis I. Private Tombs*. Oxford: Griffith Institute-Ashmolean Museum Oxford.

Potts, D.T. 2012. Fish and Fishing. In: D.T. Potts, ed. *A Companion to the Archaeology of the Ancient Near East.* Malden (MA): Wiley-Blackwell, pp. 220–35.

Powell, J. 1992. Archaeological and Pictorial Evidence for Fishing in the Bronze Age: Issues of Identification and Interpretation. In: R. Laffineur & J.L. Crowley, eds. *Εικών. Aegean Bronze Age Iconography: Shaping a Methodology. Proceedings of the 4th International Aegean Conference, University of Tasmania, Hobart, Australia 6–9 April 1992.* Aegaeum 8. Liège: Université de Liège, pp. 307–15.

Powell, J. 1996. *Fishing in the Prehistoric Aegean.* SIMA Pocketbook 137. Jonsered: Astrom Editions.

Pritchard, J. 1969. *Ancient Near Eastern Texts Relating to the Old Testament.* Princeton – New Jersey: Princeton University Press.

Reese, D.S. 1984. Shark and Ray Remains in Aegean and Cypriote Archaeology. *Opuscula Atheniensia,* XV: 188–92.

Reese, D.S. 2003. The Shells, Astragali and Fish. In: V. Karageorghis, *Excavations at Kition VI. The Phoenician and Later Levels II.* Nicosia: A. G. Leventis Foundation, pp. 418–28.

Reese, D.S. 2008. Organic Imports from Late Bronze Age Cyprus (with Special Reference to Hala Sultan Tekke). *Opuscula Atheniensia,* 31–32 (2006–2007): 191–209.

Roscher, W.H. 1890–1894. *Ausführliches Lexicon der Griechischen und Römischen Mythologie*. Band 2, Abteilung 1. Teubner.

Rostovtseff, M. 1933. Hadad and Atargatis at Palmyra. *American Journal of Archaeology, 37*:1, 58–63.

Sahrhage, D. 1999. *Fishfang und Fischkult im alten Mesopotamien*. Berlin: Peter Lang.

Sakellarakis, Y.A. 1992. *The Mycenaean Pictorial Style in the National Museum of Athens.* Athens: Kapon Editions.

Savoldi, E. 1996. Hieros ichthus. Sacralità e proibizione nell' epica greca arcaica *Annali della Scuola Normale Superiore di Pisa. Classe di Lettere e Filosofia,* 1/1: 61–90.

Schallin, A.L. 1993. *Islands under Influence. The Cyclades in the Late Bronze Age and the Nature of Mycenaean Presence*. SIMA CXI. Jonsered: Astrom Editions.

Seguda, E.G. 2023. Hercules: Lord and Guardian of the Fresh Waters. *'Ilu, Rivista de Ciencias de las Religiones,* 28: e-87407. https://doi.org/10.5209/ilur.87407.

Shapland, A. 2022. *Human-Animal Relations in Bronze Age Crete: A History through Objects*. Cambridge: Cambridge University Press.

Singer, I. 1992. Towards the Image of Dagon, the God of the Philistines. *Syria,* 69: 431–50.

South, A. 2000. Late Bronze Burials at Kalavassos-*Ayios Dhimitrios*. In: G.K. Ioannides & S.A. Hadjistylli, eds. *Proceedings of the Third International Congress of Cypriote Studies*. Nicosia: Department of Antiquities, Cyprus, pp. 345–64.

Tsukimoto, A. 2014. "In the Shadow of Thy Wings": A Review of the Winged Goddess in Ancient Near Eastern Iconography. In: D.Y. Sugimoto, ed. *Transformation of a Goddess. Ishtar – Astarte – Aphrodite*. Fribourg: Academic Press Fribourg & Göttingen: Vandenhoeck & Ruprecht, pp. 15–32.

Vermeule, E. & Karageorghis, V. 1982. *Mycenaean Pictorial Vase Painting*. Cambridge (MA): Harvard University Press.

Vlachopoulos, A. 2003. The Late Helladic IIIC "Grotta Phase" of Naxos. In: S. Deger-Jalkotzy & M. Zavadil, eds. *LHIIIC Chronology and Synchronisms. Proceedings of the International Workshop Held at the Austrian Academy of Sciences at Vienna, May 7th and 8th 2001*. Vienna: Verlag der Österreichischen Akademie der Wissenschaften, pp. 217–34.

Vlachopoulos, A. 2008. A Late Mycenaean Journey from Thera to Naxos: The Cyclades in the Twelfth Century BC. In: N. Brodie, J. Doole, G. Gavalas & C. Renfrew, eds. *Horizon: A Colloquium on the Prehistory of the Cyclades*. Cambridge: McDonald Institute for Archaeological Research, pp. 479–91.

Webb, J.M. 1999. *Ritual Architecture, Iconography and Practice in the Late Cypriot Bronze Age*. SIMA Pocket-book 75. Jonsered: Astrom Editions.

Wright, G.R.H. 1990. Of Fishes and Men. Fish Symbols in Ancient Religion. *Journal of Prehistoric Religion,* III-IV: 30–43.

Zohar, I. & Artzy, M. 2019. The Role of Preserved Fish: Evidence of Fish Exploitation, Processing and Preservation in the Eastern Mediterranean during the Late Bronze Age (14th–13th Century BCE). *Journal of Archaeological Science Reports,* 23: 900–909.

6 Underrepresented Riches from the Sea

Mari Yamasaki

Abstract

In the Late Bronze Age, fish and molluscs played an important role in the diet and the economy of Eastern Mediterranean coastal societies. Fish-roe cakes and dried splits feature in the list of goods traded over long distances, while textiles coloured with the purple dye obtained from murex shells were considered among the most luxurious items of their time and would continue to be so for centuries to come. Accordingly, the ancient textual and iconographic sources deal with these final products as objects of value, in terms of international gift exchanges and temple donations. However, despite their undeniable importance, marine faunal resources rarely surface in the discourse of those who profited from them. Surprisingly few ancient sources explicitly deal with either fish or molluscs, nor are the activities related to their procurement and processing described in much detail, if at all. This chapter aims at investigating the ancient conceptualisation of these resources by combining analysis of the archaeological remains with textual and iconographical evidence and addresses the question as to why these riches from the sea were consistently underrepresented in the sources.

6.1 Introduction

In the Late Bronze Age, marine resources are a recurring item in the economy of Eastern Mediterranean coastal societies. Preserved fish products were traded over very long distances, while textiles coloured with the purple dye obtained from murex shells were considered among the most luxurious items of their time and would continue to be so for centuries to come (Pliny, *Nat. Hist.* IX: 133–34; Mayhoff 1875–1908). Accordingly, the ancient textual and iconographic sources deal with these final products as objects of value, suitable for inclusion in royal gift exchanges and temple donations. However, despite the well-established economic significance of dried fish splits and beautifully coloured wools, the processing of marine raw materials into these prized items is hardly ever mentioned.

6.2 Fish Trade and Purple Dye: The Evidence

There is abundant evidence for the role of aquatic resources in the Levant and Cyprus beyond the needs of household consumption. Preserved fish products and purple-dyed textiles make frequent appearances in lists of tributes, gift exchange and business correspondence, while archaeologically, fish remains and heaps of murex shells are well attested at several sites.

The Ugaritic archives offer a good selection of examples of fish trade. Tablet KTU 4.427 features a variety of fish types (lines 23–29) in a long list of victuals. The fish items consist of 1000 (shekels of) ‘ṭ, possibly umbrine; 2000 (shekels of) *alpnm*, identified as a kind of small fish sold in bulk; 200 *ˈiqnˈu*, murex snail; 100 *bnn*, a type of carp; 15 *qn ʿm ʿnm*, squid (literally “arrow with two beautiful eyes”), as well as large quantities of *ygb bqʿ*, and *ˈart* from Egypt (van Soldt 1990; de Moor 1996). Johannes C. de Moor translates *ygb bqʿ* with “split dried fish”, while he proposes to relate the Ugaritic *ˈart* with the Babylonian *erūtu*, meaning “roe, spawn” (de Moor 1996: 156–57), likely referring to salted fish-roe loaves/cakes not dissimilar to modern *bottarga*.

In RS 34.167, an Ur-Baal from Tyre offers to an Ahi Milku of Ugarit a talent of fish and purple-dyed textile, among other items, in exchange for a shipment of oil, silver and copper (Malbran-Labat 1991: 57–60). Another example is letter RS 34.167 + 175, where fish is traded together with valuable textile products (van Soldt 1990: 336). Fish trade is also attested at the highest levels of royal gift exchange, as exemplified by letter RS 88.2158, from Pharaoh Merneptah

M. Yamasaki (✉)
Polish Centre of Mediterranean Archaeology, University of Warsaw, Warsaw, Poland
e-mail: m.yamasaki@uw.edu.pl

C. Tsouparopoulou, L. Recht (eds.), *Human and Aquatic Beings: Interactions in and beyond the Eastern Mediterranean (3rd–1st Millennia BCE)*, Themes in Contemporary Archaeology, https://doi.org/10.1007/978-3-031-73643-8_6

(r. c. 1213–1203 BCE) to the king of Ugarit. In line 50, four units of *lippatu*.KU_6.MEŠ are mentioned together with other precious gifts: here the plural for fish, KU_6.MEŠ, is accompanied by *lippatu*, a word of yet unknown meaning (Lackenbacher 1995, 2001: 246). Dried fish also appears in the fictional 'Tale of Wenamun' (composed sometime in the eleventh century BCE), in which 30 baskets of dried splits are listed as gifts from Egypt to Byblos. Interestingly, the type of fish being exchanged appears to be specified only for some items in KTU 4.427 and in Merneptah's letter. In all other texts, there appears to be no need for a precise designation, indicating that the involved parties probably knew which fish was being referred to. Archaeologically, while roe-cakes leave no physical evidence, fish remains are found at many locations, especially at coastal sites. Cutmarks and evidence of fish processing were detected on a large sample from Akko on both marine and freshwater species (Zohar and Artzy 2019).

For marine fish, it is generally difficult to verify whether these fish were procured locally or whether they were imported. Interesting archaeometric evidence in this regard comes from the study of seabream remains from Southern Levantine sites. This fish is normally found in the coastal waters of the Eastern Mediterranean and, up to now, its presence at coastal sites was assumed to derive from local fishing. Studies on the oxygen isotope composition of the tooth enamel of seabream from the Canaanite region suggest that, at least in part, this fish might have been imported from the Bardawil lagoon in Egypt, as early as in the Late Bronze Age (Sisma-Ventura et al. 2015, 2018). Further studies may reveal whether similar considerations apply to other marine and euryhaline species.

In the case of the Nile perch (*Lates niloticus*), however, it is clear that its presence was always a result of trade. *Lates* is a large freshwater fish autochthonous of East Africa, and remains of this species have been discovered at several sites in the Levant and Cyprus (Table 6.1). While this paper focuses on the coastal areas, it should be noted that *Lates* finds are not limited to the coast, but are known also from several inland sites, sometimes over a hundred kilometres from the sea (Van Neer et al. 2004: fig. 1). Since this species is not naturally found in the modest rivers of these regions, its presence constitutes certain proof of a recurrent import of dried fish from Egypt (Reese et al. 1986; Van Neer et al. 2004; Linseele et al. 2013; Routledge 2015). Furthermore, Nile perch is the single most frequently identified fish species in the coastal sites of the Levant and Cyprus (Yamasaki 2023: fig. 6.1).

Even more than fish products, purple dye is certainly the best known and most studied product deriving from marine animals. The earliest attestation for the exploitation of murex shells for purple dye in the Eastern Mediterranean comes from LB II–III Sarepta, Minet el Beida and Ugarit, Tell Abu Hawam, Bates' Island, Troy and Hala Sultan Tekke (Reese 2010). With the implementation of industrial techniques for purple dye production from the glands of some molluscs of the Muricidae family, in the LBA these shells become an essential raw material—and remained so until the Middle Ages. The level of wealth represented by the murex dye industry is such that the colour itself becomes synonymous with rank and power. In the course of the Iron Age, the Levantine production centres come to be identified with the colour itself: in the *Iliad*, it is said that Queen Ecuba's mantles were imported directly from Sidon (*Il.* VI, 289–91) and the association between high status and purple is expressed several times with the image of the heroes clad in their great purple vests (*Il.* VIII, 221; X, 133–134; *Od.* VIII, 84). References to artisans skilled in purple-dye craftmanship also appear in the Old Testament (NIV 2 Chr., 2:13–14).

Several tablets are generally cited as evidence of the sheer size of the purple coloured textiles trade in the LBA, particularly in royal gift exchange and tribute payment. Purple dyed textiles are listed in the above-mentioned RS 34.167 and RS 34.167 + 175. In the syllabic texts from Ugarit, we encounter specific hues of purple dyed wool, namely *hašmanu* and *takiltu*. In PRU 3.14, a hundred shekels of *hašmanu*-wool and hundred shekels of *takiltu*-wool are part of the gifts to queen Taryelli of Ugarit, while "[. . .] hundred (shekels of) *takiltu*-wool" and "two hundred (shekels of) *ḫašmanu*-wool"

Table 6.1 Remains of *Lates niloticus* from Late Bronze Age coastal sites in the Levant and Cyprus (NISP/MNI)

Site	NISP/MNI
Tel Dor (Bartosiewicz et al., 2018)	193/–
Hala Sultan Tekke (Lernau, 2018; Rose, 1994; Reese, 2008; Lernau & Reese, personal communication)	186/–
Tell Tweini (Linseele et al., 2019)	67/–
Tel Abu Hawam (Zohar & Artzy, 2019)	18/–
Sarepta (Rose & Reese, unpublished report)	12/10
Enkomi-*Ayios Iakovos* (Reese, 2022)	11/10
Episkopi-*Bamboula* (Lernau & Reese, personal communication)	10/6
Ugarit, Maison aux Albâtres (Chahoud & Vila, 2017)	10/–
Sidon (Van Neer, 2006; Chahoud & Vila, 2011/2012, 2017)	8/–
Kition-*Kathari* (Reese, 2003)	1/1

are included in a list of precious objects together with two linen cloths and 80 shekels of glass in PRU 3.187 (van Soldt 1990: 336). In the alphabetic texts we find instead *pḫm* and *iqnu*, translating *hašmanu* and *takiltu* respectively, which possibly refer to a red and a blue colour (Dietrich and Loretz 1966: 227–32; van Soldt 1990: 341–43; Thavapalan 2016: 175; 2020: 282).[1] However, as noted by Shiyanthi Thavapalan, a major problem with the coloured wools mentioned in the texts is that, with few exceptions, the quantities are too enormous and the prices too low to accept that we are dealing with genuine murex dye (Thavapalan 2016: 178–79; 2020: 236). I will return to this aspect further below.

Regardless of the problematic interpretation of the textual material on coloured textiles, archaeologically we can be certain the shell derived dye was being produced at several locations in the Levant and at least at one site on Cyprus (Fig. 6.1). Purple dye in the Mediterranean is obtained from the hypobranchial gland of gastropods belonging to the Muricidae family, and the three species used in this industry are *Hexaplex trunculus*, *Bolinus brandaris* and *Stramonita haemastoma*. An often-cited experiment by Paul Friedländer (1909) showed that ca. 12,000 molluscs were necessary to obtain 1.4 g of dye. More recently, Zvi C. Koren proposed a revised estimate of 7 snails to produce 1 g of uniformly coloured wool (Koren 2005). In either case, the quantity of murex demanded by the purple dye industry would have been enormous. So big, in fact, to leave clear evidence in the archaeological record.

Early observations by western explorers in the eighteenth and nineteenth centuries CE recount vast murex shell mounds (up to 120 m in length) at Sidon, Tyre and Minet el Beida (Thompson 1947; Schaeffer 1951; Reese 2010). Even though some of these mounds have since been significantly downsized or obliterated by urban development, modern excavations have revealed significant accumulations of murex snails from well dated contexts. At Hala Sultan Tekke (Reese 2018), Sarepta (Reese 2010) and Tell Abu Hawam (Baruch et al. 2005), thousands of crushed murex shells were found in large accumulations, often in combination with pits and basins compatible with dye production (Reese 2010; Kalaitzaki et al. 2017). One should also consider that even the discarded shells would have been used to produce lime or as floor construction material (Matoïan and Vita 2014: 322–23). As such, the crushed murex we find in association with dying installations should be regarded as a fraction of those actually used in this industry. Much more rarely, fragments of textiles with traces of murex-derived pigment have survived in tombs, often together with other prestigious grave goods, as was the case with the LC IIIA1–2 Tomb 23 from Area 8 at Hala Sultan Tekke (Niklasson 1983; Rose 1994: 233; Reese 2007).

Despite a probable overestimation of the volume of this business prompted by the textual sources, it is nevertheless clear that murex derived purple dye was certainly produced at an industrial level at a handful of Levantine sites and at Hala Sultan Tekke, and that it undoubtedly represented an extremely prized luxury good.

6.3 Procurement and Processing of Fish Products and Purple-Dye: Gaps in the Sources

From this brief overview it is clear that both processed fish and purple dye represented a non-negligible source of wealth for the Levantine coastal cities. However, any expectation of finding descriptions of their production in the textual and iconographical material from the Levant is bound to disappoint. The Levantine sources are interestingly uninformative concerning the way fish and shells were worked into the final product, especially when we compare them with those from the neighbouring regions.

Dried fish splits were a very common way of preserving and transporting fish throughout the ancient Near East. Tablets concerned with the administration of fisheries are known from as early as the Uruk III period (Potts 2012: 222). Fishing with various implements such as hook and line, traps and net, as well as fishing related activities appear on Mesopotamian cylinder seals (Sahrhage 1998: fig. 4). Mentions of cutting the fish open for drying appear also in the Babylonian Creation Myth (*Enuma Eliš*, IV.135–38; Lambert 2013: 94–95) as the god Marduk splits the sea-dragoness Tiamat in half to create the world, an image that was probably inspired by the common sight of gutting, splitting and hanging fish to dry along the Euphrates' banks (Potts 2012: 230–33).

The most detailed depictions of fish processing are found in Egypt. Here they are attested as early as the Old Kingdom and are a recurrent theme throughout Egyptian history (Wild 1953; Brewer and Friedman 1989). They include the preparation of fish roe into loaves and cakes, and show the various phases of the preparation of dried splits. Textual and iconographic accounts of fishing, either as a recreational activity or as a profession, are not rare, and the profession of fisherman is colourfully outlined in the famous *Satire of the Trades* (Lichtheim 2006).

In the Levant and on Cyprus, fishermen are a rare subject. From Cyprus, there are two known depictions which may show human figures fishing from a ship or a boat: an unprovenanced Late Cypriot cylinder seal (Cyprus Museum, inv. n. 40), and an LC III/CG amphora from *Vathyrkakas* (Westerberg 1983: figs. 12, 16.2). In the texts from Ugarit,

[1] These are just a few examples of a much too large a corpus to be discussed in full here and for a comprehensive overview I refer the reader to the works of van Soldt (1990) and Thavapalan (2016; 2020).

fishermen find their most significant representative in the character of Qodesh wa Amrur. This figure appears in the Ugaritic Baal Cycle as the servant of Athirat, one of the chief deities of the Ugaritic pantheon (KTU 1.3.VI.9–11; KTU 1.4.II.29–36; KTU 1.4.IV.1–9; Smith and Pitard 2009).

Aside from these few instances about fishing, none of the textual or iconographical source material from this area makes any reference to fish processing. Even if it is clear that processed fish was a frequent item in interregional transactions, and that fishing was a recognised activity, there is no explicit reference to fish processing from neither the Levant nor from Cyprus, and the only evidence reporting on these activities comes from the neighbouring regions of Egypt and Mesopotamia. This absence of information concerning the production processes is even more striking in the case of purple dye. In the known texts, as already pointed out by van Soldt (1990) and more recently by Thavapalan (2016, 2020), absolutely no reference is made with regards to the harvesting and processing of the molluscs.

The only textual reference to a red-purple colour obtained from murex shells comes from three brief passages from the Ugaritic Baal Cycle and in the Tale of Aqahat. In two almost identical scenes in the Baal Cycle, the goddess Anat beautifies herself with the red of the *ʾanhbm* (murex snail), which is "extracted from the sea", before and after battle (KTU 1.3.II.2–3; KTU 1.3.III.1–2; Smith and Pitard 2009). A similar scene sees Paghat, sister of Aqahat, painting her face red with murex in her preparation to avenge her brother's death (Fronzaroli 1955). While the connection between the colour and its source is clear, there is no mention of the relationship between the *ʾanhbm* (murex) and textile dyeing.

Archaeologically, a possible recognition of the connection between the murex and the precious dye may come from the above-mentioned Tomb 23 at Hala Sultan Tekke. In this rich grave, remains of purple stained textile were found together with one intact *Bolinus brandaris* shell, positioned near the skull of the deceased (Niklasson 1983). However, David S. Reese notes the water-worn state of the shell, indicating that it could not have been used to produce the dye (Reese, personal communication).

6.4 Underrepresented Riches

In discussing the perceived value of preserved fish, fish derivates and purple dye, we are thus faced with the question: why is there such a small interest on the part of the Levantine (and by extension the Cypriot) sources in the procurement of the resources themselves and their processing?

Concerning the fish, the first thing that needs to be considered is that, although a common food item, fish was never the main source of subsistence in the Levant (Yamasaki 2023: 137). Even at a site with a relatively high quantity of fish remains such as Sidon, recent stable nitrogen isotope analyses have demonstrated a predominantly terrestrial diet, in which the contribution of fish was negligible (Schutkowski and Ogden 2011; Mosapour Negari 2016; Stantis et al. 2022). This could partly explain why fish and fishing occupy a relatively small space in the sources compared to land-based activities. Nevertheless, as seen above, processed fish products were bulk traded across the Eastern Mediterranean and their role was relevant enough to be included in royal gift exchanges.

Observing the distribution of the iconographic and textual sources on how dried splits and roe cakes were obtained, we can notice that on the one hand there is no information from neither the Levant nor Cyprus. On the other hand, we do have abundant iconographic evidence from Egypt showing the fishing and processing of a variety of Nilotic fish, among which Nile perch and catfish, and the euryhaline grey mullet, into dried splits and the preparation of the roe-cakes. This is indicative of a different interest in this type of activity, and that, at least in Egypt, fish processing was worth a durable depiction.

If we have a closer look at the exchanged items, an Egyptian provenance is referred to on more than one occasion. Egyptian Nile perch is found at multiple sites along the coast, in the hinterland, and even on Cyprus. In KTU 4.427, roe-cakes are specifically described as being "from Egypt." Pharaoh Merneptah includes a quantity of special dried fish among his gifts and baskets of fish splits are brought along by Wenamun as gifts to the king of Byblos in his fictional journey. While it is true that elsewhere fish products appear to have been exchanged also within the Levant, as is the case of the letter between Ur-Baal of Tyre and Ahi-Milku of Ugarit, it seems quite probable that the Levantine traders acted as middlemen for bulk-produced Egyptian fish. The recent oxygen isotopic evidence from the Southern Levant further strengthens this hypothesis (Sisma-Ventura et al. 2015, 2018).

There is no reason to doubt that fish processing occupied a non-negligible role in Egyptian economy, whereas for the Levant and Cyprus we know little of the socio-economic dimension of this industry. We can be sure that local fisheries did exist, and that fish was processed for preservation also by Levantine fishers. However, it is possible that Levantine elites were not as involved in the fishing and fish processing business as were their Egyptian counterparts. The reasons may range from a low productivity or inferior quality of the local fisheries to a preference for the exotic. A scarce involvement by the elites would be consistent with the way the sources from the Levant highlight the Egyptian origin of fish products rather than the way they were made.

Moving on to the next item of this analysis, the question now turns to the purple-dye. If on the one hand economic motivations can underlie the relative neglect of the sources for the fish processing industry, more difficult is to understand the absolute silence on the purple dye production.

As can be gathered from the brief review presented above, there is clear archaeological evidence that purple dye was produced at various sites in the Levant, most notably Ugarit-Minet el Beidah, Tyre, Sarepta, Sidon, Tell Abu Hawam as well as at Hala Sultan Tekke on Cyprus as early as the fifteenth century BCE, and that, from these production centres, it was exported across the Mediterranean. However, we have no written or iconographic source from any place in the ancient Near East concerned with any stage of the production, from the harvesting of the murex shells to their processing, to the dyeing of the textiles. We have to wait until Pliny the Elder in the first century CE before anyone would expressly dedicate their attention to the way this extremely luxurious good was obtained (*Nat. Hist.* IX:62; Mayhoff 1875–1908). The stress on the high value connected to the textiles dyed with this colour stands in stark contrast with the absolute silence on the provenance of this good. If not for a lack of profitability nor of prestige of its final product, then why not describe its production process through words and images?

As mentioned above, hints that there was a general awareness of a connection between the red-purple colour and the murex shells are incredibly rare. We have the three brief passages from the Ugaritic Baal Cycle and the Tale of Aqahat, and possibly the association of fragments of a purple-dyed textile with a water-worn *Bolinus* inside Tomb 23 from Hala Sultan Tekke. Additionally, as Thavapalan convincingly argues (Thavapalan 2016: 178–180; 2020: 236–37), the murex dye was likely much rarer than a first reading of the sources leads us to believe, and much of the purple/red textiles we have mentioned in the texts is actually a product of plant-based dyes. What we gather from the large quantities of coloured wools being exchanged is that these items were highly sought after, independent of the origin of the dye. This is not to say that genuine purple dye was not appreciated, on the contrary, but that a garment coloured with murex would have been an extremely expensive rarity.

If we consider the distribution of the known sites where the purple dye is attested (Fig. 6.1), we count less than ten sites for the Bronze Age. Given the enormous quantity of murex needed, the difficult production phases and the uncertain results, it is possible that the knowledge of this technique

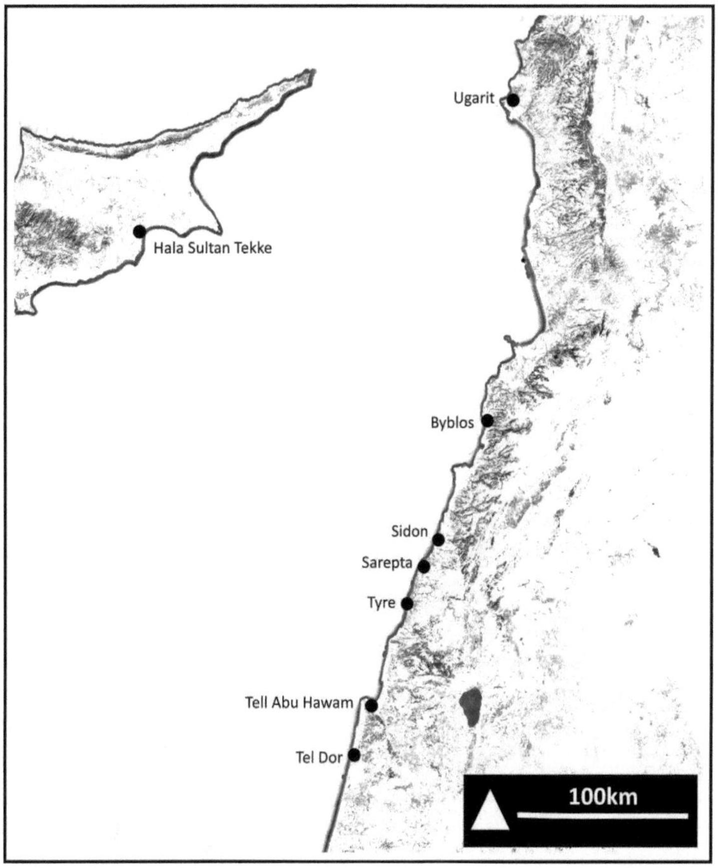

Fig. 6.1 Eastern Mediterranean sites with evidence of purple dye production. (Prepared by the author based on Google Earth images)

would have been kept as much as possible from spreading outside a limited circle of craftsmen. The limited attestations of coloured textiles with prices coherent with murex dye suggests that real murex-dyed fabric was so rare that no one, especially no one that could have left us a written account, really knew much about it except that it was very precious. Conversely, Mesopotamian literature confirms the frequent use of vegetal dyes and the instructions to obtain them. The Sippar Dye Text (BM 62788 + 82,979), dated to sometime in the seventh century BCE, but possibly referring to an older scribal tradition, provides inexpensive plant-based recipes for obtaining the most popular colours, including red, blue and purple (Reifarth and Völling 2013: 34).

Another factor that may have a role in the dearth on the dye processing itself may reside in a generic bias against activities connected to mollusc gathering or processing. The coastal installations to collect the molluscs, extract the gland, let it macerate, boil and finally obtain the dye would undoubtedly cause extremely foul odours, and were preferably located at a distance from the settlement, in such a way that the prevailing winds would carry the stench away from the houses. It may be argued that since mollusc processing must have been a particularly unattractive industry, those who were not directly involved in the production would have avoided the workshops. Nevertheless, other equally malodorous operations such as leather tanning do receive their description in the sources (van de Mieroop 1997: 26–35).

The different treatment of fishing versus shell gathering and the respective processing may find a partial explanation in the social dimension of these occupations, particularly in the categorisation of shellfish gathering as a low risk/low skill/low profit activity. An ethnographic study on the traditional fisheries in Palawan (Philippines) conducted by Susana Siar (2003) shows the differentiation of fishing tasks according to gender and age. Women and children would prevalently gather shellfish and sea cucumber in the intertidal zone, while men targeted commercial fish and octopus in the coral reef. Additionally, women would process both their own and their husbands' catch and sell it on the market. While there is a minor overlap in some of the adopted fishing techniques and targeted species, in general terms, women and men occupied different economic niches in the coastal fishing industry. This difference was reflected in the perceived value of their respective work areas. Despite both groups contributing to the household economy, with women's catch making up to 50 per cent of the income (not considering the added work of processing and selling), both female and male informants considered the income generated by women-work as subsidiary to that of men. Similarly, in another ethnographic study by Ferreira Teixeira and Silva Campos (2019) on artisanal fisheries in Brazil, it emerged that the necessary skill and labour involved in shellfish gathering is significantly downplayed in proportion to the income generation—which is also similarly downplayed against the income generated by other "manly" activities, such as fishing.

Unfortunately, we cannot reconstruct the social make-up of LBA Levantine fisheries and connected activities. The few references and representations of fishermen in the sources are all male, but as Çilingiroğlu and Çakırlar (2018) note, this may reflect the idea that fishing is a "man's job" rather than the social reality of this occupation. We can also assume that there must have existed a group of specialised shellfish gatherers that focused on the precious murex at least on a seasonal level, who left no trace in the sources. While we cannot say if this activity was performed primarily by women, men or children, we can indeed see that a lack of visibility seems to have affected this group in a way not dissimilar to what was observed with the shellfish gatherers in the Philippines and Brazil. Whether they collected edible molluscs or were involved in murex commercial harvesting, regardless of the income that their activities generated, it appears that their activity was not perceived as significant enough to be acknowledged in written or visual forms.

6.5 Conclusions

Summing up, the evidence from the Levant and Cyprus indicates a general underrepresentation of marine secondary products with notable differences between fish processing and the purple dye industry.

Fish and fishing are a subject of depictions and are mentioned, if barely, in important mythological compositions from Ugarit. Additionally, while fish processing is not directly addressed in Levantine textual and iconographical material, there are abundant sources from the neighbouring regions. One possible explanation for processing not being explicitly discussed in the Levantine sources may reside in the trade dynamics between the Levantine coastal cities and Egypt, since it appears that the former were frequently importers of Egyptian dried fish and secondary fish products. Be it for a particular prestige of Egyptian fish products or due to a relatively low productivity of local fisheries compared to the Egyptian ones is difficult to assess.

If the relative scarcity of information on the first can be explained on an economic level, the same cannot be said for the murex-based purple dye production. Even acknowledging the limited volume of the purple dye trade, this would still have consisted of a highly lucrative business. The fact that virtually no source from Egypt, the Levant and Mesopotamia discusses the method to obtain purple dye from the snails of the Muricidae family is puzzling. Multiple factors, from being a closely guarded industrial secret to a socio-economic bias against shellfish gathering, may contribute to an explanation of this apparent inconsistency between the value of the finished product and the silence on the way it was obtained.

Acknowledgements This work is part of the author's project Underwater Realms, supported by the Polish National Science Centre under the Polonez Bis 1 grant number 2021/43/P/HS3/0135, co-funded by the European Union Framework Programme for Research and Innovation Horizon 2020 under the Marie Skłodowska Curie COFUND agreement no. 945339.

Bibliography

Bartosiewicz, L., Lisk, E. & Zohar, I. 2018. Non-mammalian Vertebrate Remains. In: A. Gilboa, I. Sharon, J.R. Zorn & S. Matskevich, eds. *Excavations at Dor, Final Report. Volume IIB. Area G, The Late Bronze Age and Iron Ages: Pottery, Artifacts, Ecofacts and Other Studies*. Qedem Reports 11. Jerusalem: Hebrew University of Jerusalem, pp. 313–22.

Baruch, I., Artzy, M., Heller, J., Balensi, J. & Herrera, M.D. 2005. The Mollusc Fauna from the Late Bronze and Iron Age Strata of Tell Abu Hawam. In: D.E. Bar-Yosef Mayer, ed. *Proceedings of the 9th ICAZ Conference, Durham 2002. Archaeomalacology: Molluscs in Former Environments of Human Behaviour*. Oxford: Oxbow, pp. 132–47.

Brewer, D.J. & Friedman, R.F. 1989. *Fish and Fishing in Ancient Egypt*. Warminster: Aris & Phillips.

Chahoud, J. & Vila, E. 2011/2012. The Role of Animals in Ancient Sidon: An Overview of Ongoing Zooarchaeological Studies. *Archaeology & History in the Lebanon,* 34–35: 259–84.

Chahoud, J. & Vila, E. 2017. Exploitation of Fauna at Ras Shamra: Case Study of the "Maison aus Albâtres", Late Bronze Age, Northern Levant. In: M. Mashkour & M. Beech, eds. *Archaeozoology of the Near East 9. In Honour of Hans-Peter Uerpmann and François Poplin*. Oxford: Oxbow Books, pp. 197–216.

Çilingiroğlu, Ç. & Çakırlar, C. 2018. Toplumsal tarihin küçük bir hareketi olarak tarih öncesi Ege'de deniz kaynaklarından faydalanma. *TINA Maritime Archaeology Periodical,* 9: 49–76.

de Moor, J.C. 1996. Fishes in KTU 4.427:23-29. *Ugarit-Forschungen,* 28: 155–58.

Dietrich, M. & Loretz, O. 1966. Der Vertrag zwischen Šuppiluliuma und Niqmandu: eine philologische und kulturhistorische Studie. *Die Welt des Orients,* 3: 206–45.

Dikaios, P. 1969. *Enkomi Excavations 1948–1958: IIIa*. Mainz: Philipp von Zabern.

Ferreira Teixeira, S. & Silva Campos, S. 2019. Mollusc Gathering in Tropical Regions of Brazil. In: G. Diarte-Plata & R. Escamilla-Montes, eds. *Molluscs*. IntechOpen. https://doi.org/10.5772/intechopen.83588.

Friedländer, P. 1909. Zur Kenntnis des Farbstoffes des antiken Purpurs aus *Murex brandaris*. *Monatshefte für Chemie,* 30: 247–53.

Fronzaroli, P. 1955. *Leggenda di Aqhat*. Il Melagrano 136–37. Florence: Edizioni Fussi.

Kalaitzaki, A., Vafiadou, A., Frony, A., Reese, D.S., Drivaliari, A. & Liritzis, I. 2017. Po-Pu-Re: Workshops, Use and Archaeometric Analysis in Pre-Roman Central Eastern Mediterranean. *Mediterranean Archaeology and Archaeometry,* 17(1): 103–30. https://doi.org/10.5281/ZENODO.160967.

Koren, Z.C. 2005. The First Optimal All-murex All-natural Purple Dyeing in the Eastern Mediterranean in a Millennium and a Half. *Dyes in History and Archaeology,* 20: 136–49.

Lackenbacher, S. 1995. Une correspondance entre l'administration du pharaon Merneptah et le roi d'Ougarit. In: M. Yon, M. Sznycer & P. Bordreuil, eds. *Le Pays d'Ougarit autour de 1200 av. J.-C. Actes du Colloque International de Paris, 28 Juin-1er Juillet* 1993. Ras Shamra-Ougarit 11. Paris: ERC, pp. 77–83.

Lackenbacher, S. 2001. Une letter d'Égypte. In: M. Yon, D.Arnaud, V. Matoïan & M. Maqdissi, eds. *Etudes Ougaritiques*. Ras Shamra-Ougarit 14. Paris: ERC, pp. 239–48.

Lambert, W. G. 2013. *Babylonian Creation Myths* (Mesopotamian Civilizations 16). Eisenbrauns.

Lernau, O. 2018. Faunal Evidence. Fish Remains. In: P.M. Fischer & T. Bürge, *Two Late Cypriot City Quarters at Hala Sultan Tekke: The Söderberg Expedition 2010-2017*. Studies in Mediterranean Archaeology CXLVII. Uppsala: Astrom Editions, 563–66.

Lichtheim, M. 2006. *Ancient Egyptian Literature: A Book of Readings*. 2006 ed. Berkeley, CA: University of California Press.

Linseele, V., Van Neer, W. & Bretschneider, J. 2013. The Mysteries of the Egyptian Nile Perch (*Lates niloticus*). The Case of Tell Tweini (Syria, Middle Bronze Age-Iron Age). In: B. De Cupere, V. Linseele & S. Hamilton-Dyer, eds. *Archaeozoology of the Near East X: Proceedings of the Tenth International Symposium on the Archaeozoology of South-Western Asia and Adjacent Areas*. Ancient Near Eastern Studies 44. Leuven and Walpole (MA): Peeters, pp. 209–26.

Linseele, V., Marinova, E. De Cupere, B., van der Valk, J., Vandorpe, P. & Van Neer, W. 2019. Bronze and Iron Age Palaeo-economy in a Changing Environment. The Bioarchaeology of Tell Tweini, on the Northern Levantine Coast. In: J. Bretschneider & G. Jans, eds. *About Tell Tweini (Syria): Artefacts, Ecofacts and Landscape: Research Results of the Belgian Mission*. Leuven: Peeters, pp. 417–618.

Malbran-Labat, F. 1991. Lettres. In: P. Bordreuil & D. Arnaud, *Une Bibliothèque au sud de la ville: les textes de la 34e campagne (1973)*. Ras Shamra-Ougarit 7. Paris: ERC, pp. 27–64.

Matoïan, V. & Vita, J.P. 2014. Wool Production and Economy at Ugarit. In: C. Breniquet & C. Michel, eds. *Wool Economy in the Ancient Near East*. Oxford: Oxbow Books, pp. 310–39.

Mayhoff, K.F.T. 1875–1908. *Naturalis Historia* [Pliny the Elder]. Teubner: Leipziz.

Mosapour Negari, F. 2016. Ancient Diet Reconstruction: A Case Study of Sidon, Lebanon. *Iranian Journal of Archaeological Studies,* 6: 31–43.

Niklasson, K. 1983. Tomb 23: A Shaft Grave of the Late Cypriote III Period. In: P. Åström, E. Åström, A. Hatziantonious, K. Niklasson & U. Öbrink, *Hala Sultan Tekke 8. Excavations 1971–1979*. Studies in Mediterranean Archaeology XLV:8. Jonsered: Åströms Förlag, pp. 169–213.

Potts, D.T. 2012. Fish and Fishing. In: D.T. Potts, ed. *A Companion to the Archaeology of the Ancient Near East*. Blackwell Companions to the Ancient World. Malden (MA): Wiley-Blackwell, pp. 220–36.

Reese, D.S. 2003. The Shells, Astragali and Fish. In: V. Karageorghis, *Excavations at Kition VI. The Phoenician and Later Levels*. Nicosia: Department of Antiquities, Cyprus, pp. 413-28.

Reese, D.S. 2007. Faunal Remains from Hala Sultan Tekke Tombs. In: P. Åström & K. Nys, eds. *Tomb 24, Stone Anchors, Faunal Remains and Pottery Provenance (Hala Sultan Tekke 12)*. Studies in Mediterranean Archaeology XLV:12. Sävedalen: Åströms Förlag, pp. 50–52.

Reese, D.S. 2008. Organic Imports from Late Bronze Age Cyprus (with Special Reference to Hala Sultan Tekke). *Opuscula Atheniensia,* 31–32: 191–209.

Reese, D.S. 2010. Shells from Sarepta (Lebanon) and East Mediterranean Purple-dye Production. *Mediterranean Archaeology and Archaeometry,* 10: 113–41.

Reese, D.S. 2018. Faunal Evidence. Invertebrates. In: P.M. Fischer & T. Bürge, *Two Late Cypriot City Quarters at Hala Sultan Tekke: The Söderberg Expedition 2010–2017*. Studies in Mediterranean Archaeology CXLVII. Uppsala: Astrom Editions, pp. 532–63.

Reese, D.S. 2022. Some Comments on the Unpublished Fauna from Enkomi-*Ayios Iakovos* (Dikaios' Excavations, 1948–1958). *Conference Paper presented at Fire and Bones: A Multidisciplinary Approach in the Study of Burnt Human Remains, 3rd International Congress on Archaeological Sciences in the Eastern Mediterranean and the Middle East (ICAS-EMME 3)*. Nicosia.

Reese, D.S., Mienis, H.K. & Woodward, F.R. 1986. On the Trade of Shells and Fish from the Nile River. *Bulletin of the American Schools of Oriental Research,* 264: 79–84.

Reifarth, N. & Völling, E. 2013. Spuren aus dem Reich der Farben. Einblicke in die Kunst antiker Textilfärberei. *Antike Welt,* 44: 33–39.

Rose, M. 1994. With Line and Glittering Bronze Hook: Fishing in the Aegean Bronze Age (PhD dissertation, Indiana University, Bloomington).

Routledge, B. 2015. A Fishy Business: The Inland Trade in Nile Perch (*Lates niloticus*) in the Early Iron Age Levant. In: T.P. Harrison, E.B. Banning & S. Klassen, eds. *Walls of the Prince: Egyptian Interactions with Southwest Asia in Antiquity: Essays in Honour of John S. Holladay, Jr.* Leiden: Brill, pp. 212–33. https://doi.org/10.1163/9789004302563.

Sahrhage, D. 1998. *Fischfang und Fischkult im alten Ägypten* (Kulturgeschichte der antiken Welt 70). Philipp von Zabern.

Schaeffer, C.F.A. 1951. Une industrie d'Ugarit, la pourpre. *Les Annales Archéologiques de Syrie,* 1: 188–92.

Schutkowski, H. & Ogden, A. 2011. Sidon of the Plain, Sidon of the Sea: Reflections on Middle Bronze Age Diet in the Eastern Mediterranean. *Archaeology & History in the Lebanon,* 34/35: 213–25.

Siar, S.V. 2003. Knowledge, Gender, and Resources in Small-scale Fishing: The Case of Honda Bay, Palawan, Philippines. *Environmental Management,* 31: 569–80. https://doi.org/10.1007/s00267-002-2872-7.

Sisma-Ventura, G., Zohar, I., Sarkar, A., Bhattacharyya, K., Zidane, A., Gilboa, A., Bar-Oz, G. & Sivan, D. 2015. Oxygen Isotope Composition of Sparidae (Sea Bream) Tooth Enamel from Well-dated Archaeological Sites as an Environmental Proxy in the East Mediterranean: A Case Study from Tel Dor, Israel. *Journal of Archaeological Science,* 64: 46–53. https://doi.org/10.1016/j.jas.2015.10.004.

Sisma-Ventura, G., Tütken, T., Zohar, I., Pack, A., Sivan, D., Lernau, O., Gilboa, A. & Bar-Oz, G. 2018. Tooth Oxygen Isotopes Reveal Late Bronze Age Origin of Mediterranean Fish Aquaculture and Trade. *Scientific Reports,* 8: 14086. https://doi.org/10.1038/s41598-018-32468-1.

Smith, M.S. & Pitard, W. 2009. *The Ugaritic Baal Cycle. Volume II. Introduction with Text, Translation and Commentary of KTU / CAT 1.3-1.4.* Supplements to Vetus Testamentum 114. Leiden & New York: Brill.

Stantis, C., Maaranen, N., Kharobi, A., Nowell, G.M., Macpherson, C., Doumet-Serhal, C. & Schutkowski, H. 2022. Sidon on the Breadth of the Wild Sea: Movement and Diet on the Mediterranean Coast in the Middle Bronze Age. *American Journal of Biological Anthropology,* 177: 116–33. https://doi.org/10.1002/ajpa.24423.

Thavapalan, S. 2016. Purple Fabrics and Garments in Akkadian Documents. *Journal of Ancient Near Eastern History,* 3: 163–90. https://doi.org/10.1515/janeh-2017-0007.

Thavapalan, S. 2020. *The Meaning of Color in Ancient Mesopotamia.* Culture and History of the Ancient Near East 104. Leiden and Boston: Brill.

Thompson, D.W. 1947. *A Glossary of Greek Fishes.* London: Oxford University Press.

van de Mieroop, M. 1997. *The Ancient Mesopotamian City.* Oxford and New York: Clarendon Press and Oxford University Press.

Van Neer, W. 2006. Bronze Age Fish Remains from Sidon. *Archaeology and History in Lebanon,* 24: 86–95.

Van Neer, W., Lernau, O., Friedman, R., Mumford, G., Poblome, J. & Waelkens, M. 2004. Fish Remains from Archaeological Sites as Indicators of Former Trade Connections in the Eastern Mediterranean. *Paléorient,* 30: 101–47. https://doi.org/10.3406/paleo.2004.4775.

van Soldt, W.H. 1990. Fabrics and Dyes at Ugarit. *Ugarit-Forschungen,* 22: 321–57.

Westerberg, K. 1983. *Cypriote Ships from the Bronze Age to c. 500 BC.* Studies in Mediterranean Archaeology. Pocket Books 22. Göteborg: Åströms Förlag.

Wild, H. 1953. *Le tombeau de Ti. Fascicule II. La chapelle.* IFAO 65,2. Cairo: Institut Français d'Archéologie Orientale.

Yamasaki, M. 2023. *Conceptualizing Bronze Age Seascapes. Concepts of the Sea and Marine Fauna in the Eastern Mediterranean in the Second Millennium BCE.* Levant and Eastern Mediterranean Archaeology 2. Turnhout: Brepols.

Zohar, I. & Artzy, M. 2019. The Role of Preserved Fish: Evidence of Fish Exploitation, Processing and Long-term Preservation in the Eastern Mediterranean during the Late Bronze Age (14th–13th Century BCE). *Journal of Archaeological Science Reports,* 23: 900–909. https://doi.org/10.1016/j.jasrep.2018.12.008.

Exploring Fishing in Cyprus from the Neolithic to the Early Christian Periods

7

Maria M. Michael

Abstract

The research presented here is based on the results developed during the author's PhD research project (Michael, M. *To Fish or Not to Fish? The Case Study of Fishing Activity in Cyprus* (PhD dissertation)). University of Southampton, Faculty of Arts and Humanities, Department of Archaeology, 2022). The current chapter briefly presents the tradition of fishing on the island of Cyprus from the Neolithic to the Early Christian periods. This is accomplished by examining the archaeological finds of fishing technology and fishbone assemblages recovered at several archaeological sites on Cyprus. The iconographic and written sources and the environmental and ethnographic data are supporting classes of evidence. With the examination of the main indicators of fishing, it is possible to enable the exploration of the reasons for the presence or absence of evidence in the Cypriot maritime landscape through time. Furthermore, the parallel study of the available evidence with aspects of traditional knowledge of fishing demonstrates that the absence of evidence sometimes does not equal the absence of fishing from the archaeological record. This examination contributes towards a more holistic understanding of fishing, which involves interdependent economic, cultural, technological, and environmental aspects. Based on the reported archaeoichthyological data, it appears that fishing on Cyprus occurred more or less continuously from the Neolithic to the Early Christian periods. The systematic mapping of the archaeological sites where finds of fishing gear and fish remains have been recovered demonstrates regional and chronological patterns, which are also discussed in this chapter.

M. M. Michael (✉)
Honor Frost Foundation – Cyprus Regional Development Project, London, UK
e-mail: mmm1r18@southamptonalumni.ac.uk

7.1 Introduction

Fishing can occur as an everyday or occasional activity with fairly simple equipment and artisanal fishing methods aiming to fulfil the needs of daily subsistence; it may be solitary or communal (Bekker-Nielsen 2010: 191; Cottica and Divari 2010: 363; Trakadas 2018: 73; Michael 2022: 68–77; 2023: 1). It can be a non-industrial or large-scale activity where a group of fishers aims to catch either for food or commercial purposes (Marzano 2013: 51–88; Trakadas 2018: 73–74). To achieve this aim, fishers must acquire a cognitive knowledge of ecology (the seabed ground) and more specifically how fish species behave daily, seasonally, and annually in their life cycles, as this is the main way to understand the environment where fish species live and fishers interact with them, in order to catch them (Hela and Laevastu 1962; Acheson 1981; McNiven 2003: 330–32; O'Sullivan 2003; Cooney 2004; Westerdahl 2007: 207–208; Rohling et al. 2009: 36; Duncan 2011; Theodoropoulou 2011). It is also essential to consider the meteorology (winds, currents) and biology (fish species availability) of this environment, to highlight the seasonality of fishing, and the targeted fish species. This knowledge and the knowledge of the basic features of fishing gear seem to endure through time and may likely be passed down from father to son over generations to the present day (O'Sullivan 2001: 269; 2003: 465–66).

Fishing cannot only be characterised as a simple two-way interaction between the sea and the fisher, who procures the necessary fishing gear (equipment) to catch the different fish species that are available to them; fishing is also an activity involving interdependent technological, social, economic, cultural, biological, and environmental aspects (Bekker-Nielsen 2010: 187; Michael 2022: 68–98). As demonstrated by Bekker-Nielsen and colleagues, taking all of these factors into account is important in order to fully appreciate the complexity of fishing practices and technologies in the past (see papers in Bekker-Nielsen and Bernal-Casasola 2010).

C. Tsouparopoulou, L. Recht (eds.), *Human and Aquatic Beings: Interactions in and beyond the Eastern Mediterranean (3rd–1st Millennia BCE)*, Themes in Contemporary Archaeology, https://doi.org/10.1007/978-3-031-73643-8_7

This is here attempted in relation to Cyprus, where this has not yet been systematically studied.

Thus, this chapter attempts to examine these aspects along with direct evidence for fishing preserved in the archaeological record on Cyprus from the Neolithic to Early Christian times (9200/9000 BCE–647 CE), the fish remains (the result of fishing), and archaeological evidence of fishing technology (equipment: harpoons or spears, fish-hooks, and traps, stone, clay, and lead weights for a net or line, and structures such as fish-ponds). It might be expected that the occurrence and nature of fishing on Cyprus through time can be easily identified due to the privileged geographical position of Cyprus as an island in the eastern Mediterranean. The latter possibly encouraged the exploration of the wider environment of the Mediterranean, the navigation of the seas and coasts, and the rise of maritime connectivity within the Mediterranean basin. However, fishing and its full significance in the ancient context of Cyprus have rarely been fully acknowledged (Michael 2022: 33–62). As a result, many scholars have adopted the idea that the people on Cyprus have not engaged intensively in fishing in the past (Dikaios 1961: 290; Egoumenidou and Michaelides 2000: 112; Ionas 2001: 217; Knapp 2018: 151).

The fact that the study of fishing is a neglected subject in Cyprus seems to be due to a lack of suitable methodological approaches to record, examine and interpret archaeological evidence of fishing (Michael 2022: 62–64); as a result, the only way to gain more primary archaeological information on fishing is by raising awareness of the study of this understudied subject (Maarleveld 2010: 270). Honor Frost was the first archaeologist who addressed the problematic presentation and analysis of evidence related to fishing in the archaeological context of Cyprus, and she pointed out that a lack of solid methodological approaches could lead to misinterpretation and incorrect classification of both fish remains and evidence of fishing gear. Through her examination of three limestone fishing weights from the archaeological site of Kition-*Bamboula*, she demonstrated that the study of fishing would benefit from the development of a systematic, interdisciplinary methodology for recording the different types of archaeological evidence and combine them in a single corpus (Frost 1985: 169–71). This was achieved by the author's PhD research project (Michael 2022), which is the first attempt to gather all available evidence for fishing on Cyprus in a single body of evidence.

Consequently, this chapter aims to provide a systematic and interdisciplinary recording of available evidence of fishing on Cyprus through time. Information on the wider environmental and social context in which people on Cyprus established and developed fishing is provided, followed by a brief overview of the dataset and the methodological approach applied to the collection and analysis of the available evidence. Selected chronological and regional patterns revealed through the systematic mapping of archaeological sites where archaeoichthyological data have been recovered are discussed with the available descriptive, environmental, and ethnographic data, leading to the exploration of fishing in the Cypriot maritime landscape over time.

7.2 Geographical and Chronological Context

The study of the geographical context of maritime Cyprus is an important component for understanding and reconstructing fishing through time because it can reveal fishers' knowledge of the environment, ecology, meteorology, and biology, which mainly develops through daily interaction with and experience of them. This intangible knowledge helps improve our understanding of the relationship between fishers and their maritime environment and determine where and when fishing was established on Cyprus over time.

The island of Cyprus is in the eastern basin of the Mediterranean (Fig. 7.1). Fishing is carried out mainly within the area of the continental shelf, which is defined as the seafloor at water depths shallower than 200 m and is also narrow and at maximum extends to about 16 km from the shore (Demetropoulos 1985: 70; Department of Fisheries and Marine Research 2012: 2). Specifically, it is narrow in the north and wider in the south, while it also contains numerous rock outcrops, mainly in the northern part (Department of Fisheries and Marine Research 2012: 2). Although it is narrower in the north, ethnographic data collected by the author from the Archive of Oral Tradition and Folk Study (Cyprus Research Centre) highlight that fishing grounds on this side of the island are more fertile because the seabed is rocky and covered by seaweeds and seagrasses, which supply nutrition for many fishes (Michael 2022: 218–21). As a result, these areas can generally be characterised as desirable fishing grounds with high biodiversity and endemism, although the Mediterranean is one of the most oligotrophic ocean regions in the world because of its high salinity and temperature, especially in the eastern basin (Hela and Laevastu 1962: 18–26; Béthoux 1984: 48; Golani et al. 2006: 8–9; Coll et al. 2000: 12; Bariche 2012: 3–4). This may be the main reason why evidence of aquaculture is mainly found on the northern coast of the island, where the fishing grounds are more fertile (Michael 2022: 378–88).

Winds and currents also have a significant effect on aiding fishers in identifying and choosing fishing grounds, on sailing, or rowing to fish, and returning home without losing the catch (Marsellou 1948: 13–31; Rose 1994: 49–50; Ayodeji 2004: 49; Obied 2016: 102; Safadi 2018: 227–39). Although there is a general wind form in Cypriot coastal areas, a variety of wind patterns appears on the different sides of the island (Safadi 2016: 353–55; Department of Fisheries and Marine Research 2019: 6–7). In coastal areas, where fishing occurred, local land

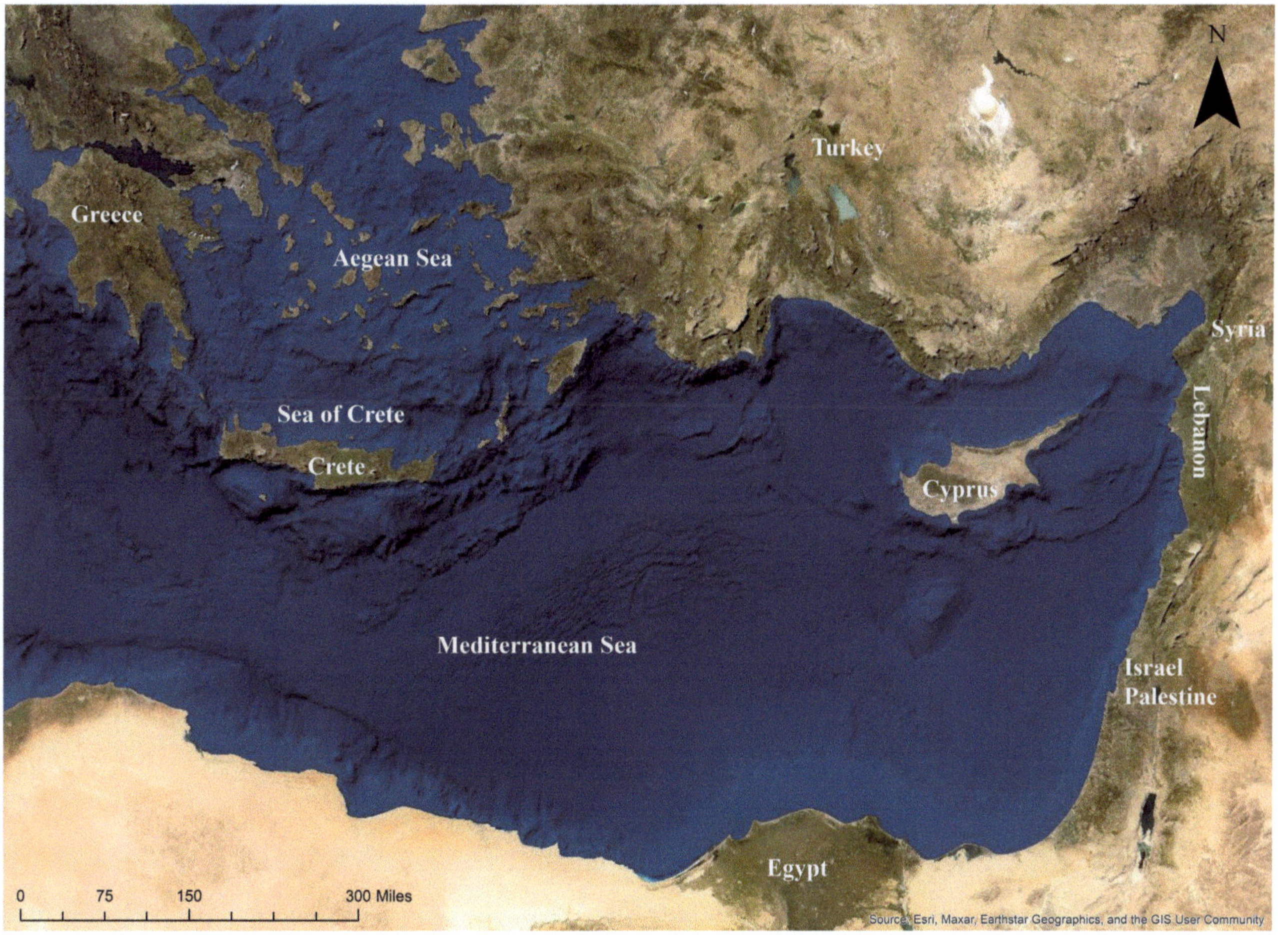

Fig. 7.1 Map of Cyprus and the eastern Mediterranean. (Produced by M. M. Michael. Source of the basemap: Esri, DigitalGlobe, GeoEye, Earthstar Geographics, CNES/Airbus DS, USDA, USGS, AeroGRID, IGN, and the GIS User Community)

and sea breezes are also superimposed on gradient winds (Meteorological Service 1986: 6). Consequently, fishers must consider these environmental conditions to achieve the successful establishment and development of fishing.

The chronological context of this study spans from the Neolithic period (tenth millennium BCE) to the Early Christian period (mid-seventh century CE) (Fig. 7.2). The archaeological finds of fishing technology and fishbone assemblages, which are the main indicators of fishing during this chronological context, are unevenly distributed, with some periods yielding more evidence than others (in particular, the Neolithic, the Late Bronze Age, and the Hellenistic and Roman periods) (Michael 2022: 257–404). As discussed in the next section (see Sect. 7.3), this phenomenon is mainly because of the lack of recording and interpretation of such finds, and the lack of a proper methodological approach. For instance, the limited or non-existent use of wet-sieving in conjunction with the absence of precise and integrated reference collections seems to be the main factor that affects the recovery of small finds, such as fish-hooks and fish remains, and the precise identification of fish species. Consequently, this evidence can be more easily analysed within its chronological and archaeological contexts, and this contextual analysis highlights exploitation patterns, ritual, and subsistence habits within Cypriot society (Michael 2022: 103–21, 257–300).

This wide chronological framework allows for the identification of patterns of continuity or/and change in fishing technology over a time span of nearly ten millennia, along with their relation to socio-economic developments on the island. For instance, fishers may have started experimenting to understand whether the use of the easily accessible raw material of copper could improve the effectiveness of their fishing gear during the Bronze Age. Also, the occurrence of copper alloys and lead fishing gear from the Bronze Age onwards demonstrates that fishers decided to adopt other raw materials in the manufacture of fishing gear, presumably due to their increased strength and resistance to corrosion (Michael 2022: 336–38, 420–22). Consequently, through this diachronic study, it is observed that the main form and function of fishing gear remained the same (continuity), but at the same time, a technological change is observed based on the material due to a desire to improve the efficiency of the fishing gear.

	Cultural Period	Period Subdivisions	Date Range
Prehistoric Periods	Late Epipalaeolithic	Akrotiri phase	ca. 11000-9000 BCE
	Neolithic period	Initial Aceramic Neolithic	ca. 9200/9000-8500/8400 BCE
		Early Aceramic Neolithic	ca. 8500/8400-7000/6800 BCE
		Late Aceramic Neolithic (Khirokitia)	ca. 7000/6800-5200 BCE
		Ceramic Neolithic (Sotira)	ca. 5200/5000-4000/3900 BCE
	Chalcolithic period (Erimi)	Early Chalcolithic	ca. 3900-3600/3400 BCE
		Middle Chalcolithic	ca. 3600/3400-2700 BCE
		Late Chalcolithic	ca. 2700-2500 BCE
	Bronze Age	Philia Phase	ca. 2450-2250 BCE
		Early Bronze Age	ca. 2500-2000 BCE
		Middle Bronze Age	ca. 2000-1650 BCE
		Late Bronze Age	
		Late Cypriot I	ca. 1650-1450 BCE
		Late Cypriot II	ca. 1450-1200 BCE
		Late Cypriot IIIA	ca. 1200-1125/1100 BCE
Historic Periods	Early Iron Age	Late Cypriot IIIB	ca. 1125/1100-1050 BCE
		Cypro-Geometric I	ca. 1050-950 BCE
		Cypro-Geometric II	ca. 950-900 BCE
		Cypro-Geometric III	ca. 900-750 BCE
	Cypriot City-Kingdoms (Iron Age)	Cypro-Archaic I	ca. 750-600 BCE
		Cypro-Archaic II	ca. 600-480 BCE
		Cypro-Classical I	ca. 480-400 BCE
		Cypro-Classical II	ca. 400-310 BCE
	Hellenistic	Hellenistic I	ca. 310-217 BCE
		Hellenistic II	ca. 217-31 BCE
	Roman	Roman	ca. 31 BCE-330 CE
	Early Byzantine period	Proto-Christian	ca. 330-647 CE

Fig. 7.2 Chronology of cultural periods in Cyprus. (Produced by M. M. Michael (sources of information: Department of Antiquities 2005–2024; Knapp 2013: 27, Table 2; Papantoniou and Kyriakou 2018: 544, Table 1))

7.3 Data and Methodology

Fish remains and evidence of fishing gear recovered from Cypriot archaeological sites constitute the main dataset of this research. Their collection was mainly conducted through an intensive desk-based study of published final reports of Cypriot archaeological sites, terrestrial or underwater, and museum inventories. Also, fieldwork was conducted for the systematic examination of the excavated archaeological finds of fishing gear identified during the desk-based study. No detailed archaeofaunal study was conducted during the fieldwork, primarily because of the restricted availability of fishbone assemblages, in conjunction with the author's own non-expert knowledge with regard to direct examination.

This evidence, which has been examined in association with environmental, biological, and ethnographic data to reconstruct fishing on Cyprus through time, has been recorded in a customised database. Distributional maps have been produced using the location of archaeological sites with archaeoichthyological data. Evidence from 239 archaeological sites dating from the Neolithic period to the Late Roman/Early Christian period has been collected and examined; fishing gear and fish remains were recovered from 74 of these sites.

7.3.1 Fish Remains

Identified and unidentified fish remains have been recovered from 54 sites distributed throughout the island and dated from the Neolithic to the Early Christian period. More specifically, 12 sites produced fishbone assemblages dated to the Neolithic period, four sites to the Chalcolithic period, 21 sites to the Bronze Age, two to the Cypro-Archaic period, and six sites to the Hellenistic/Roman periods. There are also some sites which have produced fishbone assemblages from several periods. Although the number of sites in some periods is great compared to that of other periods, this number is not always representative of the number of fish remains. For example, 21 Bronze Age sites have produced fish remains, but the quantity of these remains is much less when compared only to the Neolithic site of Cape Apostolos Andreas, Kastros, which has produced more than 6000 remains of bony fish (Garnier 1981).

The numbers of fish bones from many sites are usually very small, or/and unquantified, and it seems that the fishbone assemblages vary greatly from just one or two bones to over a thousand. For example, only 14 fish bones were recovered during the excavations from 1981 to 1989 at Kition (Cypro-Archaic I or Cypro-Classic I-II periods), but only eight were determined to be the species of Serranides, Epinephelus/Mycteroroperca (Piques 2015: 363–64). Also, it is often the case that the recovery techniques and the quantity of fish remains in a given assemblage are only provided in general terms, with or without mentioning individual specimens amongst these. For instance, an unknown quantity of fish remains was recovered at the Cypro-Geometric Tomb 79 at Salamis, and they were identified to at least three genera and species: *Diplodus annularis,* Epinephelus, and *Clarias* (Greenwood and Howes 1973). However, no quantity or identification has been provided for the fish remains recovered at the Late Bronze Age phase of Kouklia, Paphos (Halstead 1977: 263). In the aforementioned sites, no reference to the recovery techniques was included in the publication.

Researchers who study fish remains and their related activities argue that collection policy and the methodological approach of their examination and identification that archaeologists decide to apply during excavation influence their study. For instance, the quantity of fish remains at some archaeological sites is limited or their identification is not detailed because a great number of archaeologists do not employ careful recovery techniques, such as sieving (Wheeler and Jones 1989: 38–40; Van Neer et al. 2005: 132, 142; Morales-Muñiz et al. 2007: 119; Potts 2012: 221; Bernal-Casasola 2016: 193–94; Trakadas 2018: 52–53). Even where sieving has been used, the mesh size of the screens can affect the nature and quantity of a fishbone assemblage (Reitz and Wing 2008: 147–50; Ritchie 2010: 13). Smaller mesh sizes lead to the recovery of smaller fish remains; as a result, if smaller mesh size screens were not used on an excavation, then most probably a significant proportion of small fish would not be recovered and would instead be lost during the process of excavation. In the case of Cyprus, the absence of remains of small species such as *Spicara smaris* (Picarel), *Spicara maena* (Blotched Picarel), and *Boops boops* (Bogue), which have high commercial importance nowadays (FAO 2005), is probably a result of the non-existent use of sieving or/and smaller mesh sizes, and/or the depositional environment, including its acidic soil (Rose 2023: 201, 214–15).

The "selective" use of sieving by archaeologists and their experience and skills also affect the recovered assemblage (Locker 2007: 141–42; Ritchie 2010: 13; Trakadas 2018: 52–53). For instance, fish remains at the Neolithic site of Paralimni-*Nissia* are seriously underrepresented in its faunal assemblage because very little wet-sieving was carried out during the excavation (Croft 2008: 105). Moreover, unidentified fish remains were recovered at the Hellenistic site of Geronisos, although sieving was used during the excavation (Bretton-Connelly 2002: 263). Consequently, the choice of better excavation and recovery techniques can affect the recovered assemblage, which may include a larger diversity of fish remains and more species from smaller fish.

In the current research, these methodological issues cause challenges in defining the intensity of fishing during each chronological period or at each archaeological site. To overcome these issues, which create difficulties in comparisons of fish remains between archaeological assemblages, each identified fish is considered here as a unique occurrence within the chronological context in which it is recovered, while unidentified fish remains have simply been noted as present (Locker 2007: 144; Trakadas 2018: 53–54; Michael 2022: 107–12). Although the sample sizes in comparison between archaeological sites or between parts of a site are not adequate and comparable in the case of Cyprus (Reitz and Wing 2008: 113–14, 151), this approach allows at least the identification of which fish were caught. However, it does not demonstrate which species were not caught or/and which fish had a high commercial/subsistence importance within an archaeological context. On the other hand, the approach highlights the distribution of fish remains throughout the spatial and temporal context of Cyprus, and it presents the continuity of the exploitation of some fish through time (Table 7.1).

Based on the above table, it seems that more identified fish remains have been recovered from Neolithic and Bronze Age sites, while sites from later periods (Hellenistic/Roman/Early Christian) demonstrate a lack of identified fish remains. Furthermore, some identified fish species (i.e. *Argyrosomus regius* (Meagre), *Chelon ramada* (Thinlip grey mullet), *Chondrichthyes* (Sharks and rays), *Clarias* sp. (Catfish), *Dentex* sp. (Dentex), *Dicentrarchus labrax* (European seabass), *Epinephelus* sp. (Grouper), *Muraena helena* (Mediterranean moray), *Pagrus pagrus* (Red porgy), *Pagellus* sp. (Pandora), *Scomber scombrus* (Atlantic mackerel), *Scorpaena scrofa* (Red Scorpionfish), *Sparus aurata* (Gilthead seabream), *Sphyraena* sp., *Sphyraena sphyraena* (Barracuda), *Thunnus thynnus* (Bluefin tuna), *Trachurus trachurus* (Horse mackerel)) seem to have a diachronic presence at Cypriot archaeological sites, a phenomenon that may indicate long-term engagement with fishing.

7.3.2 Evidence of Fishing Technology

The other important element for the organisation of fishing is the gear used. The collection and examination of the evidence for fishing technology were challenging due to limited mentions in the excavation reports, publications, and

Table 7.1 Occurrence of identified fish in Cypriot archaeological sites through time

Family Taxa Common name	Neolithic period	Chalcolithic period	Bronze Age	Cypro-Geometric period	Cypro-Archaic period	Cypro-Classical period	Hellenistic/ Roman/ Early Christian periods
Chrondrichthyes (sharks/rays)	X: 1 site: Cape Apostolos Andreas, Kastros		X: 4 sites: Akhera / Athienou-*Pamboularin Koukkouninas* / Enkomi / Korovia-***Nitovikla***	X: 1 site: Kition	X:1 site: Athienou-*Malloura*		
Carcharhinidae							
Carcharhinus sp.	X: 1 site: *Ais Yiorkis* (Kritou Marotou)						
Dasyatidae	X: 1 site: Pareklissia-*Shillourokampos*						
Elasmobranchii Shark/ray			X: 2 sites: Hala Sultan Tekke / Idalion		X: 1 site: Salamis		
Polyprionidae							
Polyprion americanus Wreckfish			X: 1site: Hala Sultan Tekke				
Muraenidae							
Muraena helena Mediterranean moray	X: 1 site: Cape Apostolos Andreas, Kastros	X: 1 site: Kissonerga-*Mylouthkia*					
Latidae							
Lates niloticus Nile perch			X: 1 site: Hala Sultan Tekke	X: 1 site: Kition	X: 1 site: Kition	X: 1 site: Kition	
Sphyraenidae							
Sphyraena sp. *Sphyraena sphyraena* European barracuda	X:1 site: Cape Apostolos Andreas, Kastros		X: 1 site: Sotira-*Kaminoudhia*				
Carangidae	X: 1 site: *Ais Yiorkis* (Kritou Marotou)				X: 1 site: Amathus	X: 1 site: Amathus	
Seriola dumerili Greater amberjack	X:2 sites: Cape Apostolos Andreas, Kastros / Khirokitia						
Trachurus trachurus	X:2 sites: Cape Apostolos Andreas, Kastros / Kissonerga-*Mylouthkia*	X:1 site: Kissonerga-*Mylouthkia*			X:1 site: Salamis		
Cichlidae			X: 1 site: Hala Sultan Tekke				
Alosidae							
Sardina pilchardus European pilchard		X: 2 sites: Kissonerga-*Mosphilia* / Kissonerga-*Mylouthkia*					
Clupeidae	X: 1 site: Kissonerga-*Mylouthkia*	X:1 site: Kissonerga-*Mosphilia*					
Cyprinidae							
Cyprinus carpio Common carp			X:1 site: Apliki-*Karamallos*				
Labridae							
Labrus sp.				X: 1 site: Kition			

(continued)

Table 7.1 (continued)

Family Taxa Common name	Neolithic period	Chalcolithic period	Bronze Age	Cypro-Geometric period	Cypro-Archaic period	Cypro-Classical period	Hellenistic/ Roman/ Early Christian periods
Moronidae							
Dicentrarchus labrax European seabass	X: 1 site: Khirokitia	X: 1 site: Kissonerga-*Mylouthkia*	X:2 sites: Athienou-*Pamboularin Koukkounias* / Hala Sultan Tekke			X: 1 site: Amathus	
Scaridae							
Scarus sp. Parrotfish		X: 1 site: Kissonerga-*Mosphilia*					
Sparisoma cretense Parrotfish			X: 1 site: Hala Sultan Tekke				
Sciaenidae							
Argyrosomus regius Meagre	X: 1 site: Cape Apostolos Andreas, Kastros		X: 1 site: Hala Sultan Tekke			X: 1 site: Amathus	
Sciaena umbra Brown meagre	X: 1 site: Cape Apostolos Andreas, Kastros						
Umbrina cirrosa Shi drum			X:1 site: Athienou-*Pamboularin Koukkouninas*				
Sparidae	X:2 sites: Cape Apostolos Andreas, Kastros/ Khirokitia	X:2 sites: Kissonerga-*Mosphilia* / Kissonerga-*Mylouthkia*	X:2 sites: Athienou-*Pamboularin Koukkouninas* / Hala Sultan Tekke	X:1 site: Kition	X: 1 site: Salamis	X: 1 site: Amathus	
Dentex sp.	X: 2 sites: Cape Apostolos Andreas, Kastros / Khirokitia				X: 1 site: Salamis	X: 1 site: Amathus	
Diplodus annularis Annular seabream		X: 1 site: Kissonerga-*Mylouthkia*		X: 1 site: Salamis			
Diplodus sp. Seabream	X: 1 site: Cape Apostolos Andreas, Kastros						
Oblada sp. Saddled seabream	X: 1 site: Cape Apostolos Andreas, Kastros						
Pagellus sp. Pandora	X: 1 site: Cape Apostolos Andreas, Kastros		X: 1 site: Kition				
Pagrus pagrus Red porgy	X: 1 site: Cape Apostolos Andreas, Kastros		X: 1 site: Kalavasos-*Mangia*	X: 1 site: Kition			
Sarpa salpa Salema	X: 1 site: Cape Apostolos Andreas, Kastros						
Sparus aurata Gilthead seabream	X: 2 sites: Cape Apostolos Andreas Kastros / Khirokitia		X:1 site: Hala Sultan Tekke	X: 1 site: Kition		X: 1 site: Amathus	
Sparus sp.			X:1 site: Maroni-*Tsaroukkas*				
Spondyliosoma cantharus Black seabream			X: 1 site: Hala Sultan Tekke				

(continued)

Table 7.1 (continued)

Family Taxa Common name	Neolithic period	Chalcolithic period	Bronze Age	Cypro-Geometric period	Cypro-Archaic period	Cypro-Classical period	Hellenistic/ Roman/ Early Christian periods
Merlucciidae							
Merluccius merluccius European hake	X: 1 site: Kissonerga-*Mylouthkia*						
Gadidae							
Micromesistius poutassou Blue whiting		X: 1 site: Kissonerga-*Mylouthkia*					
Lophiidae							
Lophius piscatorius Anlgler				X: 1 site: Kition			
Mugilidae	X: 2 sites: Cape Apostolos Andreas, Kastros / Khirokitia		X:2 sites: Athienou-*Pamboularin Koukkouninas* / Hala Sultan Tekke	X: 1 site: Kition		X: 1 site: Amathus	
Chelon ramada Thinlip grey mullet		X: 1 site: Kissonerga-*Mosphilia*	X: 1 site: Hala Sultan Tekke				
Mugil cephalus Flathead grey mullet			X:2 sites: Apliki-*Karamalos* / Hala Sultan Tekke				
Mullidae							
Mullus sp. Mullet			X: 1 site: Hala Sultan Tekke				
Scorpaenidae							
Scorpaena scrofa Red scorpionfish	X: 1 site: Kissonerga-*Mylouthkia*			X: 1 site: Kition			
Serranidae	X:3 sites: Khirokitia / Kissonerga-*Mylouthkia* / Pareklissia-*Shillourokambos*	X:2 sites: Kissonerga-*Mosphilia* / Kissonerga-*Mylouthkia*	X:2 sites: Hala Sultan Tekke / Kition				
Epinephelus aeneus Grouper			X: 1 site: Hala Sultan Tekke				
Epinephelus marginatus Grouper (archaeological report: *Epinephelus gigas*)						X: 1 site: Amathus	
Epinephelus sp. Grouper	X: 4 sites: *Ais Yiorkis* (Kritou Marotou) / Cape Apostolos Andreas, Kastros / Kalavasos-*Tenta* / Khirokitia	X:1 site: Kissonerga-*Mosphilia*		X: 2 sites: Kition / Salamis	X: 1 site: Kition	X: 2 sites: Amathus / Kition	
Triglidae		X: 1 site: Kissonerga-*Mosphilia*					
Pleuronectidae							
Platichthys flesus European flounder	X: 1 site: Kissonerga-*Mylouthkia*						

(continued)

Table 7.1 (continued)

Family Taxa Common name	Neolithic period	Chalcolithic period	Bronze Age	Cypro-Geometric period	Cypro-Archaic period	Cypro-Classical period	Hellenistic/ Roman/ Early Christian periods
Scombridae	X:2 sites: *Ais Yiorkis* (Kritou Marotou) / Kissonerga-*Mylouthkia*	X:1 site: Kissonerga-*Mylouthkia*					
Euthynnus alletteratus Little tunny	X: 2 sites: Cape Apostolos Andreas Kastros / Kissonerga-*Mylouthkia*						
Scomber scombrus Atlantic mackerel	X:2 sites: Khirokitia / Kissonerga-*Mylouthkia*	X:1 site: Kissonerga-*Mosphilia*					
Thunnus alalunga Albacore	X: 1 site: Kissonerga-*Mylouthkia*						
Thunnus thynnus Atlantic bluefin tuna	X: 1 site: Cape Apostolos Andreas, Kastros		X:2 sites: Hala Sultan Tekke / Maa-*Palaeokastro*	X: 1 site: Kition			
Clariidae							
Clarias sp.				X: 1 site: Salamis		X:1 site: Kouklia-*Hadjiabdoullah Palaepaphos*	
Ariidae							
Netuma thalassina Giant sea catfish (archaeological report: *Arius thalassinus*)							X: 1 site: Kition-*Bamboula*
Balistidae							
Balistes capriscus Grey triggerfish (archaeological report: *Balistes carolinensis*)	X:1 site: Cape Apostolos Andreas, Kastros						
Zeidae							
Zeus faber John Dory		X:1 site: Kissonerga-*Mosphilia*					

Produced by M. M. Michael and taxonomical order by G. Fyttis, Marine Biologist, University of Cyprus

inventories, often without a detailed description or any analysis regarding their context and their importance at the site from which they were recovered (Michael 2022: 19–22, 50–61). Their descriptions are often limited to the material they were made of, their basic attributes, and their potential use (Michael 2022: 112–21). Many of the objects are also catalogued under miscellaneous with limited or no contextual information (Gabriel et al. 2005: 508); as a result, it is difficult to examine and interpret them, especially as they are rarely photographed and sometimes cannot be found within museum storerooms. Some of them are also described as multifunctional due to the inability of their excavators to distinguish their specific use. For instance, 85 perforated limestone shaped discs retrieved from the Late Ceramic Neolithic site Kantou-*Koufovounos* have been interpreted as spindle whorls, loomweights, and net or line weights (Mantzourani 2019: 317–18). Consequently, it is difficult to define their accurate use from the published photos and restricted contextual information.

Despite these limitations, evidence of fishing gear has been recorded at 48 sites dated to different chronological periods: nine Neolithic sites, two Chalcolithic sites, 12 Bronze Age sites, two Cypro-Geometric sites, two Cypro-Archaic sites, two Cypro-Classical sites, 18 Hellenistic/Roman/Early Christian sites, and one site with an unknown chronological context. Fishing gear includes a wide range of tools: fish-hooks, gorges, stone, lead, and clay weights for lines and nets, needles, traps, and harpoons, or spears. In the Cypriot context, this evidence mainly consists of fish-hooks, and stone and lead weights, whose

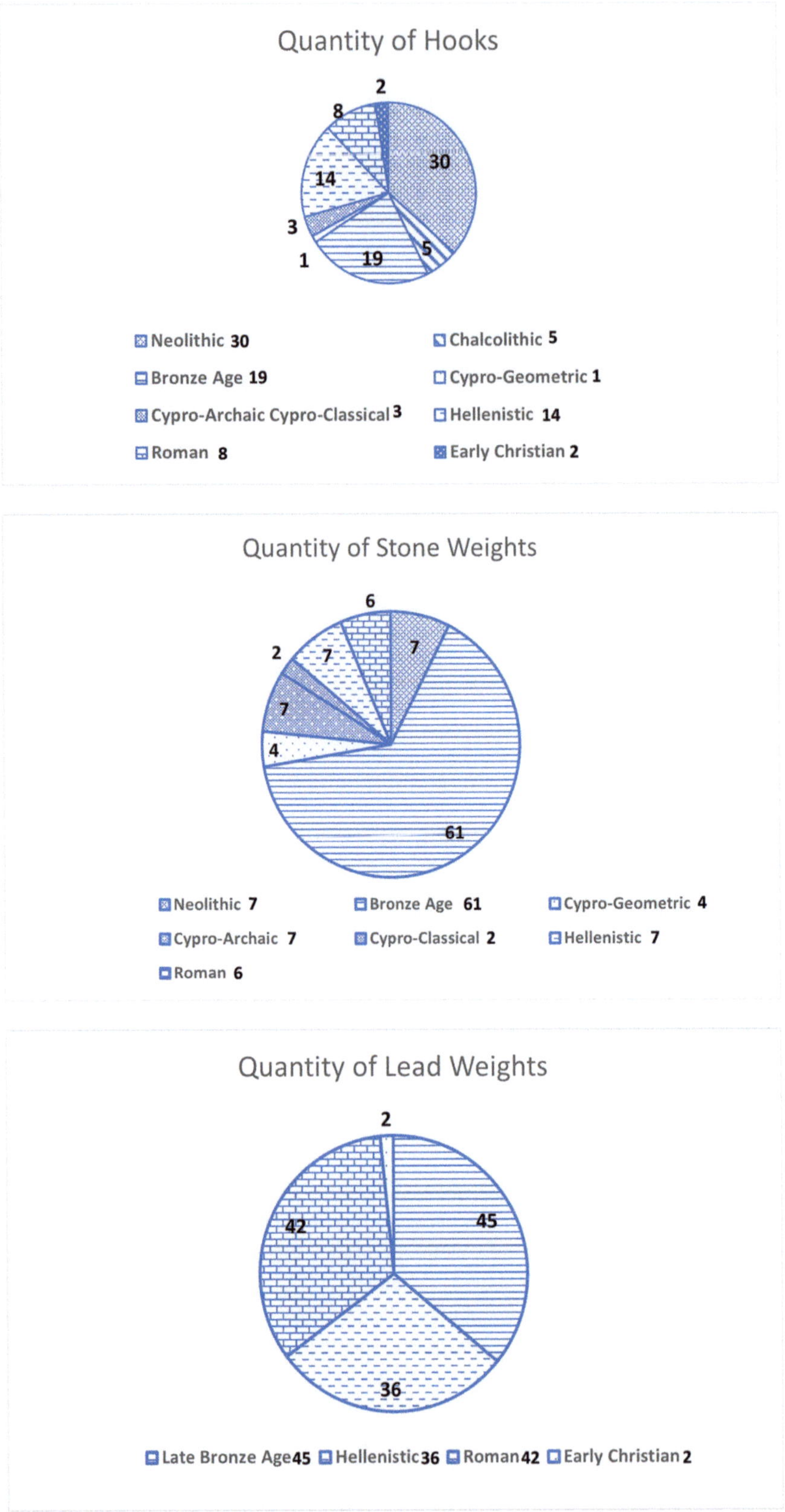

Fig. 7.3 Pie charts presenting the quantity of different identified types of fishing gear (fish-hooks, stone weights, and lead weights) recovered in Cypriot archaeological sites through time. Based on these pie charts, it is obvious that the Neolithic, Bronze Age, and later periods (Hellenistic/Roman) have produced more prominent evidence, the analysis of which can assist in the reconstruction of fishing methods and the better understanding of fishing activity. (Produced by M. M. Michael)

quantity differs from period to period (Fig. 7.3; Michael 2022: 257–300; 2023: 4–7). These quantities must be considered cautiously, and they cannot be used as indicators of the intensity of fishing on Cyprus through time. As already mentioned, the limited amount of evidence in some periods may be the result of methodological approaches, such as the use of sieving during excavation, and/or the limited experience of some archaeologists in recognising it archaeologically (Bernal-Casasola 2016: 201–202; Michael 2023: 4). For example, the function of net clay weights can be difficult to identify because they have the same shape as loomweights or spindle whorls, and as a result, they are often connected with textiles, especially if they were recovered within terrestrial, archaeological contexts (Powell 1996: 106; Cottica and Divari 2010: 347; Giner 2010: 77–78; Michael 2022: 8, 56, 277, 339). On Cyprus, it is considered that fishing gear can commonly be found at coastal sites, as fishing was an activity for people living at coastal settlements (Egoumenidou and Michaelides 2000: 114); archaeologists excavating a terrestrial site may therefore miss or incorrectly interpret evidence related to fishing because they do not expect to recover it. This can lead to misunderstandings and incorrect interpretations of the finds, and in turn, it can affect the quantity of recovered finds related to fishing.

The fish-hooks were made of bone, copper, and bronze. The earliest examples have a half-circular shape, and the latest a "J" shape (Fig. 7.4). They are usually barbed, and at the top of their shanks, there is usually an eye or grooves where the line is tied. In terms of size, fish-hooks can be very small (<25 mm), small (25–40 mm) and medium (40–80 mm) (Bernal-Casasola 2010: 89–90; Michael 2023: 4). Bone and bronze gorges have also been recovered; these were used in place of hooks. The recovery of hooks indicates the use of fishing lines.

Fishing nets are not preserved in the archaeological record of Cyprus because of their perishable materials (Leach 2006: 86, 92; Giner 2010: 58–61; Thomas 2010; Bernal-Casasola 2016: 203; Michael 2023: 5). Only the needles used to manufacture them, and the weights lined on their edges are preserved. The needles are mainly made of bronze with tongs at both ends (Fig. 7.5). The weights, which were used to anchor the net or to help it sink, are mainly made of clay, stone, and lead, and they come in a variety of sizes, shapes, and forms. The most common type of net weight is the folded rectangular lead weight, which was bent over the ground line of the net, while the weights with a more solid body formed from a lead mass and a groove, a hole, or a ring for a line attachment were mainly used as line weights, but they could also be used as net weights (Galili et al. 2002: 183–84). Although some shapes of weights could suggest their use as a fishing weight, this variety of shapes and sizes can make it difficult to come to a final conclusion about their use. To overcome this issue, scholars state that fishing weights are heavier than loomweights, and their weights usually range from 2 to 7 kg, while the weights of pebbles used as weights on the footrope of a net range between 20 and 243 g, and their length ranges between 37 to 100 mm (Frost 1991: 235; Rosenberg et al. 2016: 460). The large stone weights with straight perforations, which were used on the net edges to anchor it or as the extremities of set-long lines, weighed about 10 kg (Frost 1991: 235).

Beside the fish-hooks and stone and lead weights, the Late Bronze Age produced one harpoon and one trident, which may or may not be related to fishing (Niklasson 1983: 174, 206; Courtois 1984: 18; Rose 1994: 139, 167–68; Lindqvist 2016: 241). Three sites with remains of fish-ponds dating to later periods (Hellenistic, Roman and Early Christian) are also related to fishing, as they are evidence of aquaculture (Quilici 1971: 13–34; Nicolaou and Flinder 1976; Du Plat Taylor 1980: 154). They are rock-cut basins built entirely on the coast and filled with salt-water in which live fish were kept (Fig. 7.6). Finally, the parallel examination of ethnographic data with the iconographic and written sources (Michael 2022: 237–40) indicates the use of basket fish-traps, usually not preserved in the archaeological record because they are made of perishable materials such as wood.

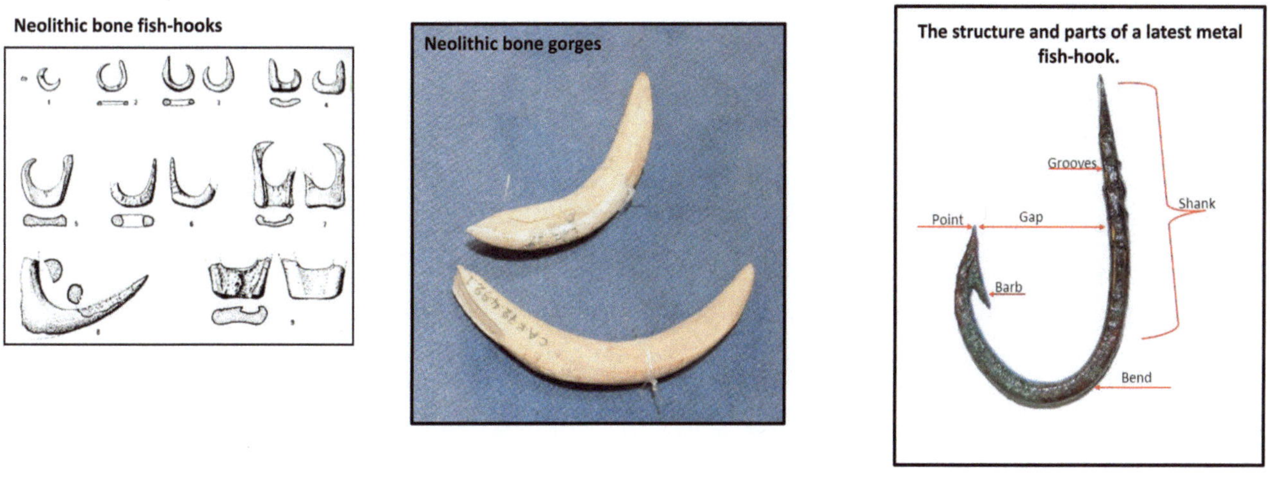

Fig. 7.4 Fish-hooks and gorge from Cyprus. (From Le Brun 1981: 203, fig. 56.1–9; Photos by M. M. Michael. By permission of the Director of the Department of Antiquities, Cyprus)

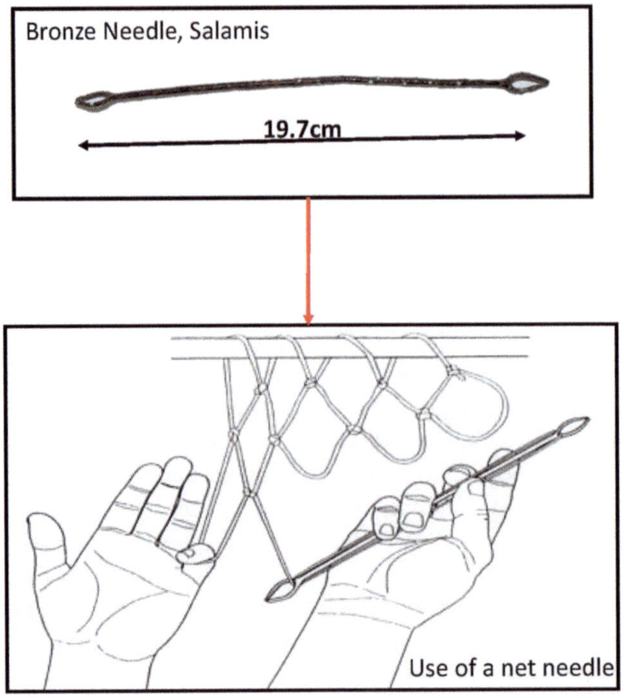

Fig. 7.5 Net needle from an archaeological site in Cyprus, and reconstruction of its use. (From Galili et al. 2013: 154, fig. 14; Photo by M. M. Michael. By permission of the Director of the Department of Antiquities, Cyprus)

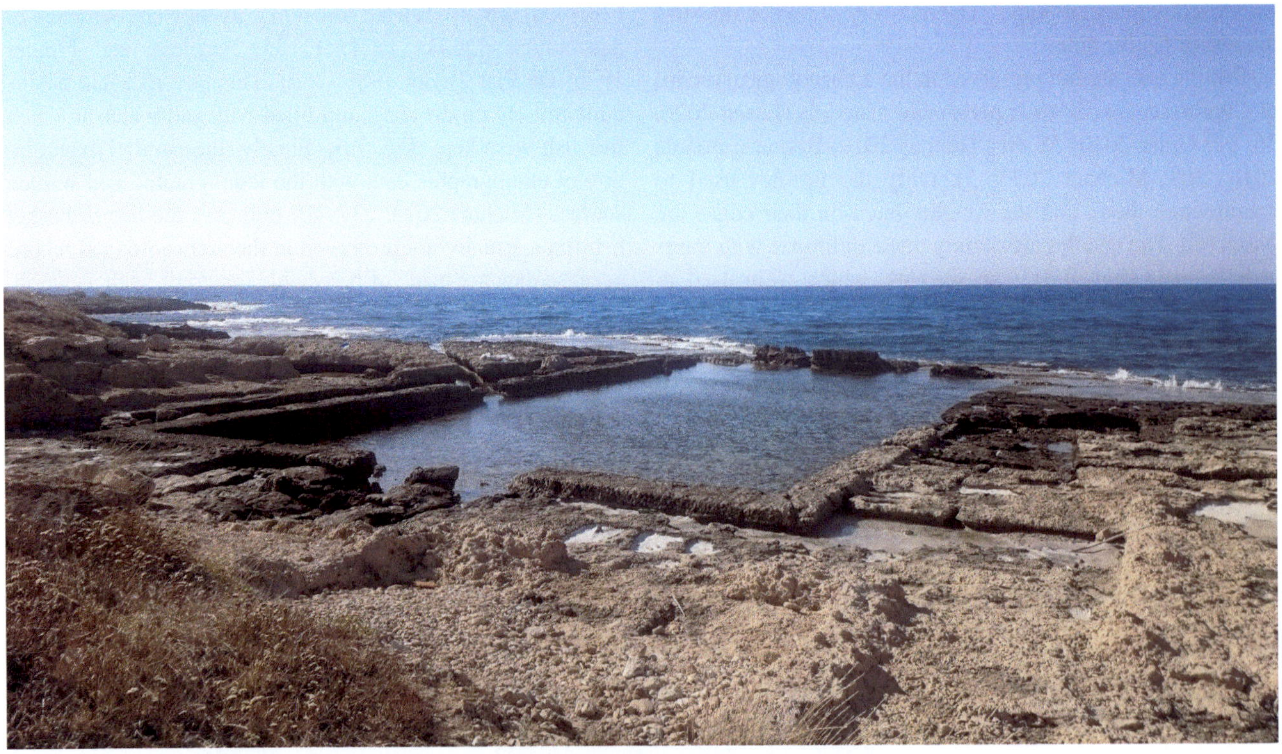

Fig. 7.6 The fish-pond at the archaeological site of Lapithos, on the north coast of Cyprus. (Photo by M. M. Michael)

7.4 Fishing in the Cypriot Archaeological Context: Evidence and Discussion

According to the available evidence, it is clear that people on Cyprus engaged in fishing throughout the period investigated here. Through the study of the primary physical and behavioural characteristics of each fish species, fishers' decision-making processes of where, when, what, and how to fish can be detected. For instance, if fish remains of *Thunnus thynnus* (Tuna), which can be fished at a maximum depth of 500 m, but seasonally migrates close to shore (Ioannou and Michaelides 2015: 94–95), were recovered at an archaeological site, they can indicate that fishers were skilled seafarers and had acquired maritime knowledge which helped them explore both inshore and offshore resources. The finds of fish-hooks and fishing weights demonstrate the use of fishing lines and nets, while the remains of fish-ponds indicate the establishment and development of aquaculture during the Roman and Early Christian periods. Consequently, the parallel study of the main indicators of fishing show that fishers on Cyprus mainly conducted daily short local-scale trips to exploit both coastal and pelagic resources, although each archaeological site has its own economic, social, technological, and physical characteristics, which influence the establishment and development of this activity (Michael 2022: 406–10).

The systematic mapping of the archaeological sites where fish remains and evidence of fishing gear have been recovered shows the temporal and regional distribution of these main indicators of fishing (Fig. 7.7). This demonstrates that

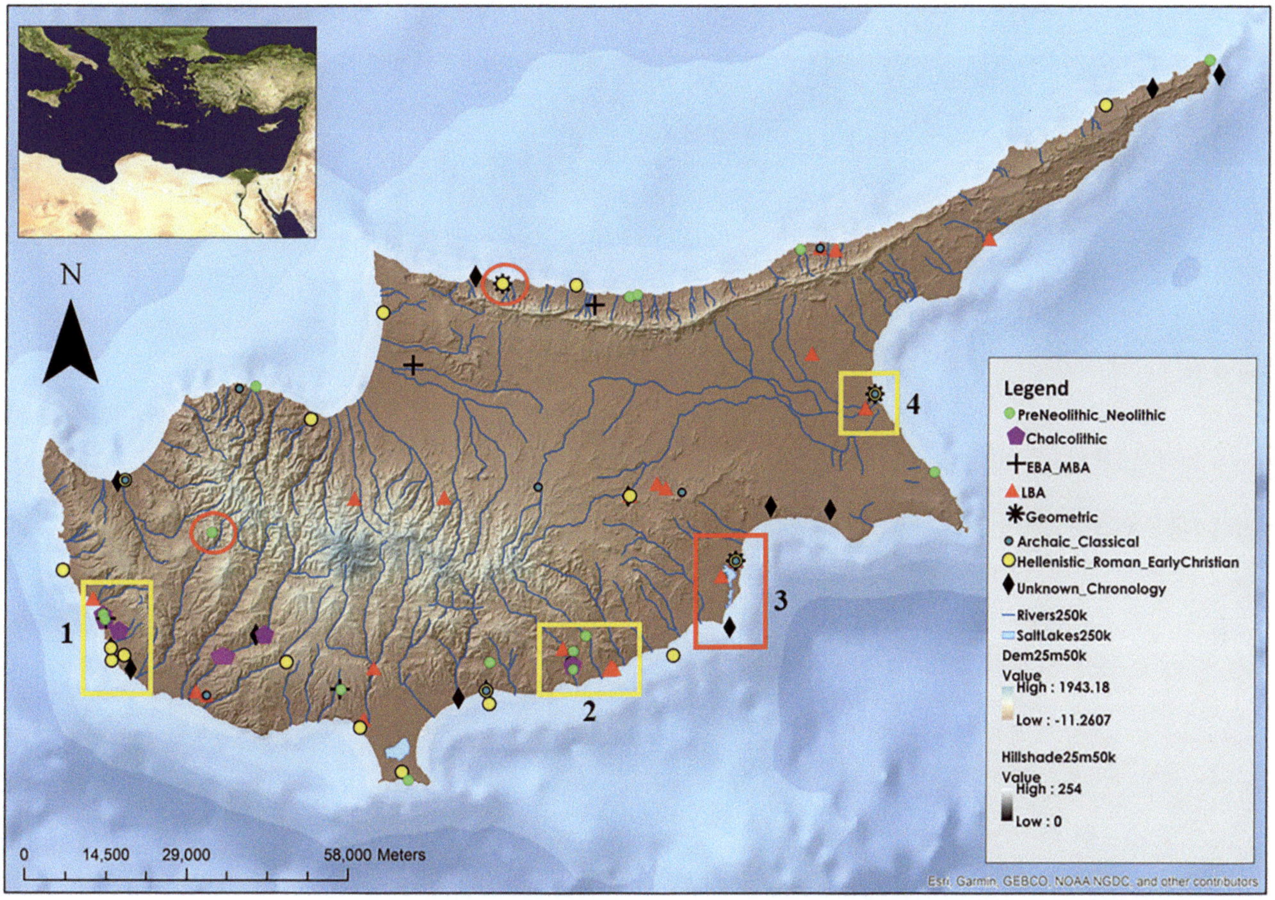

Fig. 7.7 The temporal and regional distribution of archaeological sites where archaeoichthyological evidence has been recovered. (Map produced by M. M. Michael on ArcGIS10. The green point with the red circle (*Ais Yiorkis* Kritou Marotou: Neolithic), the red rectangle (area of Hala Sultan Tekke and Kition: Late Bronze Age), and the yellow point with the red circle (Lapithos: Roman/Early Christian) indicate sites/areas discussed in the text. The numbered rectangles (yellow and red) indicate areas with long-term development of fishing (**1**. Kissonerga-*Mylouthkia* and Kissonerga-*Mosphilia*: Neolithic and Chalcolithic—Chloraka-*Palloures*: Chalcolithic—Kissonerga-*Skalia*: Bronze Age—Nea Paphos: Hellenistic and Roman, **2**. Kalavasos-*Kokkinogia*, Kalavasos-*Tenta* and Khirokitia: Neolithic—Kalavasos-*Agious*: Chalcolithic—Kalavasos-*Agios Dimitrios*, Kalavasos-*Mangia*, Maroni-*Vournes* and Maroni-*Tsaroukas*: Bronze Age—Kalavasos-*Kopetra*: Early Christian, **3**. Hala Sultan Tekke: Late Bronze Age—Kition: Geometric, Archaic and Classical, **4**. Enkomi: Late Bronze Age – Salamis: Geometric, Archaic, Classical, Hellenistic and Roman). Source for layers of Digital Elevation Model, Hillshade Coastline, Rivers and Salt Lakes: Department of Lands and Surveys, Cyprus; Source of the basemap: Esri Garmin, NCAA NGDC, and other contributors; Layer of archaeological sites produced by M. M. Michael)

fishing was known on the whole island over time, as evidence was not only recovered along the coast, the physical space where fishers mostly work, but also at inland sites. For instance, evidence of fishing has mainly been found at coastal sites during the Neolithic period, but a terrestrial site on the Troodos Mountain range, known as *Ais Yiorkis* (Kritou Marotou) (Fig. 7.7) yielded a small but indicative fishbone assemblage (Rose 2023). No fish-hooks or other finds of fishing gear have been found at *Ais Yiorkis*, but the assemblage instead included 982 marine shellfish (Rose 2023: 203), whose occurrence indicates the movement of marine resources from the coast to the uplands, as do the fish remains. The question is how these remains moved from the coast to this mountainous site; their transportation would have to be done quickly to ensure their preservation (Rose 2023: 212). One way to address this is to examine the topographical location of the site and its interaction with other sites within its wider local region.

There is evidence of local exchange of raw materials and other products during the early prehistory in Cyprus. According to Moutsiou and Agapiou (2019: 5, 8), who used least cost analysis to map out potential Aceramic Neolithic social-economic networks on the Cypriot landscape and distinguish possible routes by which obsidian circulated, rivers and their banks seem to have functioned as pathways to movement between coastal and inland sites. Based on the mapping of obsidian circulation on the island of Cyprus (Fig. 7.8), the areas from where the archaeological site of *Ais Yiorkis* obtained its obsidian supplies were the areas of Kissonerga on the western coast and of Petra tou Limniti on the northwestern coast. *Ais-Yiorkis* could be reached by following the Ezousas river from Kissonerga, or by following a

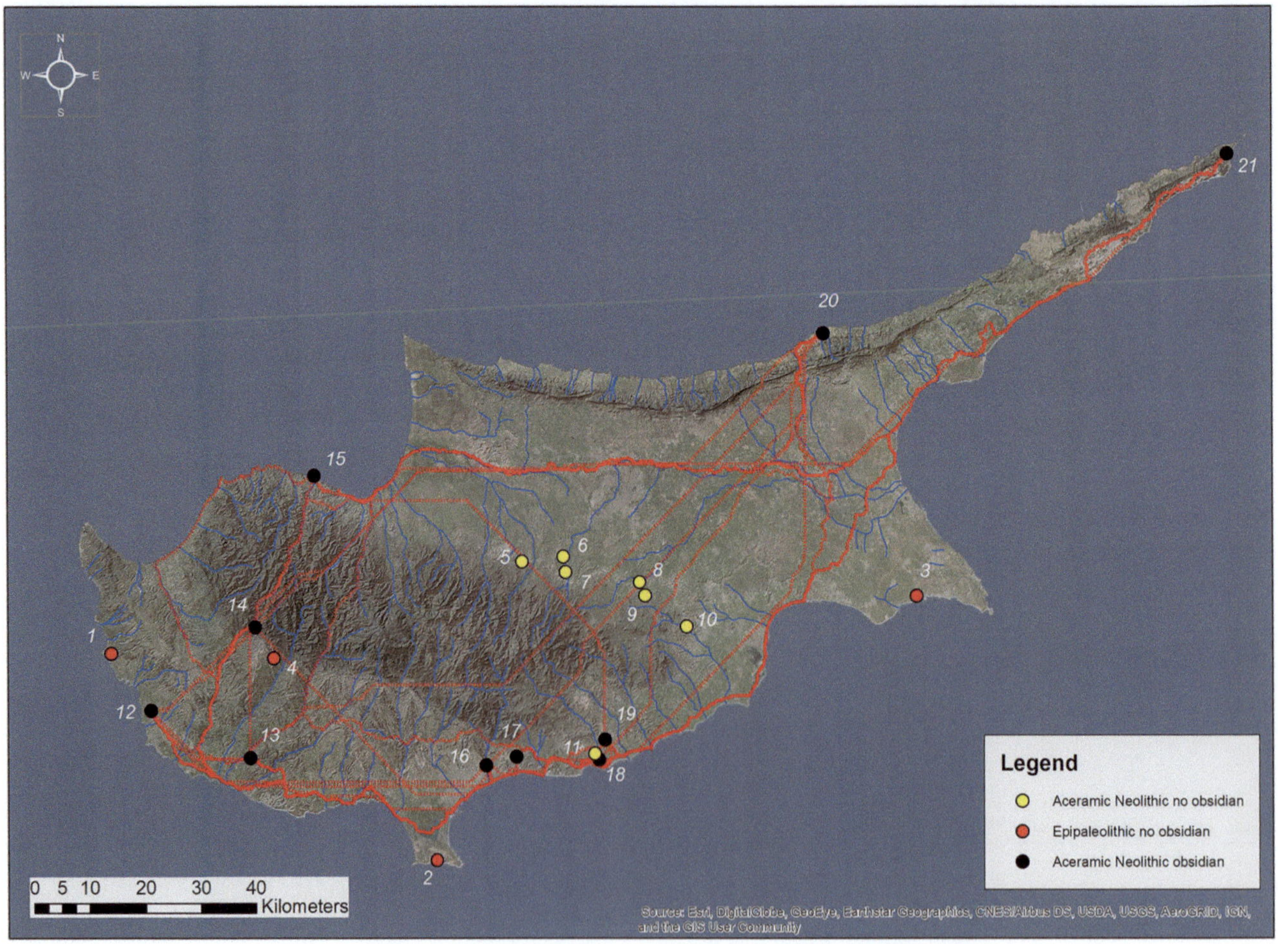

Fig. 7.8 Least Cost Paths Analysis (LCPA) of Aceramic Neolithic obsidian circulation on the island of Cyprus. Sites: 1. Akamas-*Aspros*, 2. Akrotiri-*Aetokremmos*, 3. Nissi Beach, 4. Vretsia-*Roudias*, 5. Agrokipia-*Paleokamina*, 6. Asprokremmos, 7. Politiko-*Kelaidoni*, 8. Pera Chorio-*Moutti*, 9. Alambra-*Spileos* and *Koudourka*, 10. Ayia Anna-*Perivolia*, 11. Mari, 12. Kissonerga-*Mylouthkia*, 13. Choletria-*Ortos*, 14. *Ais Yiorkis* Kritou Marotou, 15. Petra tou Limniti, 16. Parekklisia-*Shillourokambos*, 17. Ayios Tychonas-*Klimonas*, 18. Kalavasos-*Tenta*, 19. Khirokitia-*Vouni*, 20. Akanthou-*Arkosyko*, 21. Cape Apostolos-*Andreas Kastros*. (From Moutsiou and Agapiou 2019: 8, fig. 5)

river south from Petra tou Limniti into the Troodos. These routes may also be followed for the transportation of fish supplies to the site of *Ais Yiorkis*. However, the Troodos Mountain range may have been an obstacle to the transportation of fish from the northwestern coast, and as a result, the area of Kissonerga seems to be preferable as the source of fish supplies (Michael 2022: 319). The occurrence of the Scombridae family in Kissonerga's fishbone assemblage may confirm that there were local exchanges between the two sites, because this fish species has also been identified in the *Ais Yiorkis* fishbone assemblage (Céron-Carrasco 2003: 81–82; Rose 2023: 204). On the other hand, the distance from this site to *Ais Yiorkis* is about 20 km, and thus sufficient preservation such as drying would have been needed to accomplish this transportation, but no evidence has been recovered (Rose 2023: 201, 212–14, 216).

The regional and temporal distribution of archaeoichthyological evidence also demonstrates a long-term engagement with fishing in some areas of Cyprus, despite the absence of evidence from some periods (Fig. 7.7). Their long-term occurrence within the same local context indicates that fishing was not only an "emergency resource" that the inhabitants of Cyprus would choose when the agricultural supplies were not sufficient, as previously suggested (Dikaios 1961: 290; Michaelides 1998; Egoumenidou and Michaelides 2000: 112; Ionas 2001: 217; Knapp 2018: 151), but that it was an essential activity that was passed down over generations. For instance, a long-term engagement with fishing can be observed in the wider area of the archaeological sites of Hala Sultan Tekke and Kition, as evidence related to fishing has been recovered from the Late Bronze Age to Roman periods within this local area (Fig. 7.7).

The Late Bronze Age site of Hala Sultan Tekke has yielded representative evidence of fishing gear (stone fishing weights, pebble fishing weights, lead fishing weights including sinkers folded rectangular, bronze fish-hooks, and a potential trident) and identified and unidentified fish remains (see Table 7.1 for those identified) revealing shell gathering and the exploitation of coastal and pelagic resources caught in an area from 1 m to 300 m depth (Michael 2022: 351–53). This means that fishing was mainly conducted within the areas of the continental shelf and the lagoon, which existed and was used as the harbour of the site during the second millennium BCE (Gifford 1978: 166–69; Devillers et al. 2015: 75–78); today, it forms the Salt Lake after its erosion and siltation during the later second millennium BCE (Fig. 7.7).

Furthermore, the recovery of imported fish species (*Lates niloticus, Clarias* sp.) reveals the existence of potential exchange patterns (Van Neer et al. 2005: 136, 142, 148–49; Fischer and Bürge 2018: 605–606; Millek 2019: 227–28, 339, 362; Fischer and Bürge 2020: 73, 94–95, 98; Michael 2022: 353–56). Articulated fish vertebrae of a *Lates niloticus* (Nile perch) were identified close to one of the skeletons within a "luxury" tomb of 37 individuals, and a Plain White wheel-made jug filled with fish bones was found in the same tomb (Fischer and Bürge 2020: 73, 94–95, 98). The occurrence of fish remains within ceramic vessels could indicate the offering and consumption of dried, salted or smoked fish because processed fish products are mainly recovered from ceramic vessels (Pulak 1988: 5; Mylona 2020: 185–86); their deposition within a mortuary context may be connected with ritual beliefs (Kofel 2020: 107; Mylona 2020: 185–86).

The Late Bronze Age phase at Kition mainly yielded unidentified fish remains; as a result, it is difficult to extract any exploitation or exchange patterns (Reese 1985). However, evidence related to fishing has been recovered in Kition's Cypro-Geometric, Cypro-Archaic, Cypro-Classical, and Hellenistic phases (Desse 1993; Reese 2003). The fishbone assemblage from the Cypro-Geometric phase includes a variety of fish species (see Table 7.1), despite the fact that only 60 fish bones have been recovered. Based on the identified fish species, it seems that coastal and pelagic resources were exploited, as is the case at Late Bronze Age Hala Sultan Tekke. Furthermore, their recovery in a ritual context may also indicate offerings (Michael 2022: 282). The recovery of *Lates niloticus* remains (Nile perch) in the Cypro-Archaic and Cypro-Classical phases of the site suggests that the trade of imported fish which was introduced during the Late Bronze Age in this region may have continued. This is supported by the presence of imported fish from the Hellenistic phase at Kition (Desse 1993; see Table 7.1). It thus seems that the inhabitants of this region were diachronically engaged in fishing.

The examination of the Roman/Early Christian evidence of fish-ponds at the archaeological site of Lapithos (Fig. 7.7) in parallel with environmental and ethnographic data collected by the author from fishers' interviews archived at the Archive of Oral Tradition and Folk Study (Cyprus Research Centre), also demonstrates that fishers acquired knowledge of the physical characteristics of the coast, ecology (seabed ground), meteorology (winds, currents, tides, etc.), and biology (availability of fish species) through their daily interaction with maritime environments (Michael 2022: 379–88). According to ethnographic data, Cypriot fishers during the 19th and 20th centuries preferred to fish in rocky areas such as the northern coast, because the water was cleaner and there were many fish that fed on the seaweeds. Furthermore, the quality of fish was better, as even today Cypriots characteristically mention that "the best and sweeter fish is the fish from the north". Consequently, it seems that the decision of the placement of fish-pond constructions on the northern coast was not accidental and was made by considering the fertility and quality of fishing grounds in this area.

Furthermore, the fact that the north coast experiences predominantly north-westerly to north-easterly winds, which create a current tending toward land, must have influenced the decision to create fish-ponds in this part of the island (Nicolaou and Flinder 1976: 138–39; Michael 2022: 384), as this environmental condition contributed to the continuous renewal of water inside the pond (Columella, De Re Rustica, 8.17; Marzano 2013: 226–27). Consequently, the examination of fish-ponds highlights the relationship between fishers and their maritime environment, and how the characteristics of this environment influenced the establishment and development of fishing on Cyprus through time. However, the physical and socio-economic characteristics of each area/site differ, and therefore each must be considered separately to define and reconstruct fishing with its individual context to then consider the occurrence and nature of fishing on Cyprus through time, and reconstruct it based on the local-scale patterns.

7.5 Conclusions

To sum up, the evidence presented in this chapter leaves no doubt that fishing occurred in the archaeological context of Cyprus throughout a wide chronological range from the Neolithic to the Early Christian periods. Fishing was an activity conducted in the sea and on the coastline, but it was also known on adjacent land, as fish were transported to inland sites; it contributed to the sustenance, ritual, and economic aspects of Cypriot society through time. The systematic examination of identified fishbone assemblages and evidence of fishing gear suggests that each archaeological site demonstrates its own "microcosm" due to the different prevalent economic, social, technological, and geographical conditions. The chronological and regional distribution of evidence of fishing gear and fish remains highlights the importance of fishing within Cypriot society, as fishing seems to seep into economic, ritual, and cultural aspects of society diachronically. The understanding of this 'penetration' in conjunction with the comprehension of the interaction of fishers with the different aspects of their society and maritime environment leads to a better understanding of fishing on Cyprus through time.

Bibliography

Acheson, J. 1981. Anthropology of Fishing. *Annual Review of Anthropology,* 10: 275–316.

Ayodeji, K. 2004. Fishing Equipment and Methods in the Roman World (PhD dissertation, Royal Holloway, University of London).

Bariche, M. 2012. *Field Identification Guide to the Living Marine Resources of the Eastern and Southern Mediterranean.* Rome: Food and Agriculture Organisation of the United Nations.

Bekker-Nielsen, T. 2010. Fishing in the Roman world. In: T. Bekker-Nielsen & D. Bernal-Casasola, eds. *Ancient Nets and Fishing Gear. Proceedings of the International Workshop on "Nets and Fishing Gear in Classical Antiquity: A First Approach" Cadiz, November 15–17, 2007.* Aarhus: Aarhus University Press, pp. 187–204.

Bekker-Nielsen, T. & Bernal-Casasola, D. eds. 2010. *Ancient Nets and Fishing Gear. Proceedings of the International Workshop on "Nets and Fishing Gear in Classical Antiquity: A First Approach", Cadiz, November 15-17, 2007.* Aarhus: Aarhus University Press.

Bernal-Casasola, D. 2010. Fishing Tackle in Hispania: Reflections, Proposals and First Results. In: T. Bekker-Nielsen & D. Bernal-Casasola, eds. *Ancient Nets and Fishing Gear. Proceedings of the International Workshop on "Nets and Fishing Gear in Classical Antiquity: A First Approach" Cadiz, November 15–17,* 2007. Aarhus: Aarhus University Press, pp. 83–137.

Bernal-Casasola, D. 2016. Garum in Context: New Times, Same Topics in the Post – Ponsichian Era. In: T. Bekker-Nielsen & R. Gertwagen, eds. *The Inland Seas. Towards an Ecohistory of the Mediterranean and the Black Sea.* Stuggart: Franz Steiner Verlag, pp. 187–214.

Béthoux, J. 1984. Paléo-hydrologie de la Méditerranée au cours des derniers 20000 ans. *Oceanologia Acta,* 7: 43–48.

Bretton-Connelly, J. 2002. Excavations on Geronisos (1990–1997): First Report. *Report of the Department of Antiquities, Cyprus*: 245–68.

Céron-Carrasco, R. 2003. Fish Remains. In: E. Peltenburg ed. *The Colonisation and Settlement of Cyprus: Investigations at Kissonerga-Mylouthkia, 1976-1996 (LAP III.1).* Studies in Mediterranean Archaeology LXX: 4. Sävedalen: Astrom Editions, pp. 81–82.

Coll, M., Piroddi, C., Steenbeek, J., Kaschner, K., Ben Rais Lasram, F., Aguzzi, J., Ballesteros, E., Bianchi, C. N., Corbera, J., Dailianis, T., Danovaro, R., Estrada, M., Froglia, C., Galil, B.S., Gasol, J. M., Gertwagen, R., Gil, J., Guilhaumon, F., Kesner-Reyes, K., Kitsos, M.S., Koukouras, A., Lampadariou, N., Laxamana, E., López-Fé de la Cuadra, C. M., Lotze, H.K., Martin, D., Mouillot, D., Oro, D., Raicevich, S., Rius-Barile, J., Saiz-Salinas, J.I., San Vicente, C., Somot, S., Templado, J., Turon, X., Vafidis, D., Villanueva, R. & Voultsiadou, E. 2000. The Biodiversity of the Mediterranean Sea: Estimates, Patterns, and Threats. *PLoS ONE* 5.8. https://journals.plos.org/plosone/article?id=10.1371/journal.pone.0011842

Columella, *De Re Rustica,* in *Columella on Agriculture, Books 5–9,* with an English 1954 translation by E.S. Forster & E.H. Heffner (LCL 407). Cambridge (MA 1954).

Cooney, G. 2004. Introduction: Seeing Land from the Sea. *World Archaeology,* 35.3: 323–28.

Cottica, D. & Divari, L. 2010. Spheroid Clay Weights from the Venetian Lagoon. In: T. Bekker-Nielsen & D. Bernal-Casasola, eds. *Ancient Nets and Fishing Gear. Proceedings of the International Workshop on "Nets and Fishing Gear in Classical Antiquity: A First Approach" Cadiz, November 15–17, 2007.* Aarhus: Aarhus University Press, pp. 347–63.

Courtois, J.C. 1984. *Alasia III. Les objets des niveaux stratifiés d'Enkomi. Fouilles C.F.A. Schaeffer (1947–1970). Mission Archéologique d'Alasia, Tome VI* Mémoire no. 33. Paris: Éditions Recherche sur les Civilisations.

Croft, P. 2008. Appendix A: Animal Remains from Paralimni-Nissia. In: P. Flourentzos, *The Neolithic Settlement of Paralimni.* Nicosia: Department of Antiquities, pp. 101–18.

Demetropoulos, A. 1985. Report: Cyprus Fisheries. *Marine Policy,* 9(1): 69–72.

Department of Antiquities, Republic of Cyprus, Ministry of Transport, Communications and Works. 2005–2024. Chronological Table. Available at http://www.mcw.gov.cy/mcw/da/da.nsf/DMLchtable_en [accessed 13 May 2020].

Department of Fisheries and Marine Research, Cyprus. 2012. Initial Assessment of the Marine Environment of Cyprus. Part I – Characteristics. Implementation of Article 8 of the Marine Strategy Framework-Directive (2008/56/EC). http://www.moa.gov.cy/moa/dfmr/dfmr.nsf/All/224D439CDB81950042257E6E0037F444/$file/1-INITAIL%20ASSESMENT.pdf?OpenElement [accessed 15 April 2022].

Department of Fisheries and Marine Research, Cyprus. 2019. The National Contingency Plan for Oil Pollution Combating. Appendix 4: Cyprus NCP v.1.02: Cyprus Geography. Map of Sensitive Resources. Available at http://www.moa.gov.cy/moa/dfmr/dfmr.nsf/All/EFC47876B89A5BB5422583E500414E9E/$file/1.6%20APPENDIX%204%20MAP%20OF%20SENSITIVE%20RESOURCES.pdf [accessed 26 January 2022].

Desse, J. 1993. Les vestiges osseux du basin 417. In: J.F. Salles & O. Callot. Chapitre IV: Le Chantier Sud. In: J.F. Salles, ed. *Kition-Bamboula IV: Les niveaux Hellénistiques*. Paris: Éditions Recherche sur les Civilisations, pp. 103–105.

Devillers, B., Brown, M. & Morhange, C. 2015. Paleo-environmental Evolution of the Larnaca Salt Lakes (Cyprus) and the Relationship to Second Millennium BC Settlement. *Journal of Archaeological Science,* 1: 73–80.

Dikaios, P. 1961. *Sotira.* Philadelphia: The University Museum, University of Pennsylvania.

Duncan, B. 2011. "What Do You Want to Catch?": Exploring the Maritime Cultural Landscapes of the Queenscliff Fishing Community. In: B. Ford, ed. *The Archaeology of Maritime Landscapes*. New York: Springer, pp. 267–89.

Du Plat Taylor, J. 1980. Excavations at Ayios Philon, tthe Ancient Carpasia. Part I: The Classical to Roman Periods. *Report of the Department of Antiquities Cyprus*: 152–216.

Egoumenidou, E. & Michaelides, D. 2000. Gathering, Hunting, Fishing. The Procurement of Food from the Non-domesticated Animal Kingdom in Cyprus through the Ages. In: P. Lysaght, ed. *Food from Nature. Attitudes, Strategies and Culinary Practices. Proceedings of the 12th Conference of the International Commission for Ethnological Food Research, Umea and Frostviken, Sweden, 8-14 June 1998.* Uppsala: The Royal Gustavus Adolphus Academy for Swedish Folk Culture, pp. 111–20.

FAO (Food and Agriculture Organisation of the United Nations). 2005. Fishery Country Profile: 1–8. Available at http://www.fao.org/fi/oldsite/FCP/en/CYP/profile.htm [accessed 1 June 2020].

Fischer, P.M. & Bürge, T. 2018. Discussion and Conclusions. In: P.M. Fischer & T. Bürge, *Two Late Cypriot City Quarters at Hala Sultan Tekke. The Söderberg Expedition 2010-2017.* Uppsala: Astrom Editions, pp. 603–16.

Fischer, P.M. & Bürge, T. 2020. The New Swedish Cyprus Expedition 2019: Excavations at Hala Sultan Tekke (The Söderberg Expedition). *Opuscula,* 13: 73–111.

Frost, H. 1985. Fishing Tackle: Three Limestone Weights. In: M. Yon & A. Caubet, eds. *Kition-Bamboula 3. Le Sondage L-N 13 (bronze récent et géométrique I).* Paris: Recherche sur les civilisations, pp. 169–73.

Frost, H. 1991. Anchors Sacred and Profane. Ugarit-Ras Shamra, 1986: The Stone Anchors Revised and Compared. In: M. Yon, ed. *Ras Shamra-Ougarit VI. Arts et industries de la pierre*. Paris: Éditions Recherche sur les Civilisations, pp. 355–410.

Gabriel, O., Lange, K., Dahm, E. & Wendt, T. 2005. *Fish Catching Methods of the World.* Oxford: Blackwell Publishers.

Galili, E., Rosen, B. & Sharvit, J. 2002. Fishing-gear Sinkers Recovered from an Underwater Wreckage Site, off the Carmel Coast, Israel. *The International Journal of Nautical Archaeology,* 31(2): 182–201.

Galili, E., Zemer, A. & Rosen, B. 2013. Ancient Fishing Gear and Associated Artifacts from Underwater Explorations in Israel – A Comparative Study. *Archaeofauna,* 22: 145–66.

Garnier, J. 1981. Appendice V: Les poissons du Cap Andreas Castros. In: A. Le Brun, *Un site néolithique précéramique en Chypre: Cap Andreas-Kastros*. Mémoire no. 5. Paris: Editions A.D.P.F. Recherche sur les grandes civilisations, pp. 93–94.

Gifford, A.J. 1978. Paleogeography of Archaeological Sites of the Larnaca Lowlands, Southeastern Cyprus (PhD dissertation, University of Minnesota).

Giner, A.C. 2010. Fishing Nets in the Ancient World: The Historical and Archaeological Evidence. In: T. Bekker-Nielsen & D. Bernal-Casasola, eds. *Ancient Nets and Fishing Gear. Proceedings of the International Workshop on "Nets and Fishing Gear in Classical Antiquity: A First Approach" Cadiz, November 15–17, 2007.* Aarhus: Aarhus University Press, pp. 55–82.

Golani, D., Öztürk, B. & Başusta, N. 2006. *Fishes of the Eastern Mediterranean*. Istanbul: Turkish Marine Research Foundation.

Greenwood, P.J. & Howes, G.J. 1973. Appendix 8: Fish Remains. In: V. Karageorghis, *Excavations in the Necropolis of Salamis, III/5 (Text)*. Nicosia: Department of Antiquities, pp. 259–68.

Halstead, P. 1977. A Preliminary Report on the Faunal Remains from Late Bronze Age Kouklia, Paphos. *Report of the Department of Antiquities, Cyprus*: 261–75.

Hela, I. & Laevastu, T. 1962. *Fisheries Hydrography. How Oceanography and Meteorology Can and Do Serve Fisheries.* London: Fishing News (Books) Ltd

Ioannou, G. & Michaelides, N. 2015. *Τα 100 σημαντικότερα είδη ψαριών των κυπριακών θαλασσών*. Λευκωσία: Γραφείο Τύπου και Πληροφοριών για το Τμήμα Αλιείας και Θαλάσσιων Ερευνών.

Ionas, I. 2001. *Παραδοσιακά Επαγγέλματα της Κύπρου*. Λευκωσία: Δημοσιεύματα του Κέντρου Επιστημονικών Ερευνών XXVII.

Knapp, A.B. 2013. *The Archaeology of Cyprus: From Earliest Prehistory through the Bronze Age*. Cambridge University Press.

Knapp, A.B. 2018. *Seafaring and Seafarers in the Bronze Age Eastern Mediterranean*. Sidestone Press.

Kofel, D. 2020. A Note on the Vessel Contents in CQ4 and Tomb RR. In: P.M. Fischer & T. Bürge, The New Swedish Cyprus Expedition 2019: Excavations at Hala Sultan Tekke (The Söderberg Expedition). *Opuscula,* 13: 105–107.

Le Brun, A. 1981. *Un site néolithique précéramique en Chypre: Cap Andreas-Kastros*. Mémoire no. 5. Paris: Editions A.D.P.F. Recherche sur les grandes Civilisations.

Leach, F. 2006. Fishing in Pre-European New Zealand. *Archaeofauna,* 15: 1–266.

Lindqvist, A. 2016. A Study of Fishing Methods Used at Hala Sultan Tekke during the Late Bronze Age. In: G. Bourogiannis & C. Muhlenbock, eds. *Ancient Cyprus Today: Museum Collections and New Research*. Uppsala: Astrom Editions, pp. 239–48.

Locker, A. 2007. In Piscibus Diversis: The Bone Evidence for Fish Consumption in Roman Britain. *Britannia,* 38: 141–80.

Maarleveld, T. 2010. "Fish and Chips of Knowledge": Some Thoughts on the Biases of the Archaeological Record. In: T. Bekker-Nielsen & D. Bernal-Casasola, eds. *Ancient Nets and Fishing Gear. Proceedings of the International Workshop on "Nets and Fishing Gear in Classical Antiquity: A First Approach" Cadiz, November 15–17, 2007.* Aarhus: Aarhus University Press, pp. 257–75.

Mantzourani, E. 2019. *Ανασκαφή του Νεολιθικού Οικισμού Καντού-Κουφόβουνου στην Κύπρο. 2 Τόμοι.* Λευκωσία: Τμήμα Αρχαιοτήτων Κύπρου.

Marsellou, N.T. 1948. *Το ψάρεμα και τα Ψάρια. Τέχνη-Θεωρία-Πρακτική Εφαρμογή με Λεπτομερή Μελέτη περί του καιρού. Πρακτική Προγνωσις και Σημερινή Εξέλιξις της Μετεωρολογικής Επιστήμης.* Αθήνα: N. Τιλπερόγλου.

Marzano, A. 2013. *Harvesting the Sea. The Exploitation of Marine Resources in the Roman Mediterranean*. Oxford University Press.

McNiven, I. 2003. Saltwater People: Spiritscapes, Maritime Rituals and the Archaeology of Australian Indigenous Seascapes. *World Archaeology,* 35.3: 329–49.

Meteorological Service, Ministry of Agriculture and Natural Resources. 1986. *A Study of the Surface Winds in Cyprus.* Meteorological Paper Series no. 8. Nicosia.

Michael, M. 2022. To Fish or Not to Fish? The Case Study of Fishing Activity in Cyprus (PhD dissertation, University of Southampton, Faculty of Arts and Humanities, Department of Archaeology).

Michael, M. 2023. Fishing Activity in Cyprus through Time: An Archaeologically Neglected Aspect in Cypriot Society. *Journal of Archaeological Science: Reports,* 49: 1–10.

Michaelides, D. 1998. Food in Ancient Cyprus. In: P. Lysaght, ed. *Food and the Traveller. Migration, Immigration, Tourism and Ethnic Food. Proceedings of the 11th Conference of the International Commission for Ethnological Food Research, Cyprus, June 8-14 1996*. Nicosia: Intercollege Press, pp. 22–43.

Millek, J.M. 2019. *Exchange, Destruction, and a Transitioning Society. Interregional Exchange in the Southern Levant from the Late Bronze Age to the Iron I.* Tübingen: Tübingen University Press.

Morales-Muñiz, A., Antipina, E., Antipina, A. & Roselló, E. 2007. An Icthyoarchaeological Survey of the Ancient Fisheries from the Northern Black Sea. *Archaeofauna,* 16: 117–72.

Moutsiou, T. & Agapiou, A. 2019. Least Cost Pathway Analysis of Obsidian Circulation in Early Holocene-Early Middle Holocene Cyprus. *Journal of Archaeological Science Reports,* 26.101881: 1–11. https://doi.org/10.1016/j.jasrep.2019.101881

Mylona, D. 2020. Marine Resources and Coastal Communities in the Late Bronze Age Southern Aegean: A Seascape Approach. *American Journal of Archaeology,* 124(2): 179–213.

Nicolaou, K. & Flinder, A. 1976. Ancient Fish-tanks at Lapithos, Cyprus. *The International Journal of Nautical Archaeology,* 5(2): 133–41.

Niklasson, K. 1983. Tomb 23: A Shaft-grave of the Late Cypriote III Period. In: P. Åström, E. Åström, A. Hatziantoniou, K. Niklasson & U. Öbrink, *Hala Sultan Tekke 8. Excavations* 1971–1979. Studies in Mediterranean Archaeology XLV: 8. Göteborg: Astrom Editions, pp. 169–87.

Obied, C. 2016. Rethinking Roman Perceptions of Coastal Landscapes: A Case-Study of the Levant (PhD dissertation, University of Southampton, Faculty of Arts and Humanities, Department of Anthropology).

O'Sullivan, A. 2001. *Foragers, Farmers and Fishers in a Coastal Landscape*. Dublin: The Royal Irish Academy.

O'Sullivan, Aidan. 2003. Place, Memory, and Identity among Estuarine Fishing Communities: Interpreting the Archaeology of Early Medieval Fish Weirs. *World Archaeology,* 35(3): 449–68.

Papantoniou, G. & Kyriakou, N. 2018. Sacred Landscapes and Territoriality of Iron Age Cypriot Polities: The Applicability of GIS. *American Journal of Archaeology,* 122(4): 541–77.

Piques, G. 2015. Les poisons. In: A. Gardeisen. Chapitre 7: La Faune (Fouilles 1981–1989). In: A. Caubet, S. Fourrier & M. Yon, eds. *Kition-Bamboula VI: Le Sanctuaire Sous la Colline*. Travaux de la Maison de l'Orient et de la Mediterranée no. 67. Lyon: Leon Levy Foundation, pp. 363–64.

Potts, D.T. 2012. Fish and Fishing. In: D.T. Potts, ed. *A Companion to the Archaeology of the Ancient Near East*. Chichester: Wiley-Blackwell, pp. 220–35.

Powell, J. 1996. *Fishing in the Prehistoric Aegean*. Studies in Mediterranean Archaeology and Literature Pocket-book 137. Jonsered: Astrom Editions.

Pulak, C. 1988. The Bronze Age Shipwreck at Ulu Burun, Turkey: 1985 Campaign. *American Journal of Archaeology,* 92: 1–37.

Quilici G.S. 1971. Prima perlustrazione topografica. In *Consiglio Nazionale delle Ricerche, Studi Ciprioti E Rapporti di Scavo, Consiglio Nazionale delle Ricerche Istituto per gli Studi Micenei ed Egeo-Anatolici. Biblioteca di antichita cipriote Vol. 1.* Rome: Edizioni dell'Ateneo.

Reese, D. 1985. Shells, Ostrich Eggshells and Other Exotic Faunal Remains from Kition. In: V. Karageorghis, *Excavations at Kition V/II. The Pre-Phoenician Levels*. Nicosia: Department of Antiquities, pp. 340–415.

Reese, D. 2003. The Shells, Astragali and Fish. In: V. Karageorghis, *Excavations at Kition. VI. The Phoenician and Later Levels Part II.* Nicosia: Department of Antiquities, Cyprus, pp. 413–33.

Reitz, J.E. & Wing, E.S. 2008. *Zooarchaeology*. 2nd edition. Cambridge: Cambridge University Press.

Ritchie, C.K. 2010. The Ertebølle Fisheries of Denmark, 5400-4000 B.C. (PhD dissertation, University of Wisconsin-Madison, Department of Anthropology).

Rohling, E., Abu-Zied, R., Casford, J., Hayes, A. & Hoogakker, B. 2009. The Marine Environment: Present and Past. In: J. Woodward, ed. *The Physical Geography of the Mediterranean*. Oxford: Oxford University Press, pp. 33–67.

Rose, M. 1994. With Line and Glittering Bronze Hook: Fishing in the Aegean Bronze Age (PhD dissertation, Indiana University).

Rose, M. 2023. The *Ais Giorkis* Fish Remains. In: A.H. Simmons, ed. *From the Mountains to the Sea; Ais Giorkis, an Early Neolithic Hamlet in the Troodos Foothills of Western Cyprus*. Nicosia: Astrom Editions, pp. 201–16.

Rosenberg, D., Agnon, M. & Kaufman, D. 2016. Conventions in Fresh Water Fishing in the Prehistoric Southern Levant: The Evidence from the Study of Neolithic Beisamoun Notched Pebbles. *Journal of Lithic Studies,* 3(3): 457–78.

Safadi, C. 2016. Wind and Wave Modelling for the Evaluation of the Maritime Accessibility and Protection Afforded by Ancient Harbours. *Journal of Archaeological Science Reports,* 5: 348–60.

Safadi, C. 2018. The Maritime World of the Early Bronze Age Levant through Space and Time (PhD dissertation, University of Southampton, Faculty of Arts and Humanities, Department of Archaeology).

Theodoropoulou, T. 2011. Fishing Together, Fishing on Its Own: Fish Exploitation Patterns at the Neolithic Alepotrypa Cave (Diros, Greece) and Aegean Prehistoric Fishing Traditions. *International Journal of Osteoarchaeology,* 29: 395–406.

Thomas, R. 2010. Fishing Equipment from Myos Hormos and Fishing Techniques on the Red Sea in the Roman Period. In: T. Bekker-Nielsen & D. Bernal-Casasola, eds. *Ancient Nets and Fishing Gear. Proceedings of the International Workshop on "Nets and Fishing Gear in Classical Antiquity: A first approach" Cadiz, November 15–17, 2007.* Aarhus: Aarhus University Press, pp. 139–60.

Trakadas, A. 2018. *In Mauretaniae Maritimis: Marine Resource Exploitation in a Roman North African Province.* Stuttgart: Franz Stelner Verlag.

Van Neer, W., Zohar, I. & Lernau, O. 2005. The Emergence of Fishing Communities in the Eastern Mediterranean Region: A Survey of Evidence from Pre- and Protohistoric Periods. *Paléorient,* 31(1): 131–57.

Westerdahl, C. 2007. Fish and Ships: Towards a Theory of Maritime Culture. *Deutsches Schiffahrtsarchiv,* 30: 191–236.

Wheeler, A. & Jones, A.K.G. 1989. *Fishes*. Cambridge Manuals in Archaeology. Cambridge University Press.

Cockles and Carps: Human-Aquatic Being Relations in the Mid to Late Third Millennium BCE Tombs of Ur

8

Lærke Recht and Christina Tsouparopoulou

Abstract

This chapter explores the intricate dynamics between human and aquatic beings within the third millennium BCE tombs of ancient Ur, through comprehensively examining fauna, fishing tools and visual representations from the tombs. Despite recent advancements in Human-Animal Studies, interactions involving aquatic species remain relatively understudied. Fish and molluscs, though underrepresented in physical remains, played diverse roles in funerary practices, embodying an ideology of abundance. Molluscs, primarily obtained through extensive trade, underwent substantial transformations, often rendering their origins unrecognisable. Bivalve shells, intertwined with pigments, emerge as pivotal elements in complex networks of personal identity, mortality, and posthumous transformation. Conversely, artefacts like cylinder seals and various forms of jewellery bear the imprint of meticulous craftsmanship, resulting in finished products that often bear scant resemblance to their antecedents. This chapter contributes to a deeper understanding of the dynamic interactions between human and aquatic beings in ancient Mesopotamia, shedding light on cosmologies and religious practices of this area.

Supplementary Information: The online version contains supplementary material available at https://doi.org/10.1007/978-3-031-73643-8_8.

L. Recht (✉)
Institut für Antike, University of Graz, Graz, Austria
e-mail: rechtl@tcd.ie

C. Tsouparopoulou
Institute of Arcaheology, UKSW Warsaw, Warszawa, Poland

Department of Archaeology, Durham University, Durham, UK

Abbreviations

BM	British Museum, London
IM	National Museum of Iraq, Baghdad
Penn	University of Pennsylvania Museum of Archaeology and Anthropology
RLA	Reallexikon der Assyriologie und Vorderasiatischen Archäologie
ETCLS	Black, J.A., G. Cunningham, E. Fluckiger-Hawker, E Robson, and G. Zólyomi. 1998 *The Electronic Text Corpus of Sumerian Literature* (http://www-etcsl.orient.ox.ac.uk/). Oxford

8.1 Introduction

This chapter examines the role of aquatic beings in the mid to late third millennium BCE tombs at ancient Ur in modern-day southern Iraq. We explore relations between aquatic beings (fish and molluscs) and humans as reflected in the fauna, fishing equipment, worked objects and visual evidence found in the tombs. These reveal complex and heterogeneous relations that tied people to the aquatic world through direct and indirect interaction. Research based in or inspired by the field of Human-Animal Studies and multispecies approaches have made their way to archaeology especially in the last decades, offering new insights and perspectives on the role of nonhuman animals in the past (e.g. Sykes 2015; Boyd 2017; Pilaar Birch 2018; Recht 2022, see also further references in the introduction of this volume). However, relations between humans and aquatic species remain underrepresented, despite the central role of water and its inhabitants. The aim in this chapter is precisely to focus on the dynamics of these interactions using the material culture of the tombs at Ur as a case study and as an indication of the variety of relations.

Since their excavation and publication in the 1920s and 1930s by Leonard Woolley, the tombs at Ur have received

C. Tsouparopoulou, L. Recht (eds.), *Human and Aquatic Beings: Interactions in and beyond the Eastern Mediterranean (3rd–1st Millennia BCE)*, Themes in Contemporary Archaeology, https://doi.org/10.1007/978-3-031-73643-8_8

much attention from scholars and have reached worldwide fame. The excavations cemented the importance of the site especially in the third millennium BCE, and Woolley and his colleagues extensively published the results in an exemplary manner of detail for the time in a series of 'Ur Excavations' volumes (Hall and Woolley 1927; Woolley 1934, 1939, 1955, 1962, 1965, 1974; Legrain 1936, 1951; Woolley and Mallowan 1976) along with a string of articles in the London-based *The Antiquaries Journal*. The sacred precinct with its temples, ziggurat, the *giparu* (residence of the high priestess), the so-called mausoleum, the tombs, and the early second millennium BCE domestic quarters have become mainstay in studies of ancient Mesopotamia, and Ur continues to be a centre of discussion (see, most recently, papers in Frame et al. 2021). The tombs, and in particular the so-called Royal Cemetery, have sparked many specialised studies (examples from the last 25 years include Winter 1999; Reade 2001; Dickson 2006; Gansell 2007; Baadsgaard et al. 2011, 2012; Vidale 2011; Vogel 2014, 2015). Lately, the UrOnline project has launched, which is an effort to digitally reunite all artefacts from Woolley's excavations with their records in one virtual place.

The analysis presented in this chapter is based on legacy data: Woolley's published records (1934 volume), the information on UrOnline and Woolley's field notes.[1] The majority of tombs included here are dated to the Early Dynastic III period and were originally assigned this date by Woolley. However, in subsequent studies, the dates of some individual tombs have been re-evaluated based on finds, stratigraphy and internal spatial relations (Nissen 1966; Pollock 1985). Consequently, some of the tombs included here are rather Akkadian (especially from the earlier part of the period), and we here follow the dates suggested in Pollock's study, where available.

8.2 Fish, Fishing Gear and Feasting

Woolley frequently found animal remains in the tombs of the cemetery. Some parts or joints of animals appear to have been part of food offerings, but other animals were likely sacrificed as part of the funerary proceedings and placed in a manner that suggests a carefully constructed role (Recht 2011). This is, for example, the case with the bovines placed with wheeled vehicles or a sledge, and sometimes even still wearing their equipment. Beside the great number of molluscs and worked mollusc items mentioned below, only a few Early Dynastic III-Akkadian tombs contained remains of aquatic species.[2] Thus, PG 468 contained 'fish bones', and in PG 1054, a vertebra of a fish was found in a saucer (Woolley 1934: 410).[3] The size suggested a fish of 25–30 cm in length. Finally, shark vertebrae were identified in PG 1232; Woolley passes on a note from J.R. Norman from the British Museum (who made the identification) that sharks will swim into big rivers from the sea and can be found in the Euphrates even today (1934: 410; for fish remains from the Ur tombs, see also Greenfield, this volume). So although these remains may be somewhat surprising, it need not imply that the shark had been imported. There is no further contextual information about these finds, but their scattered nature and the example from PG 1054, with the fish bone inside a vessel, suggest that the fish remains were from meals or food offerings.

Fish were extensively exploited as a food resource, and the rivers and canals abounded with edible versions. The later third millennium BCE Ur III archives demonstrate some of the extent to which fishing had become institutionalised (Englund 1990; administrators and foremen of fishermen are known already in the Archaic professions list of the Uruk III period at Uruk), with very large amounts collected, often in baskets of various sizes (Borelli 2021). The catch was not only intended for human consumption, but also frequently offered to deities (van Buren 1948). One of the earliest known examples comes from Ubaid-period Eridu (c. 6000–4000 BCE), where ample quantities of fish bones were found in prehistoric Temples VII and VI (Safar et al. 1981: 104, 107, 110). The site also provides an early precedent for the use of fish as a food offering in graves, as for example Grave 106, which contained a dish and cup with fish bones (Safar et al. 1981: 134). Continued engagement with fish as part of religious practices can be found in deposits and structures at Tello/Girsu, where the so-called 'Maison des Fruits', a sacred building on Tell K, may have been associated with water and fish (Tsouparopoulou 2024).

That fish were indeed considered appropriate food for deities and important celebrations is demonstrated by one of the most famous objects to come to light from the cemetery: the so-called Standard of Ur. The standard, whose exact function is still not known, is a ca. 22 cm high and 50 cm long rhomboid box with inlaid decoration on four sides, found in PG 779 (BM 121201). On the two long sides, there are elaborate scenes in three registers related to battle (the 'War' side) and feasting with a procession of what appears to be the loot from the battle (the 'Peace' side). The latter includes, in the second register, a man carrying two fish hanging from each of his hands (Fig. 8.1). As reconstructed,

[1] UrOnline is a collaborative project that digitally presents the finds from Ur, especially those from the British Museum and the Penn Museum. This also includes scans of Woolley's original field notes. The website of the project can be found at www.ur-online.org (last accessed 7 October 2022).

[2] These may, however, be underrepresented since most fish remains are not easily recovered by hand, but require wet or dry sieving.

[3] PG 610 (dated to Ur III by Nissen but not included in Pollock's study) also contained 'the bony spine of a perch-like fish', but Ellison and colleagues consider it more likely to be from a local freshwater fish (Ellison et al. 1978: 176).

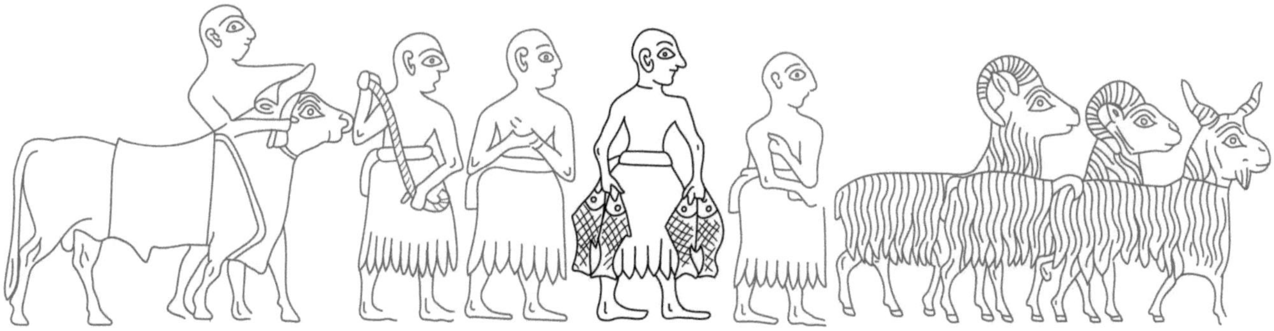

Fig. 8.1 Drawing of part of the 'Peace' side of the Standard of Ur. Drawn by L. Recht, after BM 121201

he is located roughly in the centre, with additional men accompanying livestock (sheep, cattle) before and behind, and like the other figures, he wears a belted skirt (*kaunakes*), has a bare chest, shaved head and no beard.

The four fish are generically rendered so that identification of a specific species is not possible. They hang tail down. They are here clearly part of the overall assemblage of goods that were probably part of the loot as a result of conflict, presumably with another city-state. The banquet scene depicted in the top register is perhaps the victory feast, and the rows of goods and animals may have been displayed and presented as part of the celebrations. It is reasonable to here interpret the fish as intended for consumption at the feast. Unlike the depicted mammals, these fish would have already been dead and could not simply be added to temple or palace flocks. Therefore, it is likely that they were promptly consumed. In this context, they thus symbolise both wealth and abundance in a general sense, and a reference to some of the specific food enjoyed as part of a banquet.

Banquet scenes are a common motif in Early Dynastic visual evidence and are also found on plaques and the miniature medium of seals. One such seal comes from PG 1749 (U.14473A, Woolley 1934: Pl. 200 no. 102). It has a typical two-register composition with seated figures drinking and toasting in the top register, and a row of figures bringing supplies in the lower register, including one carrying a bunch of fish in a manner similar to that seen on the standard.[4]

In contrast to the shells, it seems likely that the majority of fish exploited at Ur came from local freshwater species procured from the Euphrates.[5] Various species of carps were particularly common (Sahrhage 1999: 50–59, Sumerian 'suhur'), and are reported as extensively fished in the written records of the third millennium BCE (*RLA*, 'Fischerei'), as well as in faunal assemblages (e.g. in contemporary and nearby Abu Tbeirah, Alhaique et al. 2021: 70–72). It is thus also possible that carps were in fact the original inspiration for many of the generic representations of fish that can be found in the visual evidence. Evidence of local fishing in the form of tools and equipment can be found not only in the written sources, but also in the material culture of Ur and nearby sites. The digital catalogue at UrOnline lists 50 objects of equipment related to fishing (fishhooks, harpoons/spears, tridents).[6] Several harpoons (PG 580, PG 755, PG 1078) and one fishhook (PG 1312) come from the Early Dynastic graves. In PG 755, four harpoons were found together, suggesting that they were a specialised set placed in the tomb (no. 58 on the plan in Woolley 1934: 157). It is not clear if they reflect the identity or activities of the deceased. PG 755 is believed to be the tomb of a royal person. A seal is inscribed with the name 'Meskalamdug', and PG 1054 also has a seal referring to a 'King Meskalamdug'.[7] He may also have been king of Kish, if the Meskalamdug mentioned on an inscription on a lapis lazuli bead from the so-called 'Treasure of Ur' found in Mari is the same man (Dossin 1968; cf. translation in Aruz 2003: 143). This was a tomb with a wealth of grave goods, and there were also other 'sets' of tools and weapons, so it may be that rather than specifically referring to Meskalamdug as a fisherman, the set of harpoons is a nod to one of his many responsibilities as a ruler, which would include overseeing fishing, fishermen and the distribution of their catch.

All the harpoons found were of copper/copper alloy and could therefore be used for fishing, in contrast to some of the other items in Meskalamdug's tomb, such as a gold dagger.

[4] A sealing depicting a person carrying hanging fish in each hand also comes from the cemetery area, but from a pit rather than a specific tomb (Legrain 1936: Pl. 16 no. 302, U. 14599). Examples from other sites of a similar figure, either nude with only a belt or wearing a skirt, are mentioned in van Buren (1948) and Tsouparopoulou (2024).

[5] Some imports may still have occurred: Ellison et al. (1978) report a tunny fish, possibly from the Gulf or the Mediterranean, albeit from a much later period (Neo-Babylonian context). Fish and fishing tend to be understudied in general in the Near East - for a recent paper on this topic for the Levant, see Lernau et al. 2021.

[6] http://www.ur-online.org/term/968/, accessed 13/9/2022.

[7] Reade suggests PG 755 belongs to a prince, the son of Meskalamdug, and that PG 789 is the tomb of king Meskalamdug himself (Reade 2001).

Attached to a ca. 2 m long wooden handle, similar tools are reported as still in use today in the rivers and marshlands of southern Mesopotamia (Jawad 2006). Modern fishermen usually stand in a canoe at night and attract fish with a light and by throwing food in the water; the catch is species of barbel, which were also common in the third millennium BCE.

The one hook from PG 1312 (with more found in other contexts at Ur) testify to line-and-hook type of fishing. Certainly other types of fishing would have been common, as for example with use of bare hands, a trap, or a net with net sinkers; the latter do preserve in other contexts since they might be made of stone or metal, but these are not recorded in the assemblages of the Early Dynastic and Akkadian tombs (for reviews of fishing and types of fishing in Mesopotamia, see Sahrhage 1999: 97–116; Potts 2012; *RLA* 'Fischerei').

8.3 Shell as Raw Material

Shells were used as a raw material for a great range of objects found in the tombs. Objects made entirely of shell include vessels, cylinder seals, rings, pendants, figures, roundels, beads and a ladle. Composite objects that have parts made of shell include dice, gaming boards, lyres, sceptres/staffs, a knife (as the handle), handles, containers and other unique items such as the so-called 'Standard of Ur' and 'Ram in a Thicket'. For the composite objects, parts of shells were most commonly worked into various types of decorative inlays (Moorey 1994: 136).

Some of the molluscs were used unaltered, as the bivalve shell containers discussed below, and very likely some of the beads, although this is not specified in the publications. The rest were almost certainly imported as raw material and worked into their final shape in Mesopotamia, possibly at Ur itself. Shells were used abundantly: a large amount of the cylinder seals from the Early Dynastic and Akkadian tombs were of shell, but even more abundant are the plaques and inlays that have been shaped from shells. The molluscan material from Ur was not identified to species by an expert. The bivalve shell objects were likely made of Arcidae and Cardiidae molluscs; based on the published photographs on UrOnline of the exterior, these are all the marine *Vasticardium* sp. (some specimens possibly representing *Vasticardium elongatum* and *Vasticardium rhegminum*).[8] The main species identified by Gensheimer (1984; see also Moorey 1994: 129–139) are: *Engina mendicaria* (often used for beads and pendants), *Fasciolaria trapezium* (especially used for inlays, while the columella was used for long and thin beads and cylinder seals), *Lambis truncata sebae* (a large shell used for containers and inlays). At Ur, we also find a few examples of mother of pearl from the black-lip oyster (*Pinctada margaritifera*), cowrie shells and *Turbinella pyrum*, its columella used for large cylinder seals.

The species are nearly all marine and appear to have been primarily imported from the Gulf (Magan and Dilmun), although the Red Sea is also be a possible source for some species. The *Turbinella pyrum* may have come from the Indus Valley (Meluhha) through the Gulf. The trade between Mesopotamia and the Gulf in the third and early second millennium BCE is well known and demonstrated both in texts and archaeologically (see e.g. Oppenheim 1954; Edens 1992 and references therein; Potts 2003; Laursen and Steinkeller 2017; Kenoyer 2008; for the evidence of contacts with the Indus Valley, which now also finds support from stable isotope studies: Kenoyer et al. 2013). This trade included the export of wool, textiles and barley to the Gulf and the import of copper, stone, semiprecious stones, shells and ivory as well as a few finished products into Mesopotamia. Edens classifies shells as a luxury commodity, and notes that the import of shells peaked in the third millennium BCE, especially in the Early Dynastic II-III period (Edens 1992). The decline in shell objects in Mesopotamian contexts after the Akkadian period is not due to a restriction in access, and thus reflects a change in their social and economic value, both in mortuary and non-mortuary contexts - their role was often taken over by stone, as exemplified by the increase in cylinder seals of harder stone at the expense of shell cylinder seals.

The vast majority of molluscan material appears to have arrived in Mesopotamia as raw material rather than as finished products. While some could have been consumed in other circumstances or before being transformed, what we find reflected in the tombs is related to their use for the range of objects we see there. As Moorey notes, shells were fairly easily worked, and probably in the same tradition and workshops as those manufacturing objects from bone and stone (Moorey 1994: 129). The quite soft material means that shells could easily be shaped using stone or bronze tools. Further, the many composite objects that combine stone, metal, shell, bitumen and ostrich eggshell demonstrate close relations between the different workshops, not to mention an association between these 'luxury' materials. A conceptual correlation between stone and shell may also be reflected in some of the Sumerian terms referring to shells, which combine the designations for stone ('na_4') and fish ('$pe\check{s}_4$') (Oppenheim 1963).

Some of the finds considered the most technologically advanced and frequently depicted in relation to the royal tombs of Ur are exactly such composite objects that include shell, as for example the inlays on the Standard of Ur, but also those of the gaming boards, lyres and the so-called 'Ram in a Thicket'. Broadly speaking, such composite objects are

[8] We thank Mari Yamasaki and David Reese for making these identifications.

especially found in the wealthier tombs of the cemetery. Other shell finds appear to have been slightly more accessible. The shell cylinder seals are listed not only for some royal tombs, but also for some tombs that do not otherwise contain a great number of grave offerings (for example, PG 231 with a dagger, a reticule, a copper amulet and a shell cylinder seal, or PG 233 with an adze, a razor, three ceramic vessels and a shell cylinder seal).

8.4 Shells and Fish in Clothing and Jewellery

Molluscs were used in a range of items related to clothing and jewellery, especially as beads and 'rings', although not as popular as other material such as gold, silver, carnelian and lapis lazuli in the royal tombs. Occasionally, small beads were made of shell (e.g. as found in PG 55, with thousands of tiny white shell beads, U.8009), and there are a few examples of mother-of-pearl 'pendants' (e.g. PG 95, U.8073). For clothing, the rather anonymously labelled 'shell rings' seem to nearly always occur as a group, suggesting bracelets, necklaces, anklets or belts: PG 143 with 30–31 (with a reconstructed length of 124 cm, according to Baadsgaard 2008: 342, Table 6.27); PG 147 with 17 shell rings (U.8146A-Q); PG 789 with 5–6 shell rings for burial 24 (U.10816F), decayed shell rings on the breast of burial 26 (U.10818D), and an undisclosed number of rings for burial 38 (U.10830E); PG 895 with 10 shell rings (U.10749A-J); PG 1237 with shell rings for burials 12/14/41 (U.12360D-G, U.12399D); PG 1266 with 16 shell rings just below the knees of the central burial (U.12137A-P); PG 1307 with 20 shell rings (U.12246A-T); and PG 1600 with 17–18 shell (U.15408). These rings are rarely illustrated or described in detail, but a few sets have been photographed (e.g. the ones from PG 1600, U.15408, http://www.ur-online.org/subject/15808/, set of 17–18 rings described as found against the front of the body in Woolley's catalogue card). These show that the rings are fairly large and flat, so unlikely to have functioned as beads or finger rings. Woolley considers them to have been part of waistbands: "Shell rings have in several cases been found lying around the waist of a body, and were certainly the decoration of a belt" (1934: 87). This identification is generally supported by the descriptions of where they are found, as well as the length reconstructed. They may have been set on a leather or fibre band that would not have preserved. The groups of shell rings appear more or less equally in very wealthy and sometimes royal tombs, and in less wealthy ones with a more limited amount and range of grave offerings (e.g. PG 147, PG 1307, PG 1600).

Apart from the shell-made pieces of jewellery and items of clothing, a few fish-shaped amulets should be mentioned here. Amulets and beads of expensive materials are common in the royal tombs and found in the thousands with certain individuals, possibly related to expressions of identity of certain queens or other high-status women (Gansell 2007; Baadsgaard 2016). The iconography of their outfits included many references to flora and fauna, as well as deities and hybrid/fantastic creatures. Among these, aquatic beings are not common, but from the tomb of Pu-abi (PG 800), we do find two gold and one lapis lazuli fish-shaped amulets (U.10944A-B, U.10945) (Fig. 8.2); another lapis lazuli fish comes from PG 1195 (U.11966). They are pierced along the length of or through their body, demonstrating that they would have been strung with other beads and amulets, perhaps as part of a bracelet or necklace.

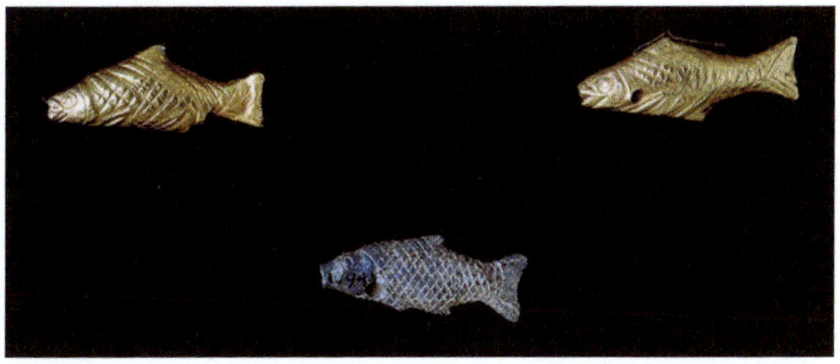

Fig. 8.2 Amulets in the form of fish, from the Early Dynastic Royal Cemetery of Ur, from Puabi's tomb. Fish amulet of lapis lazuli (U.10945; BM 121407) found with the two gold fish amulets on Pu-abi's right arm (U.10944; BM 121405; BM 121406). (©The Trustees of the British Museum)

8.5 Shells as Cylinder Seals

The use of shell as a predominant material for the cylinder seals discovered in the Early Dynastic III-Akkadian tombs, alongside lapis lazuli, has been well documented. The publications and field notes indicate the presence of 60 shell cylinder seals retrieved from 48 tombs (see Appendix 8.1, online supplementary material).[9] These seals were in most cases crafted from shell columellae of *Fasciolaria trapezium*, sourced from the Indian Ocean; a few larger examples may instead have been made from the columellae of *Turbinella pyrum* (Gensheimer 1984; Moorey 1994: 130, 136).

The preservation of shell material proved less robust compared to materials like lapis lazuli, resulting in the decay of many shell cylinder seals, and in some cases, the irreparable state of these artefacts led to their non-preservation. Consequently, numerous seals were not formally catalogued or assigned a unique U number, contrary to the customary procedure for cataloguing findings. Unfortunately, the extent of such losses remains unknown, making the presented figures likely to underestimate the actual abundance of shell cylinders within the tombs and the number of tombs that contained them.

[9] Many more are listed in tombs dated to other periods, and some that are not assigned to a specific tomb, in the catalogue by Legrain in Woolley 1934: 323–64.

Noteworthy is the remarkable uniformity observed in the imagery, materials, and seal shapes of the Early Dynastic III seals, particularly those recovered from the royal tombs of Ur. Such consistency sets them apart from seals of preceding and subsequent periods. In terms of imagery, the Early Dynastic III seals from the royal tombs predominantly depict banquet and contest scenes (Costello 2018). Those seals made of shell feature mostly contest scenes (Fig. 8.3) (nearly all of those where an image is available, see also Appendix 8.1, online supplementary material), agreeing with Pittman's observation that the contest scenes are more often on the white seals (made from shell, calcite, or gold, although examples from the latter two are rare—see Moorey 1994), while the banquet scenes are more often on the blue lapis seals (Pittman 1998: 76).

While most of the shell seals are engraved in one register, there are a few that are engraved in two registers. U.8643 (PG 357) has an upper register depicting a banquet scene, while the lower register a contest scene, U.12461 (PG 1236) has a contest scene in its upper register and a scene of soldiers, equids and a wheeled vehicle in the lower register, and U.12330 (PG1236) has two registers, apparently both with a banquet scene (Legrain in Woolley 1934: 339). There are also two more seals, U.8290 (PG 168) and U. 8513 (PG 261), where half of the circumference of the seal has one register only, the latter bearing the inscription 'An-gig-hu-lugal'. A much worn seal, U.10359, depicts a presentation scene of four figures in long robes approaching a seated figure, who seems to be wearing a flounced robe. This seal dates to the Akkadian period, as does PG 832, where it was found (Pollock 1985: 152).

The selection of shell as a material for these seals indicates a departure from representing actual animals, as the shells undergo complete transformation during the engraving process. Presumably, shell was chosen for its appealing appearance, ease of engraving, and its association with long-distance trade.

8.6 Shells as Containers

Shells themselves were used as containers. There are two main types or shapes: bivalve shell containers and *lambis* vessels. Both were made from shells imported from or through the Gulf.

8.6.1 Bivalve Shell Containers

One type consisted of bivalve shells, specifically those belonging to the Arcidae and Cardiidae families, and commonly referred to as 'cockle-shells' in Woolley's descriptions. Roughly 4–5 cm in diameter, these were largely unmodified since they already have a shape very suitable for accommodating small amounts of a thick paste (Fig. 8.4). These bivalve shells were quite common in the Early Dynastic III and Akkadian tombs at Ur. They were found in at least 125 of the tombs dated to this period, both the private and the so-called royal tombs (see Appendix 8.2, online supplementary material). Unfortunately, Woolley does not always specify the number of bivalve shells found in a tomb (frequently simply writing 'cockle shells containing paint'). It is therefore not possible to provide an exact number of bivalve shells found in the tombs, but certainly they far exceed 125; a conservative estimate based on the numbers given returns a minimum total of over 450 bivalve shell containers in the

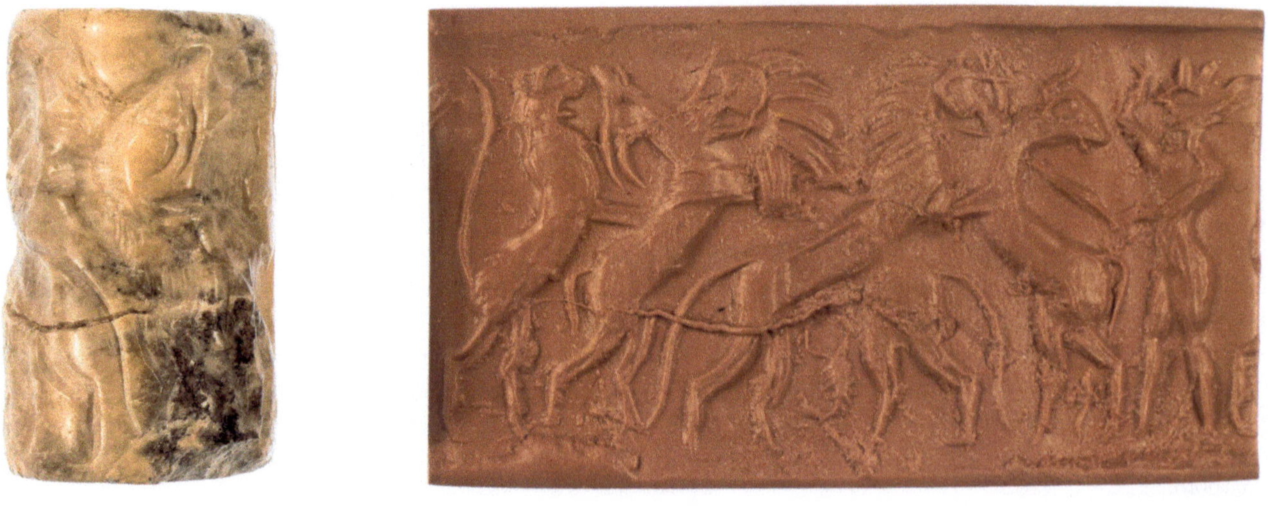

Fig. 8.3 Pale orange-brown shell cylinder seal with contest scene; badly damaged. BM 123621; U.14277. (©The Trustees of the British Museum)

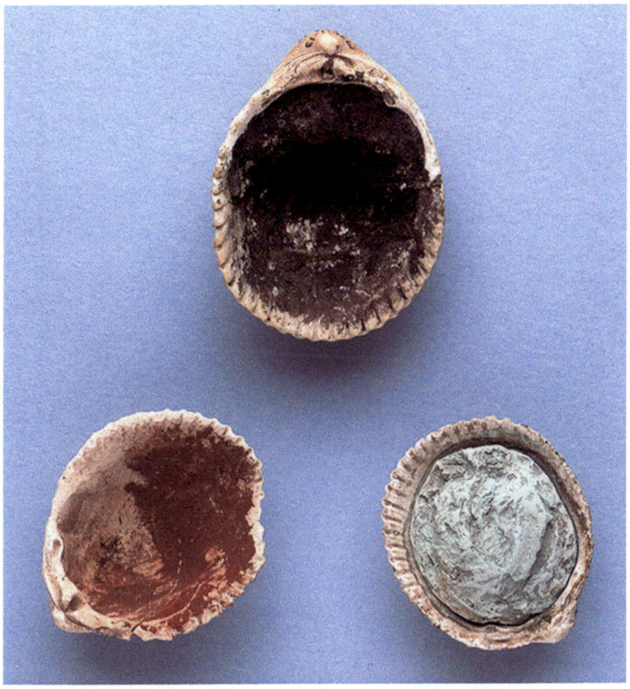

Fig. 8.4 Bivalve shells containing black, green and red paste. BM 139529; 139531; 139534. (©The Trustees of the British Museum)

tombs.[10] This, much higher, number is because they were frequently found in pairs (i.e. as the two halves of the shell placed together), and because especially some of the royal tombs contained a greater amount. For example, at least 50 instances of bivalve shell containers are recorded for PG 789, and 74 for PG 1237.

Many of the tombs are not described in detail, with only the main elements listed in tabular form. It is thus difficult to ascertain the exact placement of the bivalve shells in the tombs and their association with the deceased and other finds. However, the tombs that are described in detail, combined with comments from Woolley's field notes, suggest that bivalve shells were directly associated with individual human skeletons, rather than a general part of tomb assemblages or more exclusively related to mortuary practices, such as the *Lambis* containers discussed below. In the examples mentioned in Woolley's text, field notes, and illustrated in plates and plans, they are nearly always placed near the head, arms/mid-body or feet of a human skeleton. Further, in the death pits of PG 789 and PG 1237, they are listed as part of the assemblage of objects belonging to specific bodies of 'attendants'. Bivalve shells would thus appear to have been personal items. However, for the 'attendants' in PG 789 and PG 1237, the sets of finds become so standardised that they are unlikely to represent individual personal expressions of identity, but rather appear to be carefully constructed representations of *types* of identities created as part of the broader rituals that also involved their sacrifice. In particular for PG 1237, 'The Great Death Pit', the bivalve shells appear almost to have been mass-produced for the occasion: nearly half the bodies (33) have bivalve shells associated with them, and in all but one instance, described as containing green paint (Fig. 8.5).[11]

The bivalve shells are not exclusive to the sacrificial victims of death pits, but also occur in the main chamber and are thus associated with the presumed royal person and main deceased (for example in PG 800). Many were found in the so-called private graves - in these cases, the lack of detailed descriptions prevents us from clear associations, but it seems that most of these contained only a single deceased.

For example, PG 867 is a private grave with a single inhumation in a simple shaft grave, the body wrapped in matting and placed in a crouched position on its left side. It was accompanied by jewellery (hair rings, necklace, finger rings, bracelets), four cylinder seals, several copper bowls, a calcite vessel, a ceramic vessel and, placed by the knees (no. 12 on the plan), three bivalve shells containing black paint (Woolley 1934: 161). PG 1100 is another single inhumation private grave, this time in a coffin with the offerings inside the coffin with the deceased (Fig. 8.6). The body was placed on the right side along with jewellery, copper objects, calcite and ceramic vessels, and a bivalve shell containing black paint (no. 6 on the plan, behind the legs; Woolley 1934: 164–65). While these two graves can hardly be described as poor, they are certainly not as wealthy in offerings and construction as the royal tombs, and may therefore have belonged to individuals of a less wealthy status (in Woolley's terminology, 'middle class'). It thus seems that bivalve shells were available and meaningful to people outside 'royal' circles.

The bivalve shells were also not exclusive to adults, as at least two examples demonstrate: one of the 'attendants' in PG 1237 with bivalve shells is described as 'a very young person' (Woolley 1934: 117), and the single coffin burial of private grave PG 1133 was that of a child, buried with copper models of bivalve shells (Woolley 1934: 167–68).

We cannot confirm Woolley's contention that 'Every woman's grave of the old cemetery seems originally to

[10] This is based on the assumption that when Woolley writes 'cockle shells' in the plural, at least two shell containers were present, and that 'cockle shell' in the singular refers to only one shell (or, more precisely, one half of the shell of the live animal). The rest are based on numbers given for each tomb/group in a tomb. This gives a minimum of 454 bivalve shell containers, which is likely to be higher. Judging by Woolley's field notes, the bivalve shells were not always assigned a specific field number, and therefore are not all listed as part of tomb assemblages on UrOnline, nor do they always make it into Woolley's published descriptions.

[11] Only for U.12404E is the paint instead described as green-blue in Woolley's notes.

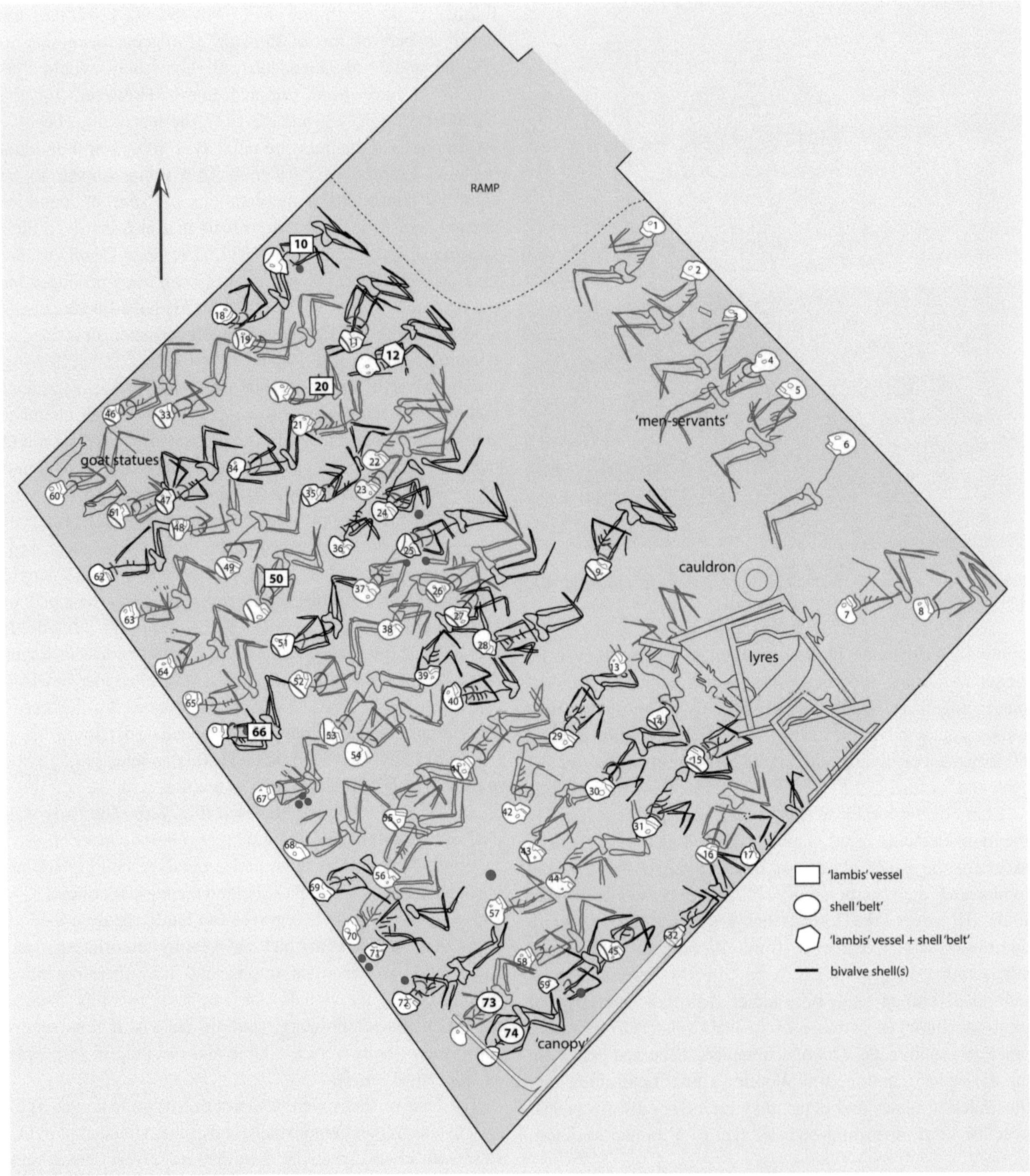

Fig. 8.5 Simplified plan of PG 1237 showing skeletons with associated bivalve shell(s) (black outline), 'lambis' vessel and/or shell rings 'belt'. (Drawn by L. Recht after Woolley 1934: pl. 71. Associations of skeletons and finds based on Woolley's description (1934: 113–20))

have contained cosmetics; such were an invariable part of the tomb furniture. The ordinary receptacle is a cockle-shell…' (Woolley 1934: 245), simply because the skeletons have not been sexed by a specialist. Bivalve shells were also found in Early Dynastic IIIB - early Akkadian A tombs at Kish; here, the shells appear to be more strongly associated with male burials. They occurred in 57 per cent of male graves, 22 per cent of female graves, and 25 per cent of young adult/child

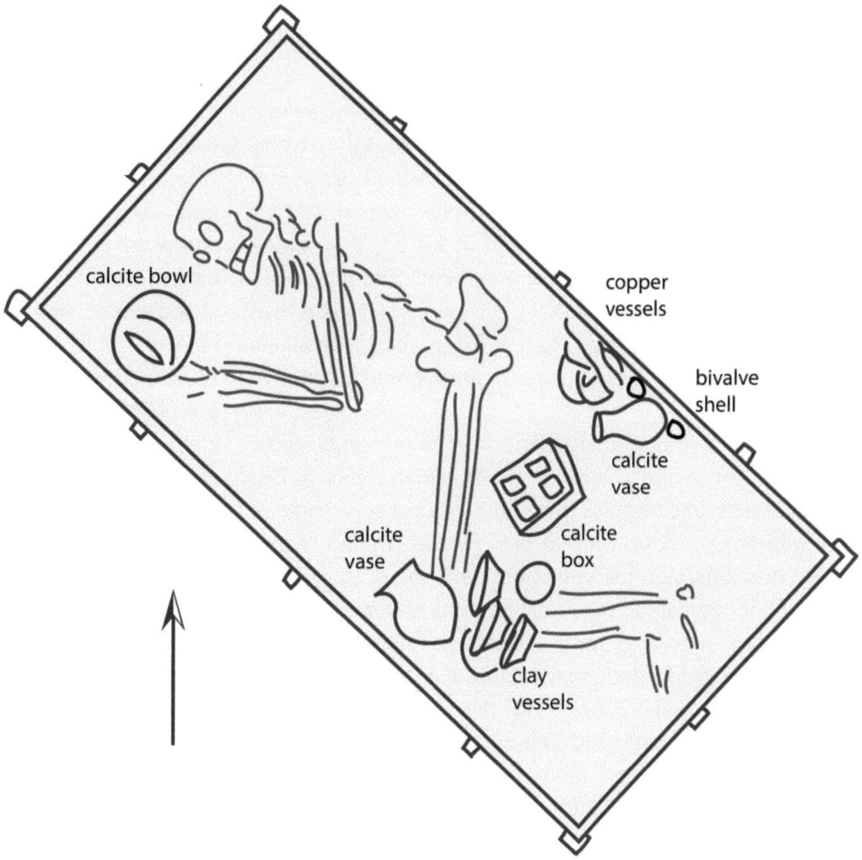

Fig. 8.6 Plan of PG 1100 showing position of skeleton and associated finds. (Drawn by L. Recht after Woolley 1934: 165, fig. 41)

graves (Moorey 1994: 134; see also Mackay 1929: 131–32).[12]

That the bivalve shell shape was significant is demonstrated by their imitation in other materials. From the Early Dynastic III tombs at Ur, we find imitations in silver, gold and copper/bronze (Appendix 8.2, online supplementary material). These metal skeuomorphs might to some extent have been upscale versions of the natural ones: the only extant gold pair was found in the main chamber of PG 800, immediately next to the head of what is presumed to be queen Pu-abi (no. 117 on Woolley's plan, pl. 36). A pair of silver bivalve shells and the few examples of unusually large bivalve shells also come from the main chamber of this wealthy tomb.

As mentioned, the shells contained mineral pigments (Bimson 1980). The vast majority are described as containing paint, and it can thus reasonably be assumed that this was the general function of the examples found. Preservation of the pigments was in many cases very good, and it is therefore possible to note some of the colours present, as found: shades of red, white, green, black, blue, brown and yellow. Green and black were by far the most common. There is also a single example from PG 1145 of a stone carved like a bivalve shell but with an internal division akin to a modern painter's palette (U.11841), emphasising the importance of the pigments, and that a range of colours was indeed used.

8.6.2 Bivalve Shells, Death and Pigments

The bivalve shells appear to be exclusively connected with pigments. Although Woolley does not always note the presence of a pigment in them, there is nothing to suggest that they were placed in the tombs as food offerings. As mentioned above, in all undisturbed recorded contexts, the bivalves are closely associated with individual human skeletons, and appear to be personal but standardised items. That is, part of the individual equipment of each body, along with jewellery, clothes and in some cases razors, cosmetic sets, headgear and tools or weapons. This situation is mirrored in Cemetery A at Kish, where bivalve shells with pigment were also found in a number of tombs - and, as Aynard points out, because the pigment was intact, it seems

[12] Shells were also quite common in the private graves at Khafajeh, for which see Delougaz et al. 1967: 137–42. Although apparently less common, they are also reported from northern Mesopotamia, for example in Early Bronze Age Tomb 1 at Tell umm el-Marra, where cockle shells filled with black pigment were found (Schwartz et al. 2003: 332 and footnote 30; note also that these were identified as marine *Trachycardium* as well as local freshwater *Unio*).

that they were prepared specifically for the tombs (Aynard 1966: 30); we do not know if the same applies to the Ur examples.

The items found with the bodies in death pits indicate that substantial effort was made to demonstrate not only wealth, access to faraway resources through long distance trade and excellent craftsmanship, but also to display their identities. The high level of standardisation (and depositional grouping within the pits) represented by the sets found with the bodies suggests that these were class or group-based rather than purely individual identities. The bivalve shell containers with their pigments were apparently an appropriate part of such a display.

Why do bivalve shells, with their range of pigments, occur in so many of the Early Dynastic III-Akkadian tombs at Ur? We would here like to point to a number of aspects, some or all of which may have been at play in each of the tombs. Beyond the skeuomorphic vessels, bivalve shells are not part of the rich iconographic repertoire of third millennium BCE Mesopotamia, despite a clear fascination with animals in many different forms and contexts (see e.g. papers in Recht and Tsouparopoulou 2021), and they are very rare in Mesopotamian literature. Certainly, part of their value would have been related to the prestigious long-distance trade with the Persian Gulf and further afield that also provided the material for many of the other luxury items deposited with the Early Dynastic III burials, and especially in the royal tombs. This may go beyond the economic aspect. Since the bivalve shells are mainly marine species, they may also have acted as a reference to or even symbol of the (salt water) sea. Although not extremely prominent in Sumerian literature, the sea does occur, especially as a physical boundary of the world to the south, but also as a cosmological boundary between the living and the dead (Verderame 2020). As Verderame has shown, the sea is also described as providing a yield (literally 'giving birth'), and are in other ways associated with female elements through being gendered female, through the goddess Nanše and, in later tradition in the *Enuma Eliš,* with Tiamat (Verderame 2020, 2022).[13] The marine creatures coming from the sea as part of the yield, especially those with a durable and pleasing material such as shell, may thus have become part of a complex network of symbols of liminal space, death and abundance.

This is to some extent supported by the pigments. Assuming that these were indeed applied to the human body and face, we may turn to the text known as *Inanna's Descent to the Netherworld.* In her descent, Inanna uses mascara to enhance her appearance: 'She placed mascara which is called "Let a man come, let him come" on her eyes' (ETCLS, lines 108–13).[14] Here, the alteration is part of a performative act of seduction. Although of a much later period, the otherwise similar descriptions noted in texts from late second millennium BCE Ugarit alert us to other uses of cosmetic pigments, where red is applied before and after battle (Yamasaki, this volume). Back in Early Dynastic Mesopotamia, the sculptural arts also suggest a connection between applied pigments and worship. Votive statues such as the well-known assemblage from Tell Asmar and other Mesopotamian sites have characteristic single eyebrows and large eyes (see examples in Aruz 2003). These have been interpreted as representing the attention paid to deities (Winter 2000; cf. Evans 2012: esp. Chapter 3). As is often noted, the eyes are clearly emphasised by their size and by consisting of inlays (usually as the only part of the statues) of shell, limestone and lapis lazuli. What is less frequently mentioned is that the eyes were also lined with black using bitumen. This likely represents use of a black pigment to enhance or alter appearance, and the statues demonstrate that it was not gender specific, at least in the context of votives.[15] As also argued by Cassin, cosmetic pigments appear to have had transformative powers that were utilised under various circumstances (*RLA*, 'Kosmetik'), and would thus be particularly suitable for liminal spaces and events, including those related to death.

We cannot reconstruct every detail surrounding the events and rituals of the funerals of the people buried in the tombs at Ur, but it is tempting to consider the bivalve shells and their pigments as part of the elaborate displays that clearly marked the events. The pigments might have been used not only to alter or enhance the appearance of the deceased (including 'attendants'), but also to hide or even somewhat delay the process of decomposition[16] during the period when the bodies and entire assemblages were visible to humans.

8.6.3 'Lambis' Shell Vessels

The second type of shell container is larger, and Woolley labels it a 'shell lamp' (or 'conch shell cut as a lamp') throughout the 1934 publication; in his field notes, they occur as 'large shell cut as lamp' or 'large shell cut as ladle' (usually 15–19 cm long) (Fig. 8.7). In this case, the *Lambis*

[13]In the *Enuma Eliš*, Tiamat is described as split into two 'like a fish for drying', when her body is used to create the world (Foster 2005: 462, line 137).

[14]https://etcsl.orinst.ox.ac.uk/section1/c141.htm; translation https://etcsl.orinst.ox.ac.uk/section1/tr141.htm

[15]That this was not a new practice can be seen on a Late Uruk period alabaster statue of a 'priest-king' from Uruk, where the eyeliner is even inlaid and very thick (Aruz 2003: 25, fig. 11b); an Early Dynastic example from Mari has eyeliner inlays of lapis lazuli (Aruz 2003: 162, no. 105), suggesting that the liner was not always black (see also Vogel 2022: 36 for various types and colours of make-up and their possible applications).

[16]There is other evidence to suggest that attempts were made to preserve the body (Baadsgaard et al. 2011, 2012).

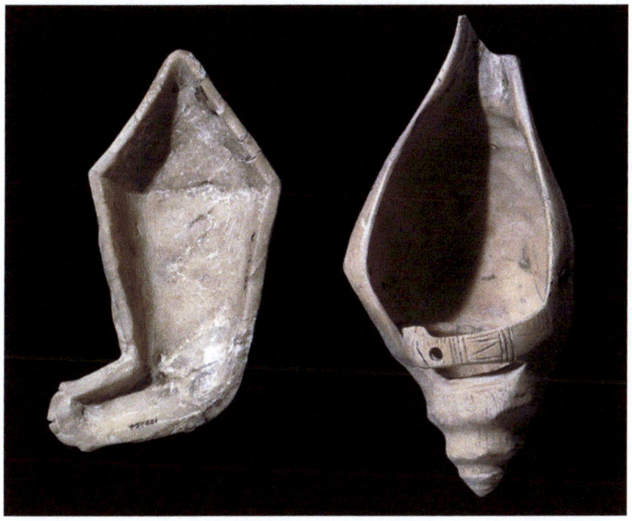

Fig. 8.7 *Lambis* vessels. Left: calcite, engraved with human-headed bovine on exterior (BM 122254; U.11795, PG 1134). Right: shell, engraved with bird head and neck, eye inlaid with lapis lazuli (BM 120861; U.8538, PG 265, probably Akkadian). (©The Trustees of the British Museum)

truncata sebae was usually used as the basis for the shape;[17] it was modified by removal of the digitations to provide a more even external surface and of the internal column to provide space for contents. In some cases, they were decorated with engraving and inlays on the exterior or even as composite vessels, as for example U.8191 from PG 143, which has an engraved bird's head and neck across the front, or U.8313 from PG 183, where the main body is made of shell, and the breast, neck and head of a duck is made of limestone, shell, lapis lazuli and bitumen (Woolley 1934: 531). In the tabular list of tombs and the tomb descriptions, Woolley mentions 45 examples of this type of shell container as present in 39 tombs in the cemetery (Appendix 8.3, online supplementary material). Like the bivalves, the lambis shape was also imitated in other materials, but even more extensively so (Woolley's Type 115). Woolley lists 23 copper/bronze, eight silver, one gold and three alabaster/white calcite examples. Some of these were also decorated and/or composite vessels, such as for example two calcite vessels with engraved human-bull heads (Woolley 1934: pl. 182).

Although not as common as the cosmetic containers, these *lambis* vessels were thus also quite common, and found in the same types of tombs: that is, both some of the extremely wealthy royal tombs and in less luxurious but still well-equipped single inhumation burials.

Throughout the publication, Woolley calls these vessels lamps, both in the shell and the skeuomorphic versions. In his discussion, he does mention the problem that there is no evidence of burning and suggests that they may instead have been used for pouring, but ultimately decides that the shape rather resembles a lamp and explains the absence of burning as the vessels being unused when deposited (Woolley 1934: 283). In his field notes, Woolley frequently described these vessels as 'ladles', but this identification does not make it into his published discussion in any detail. The metal examples have been examined by Müller-Karpe (*RLA*, 'Metallgefässe') and by Winter (1999), both of whom are not persuaded by the lamp theory and rather prefer a pouring function. As also noted by both, a *lambis* vessel may be depicted on the inlay decoration of one of the lyres from PG 789, carried by an anthropomorphised lion (Woolley 1934: pl. 105). Based on this and a detailed analysis of the contexts of the vessels as recorded in Woolley's publication and notes, Winter convincingly argues that they were part of mortuary rituals and perhaps functioned as transfer vessels, effectively funnelling liquids between other containers (conceptually not dissimilar to Woolley's ladle idea, one might note). Some of the vessels are placed as if belonging to individual skeletons, similar to the placement of the bivalve containers - as for example in PG 1237, where *lambis* vessels were associated with bodies 10, 12, 20, 50 and 66 (Woolley 1934: 117, 119–20). But they are not as consistently placed with the bodies as the bivalve shells and other items (e.g. jewellery) and thus do not give the impression of being a standard part of a personal set. This impression is strengthened by their other placements and patterns, where they occur especially in a north/northwest location and in fills above the main funerary depositions. Winter suggests that these depositional patterns and the shape of the vessel could relate to ritual performance related to libations surrounding the mortuary practices at and after the burial itself (Winter 1999).

Although the shells have been transformed to act as efficient containers, their shape is still characteristic, to the extent that they are easily recognisable in their skeuomorphic forms. Since they originally came from the sea, the connection with liquids may not be a coincidence, and they may have been considered particularly suitable for libations partly for this reason.

What is of further interest in terms of expressions of human-animal relations and more broadly of materiality, is the play with and hybridisation of animals and materials. *Lambis* shells, the abode of creatures of the sea, are turned into birds, creatures of the sky. The birds are difficult to identify more specifically, but in some cases might be birds of prey (e.g. U.8191 in PG 143), in others waterbirds (e.g. U.8313 in PG 183) - and hence embodying another reference to aquatic-related beings. The alabaster examples add additional layers of multispecies and human-thing entanglements: stone is transformed into an already

[17] Previously misidentified as *Turbinella pyrum* (Moorey 1994: 130); in a few cases, *Turbinella pyrum* (which likely came from the Indus Valley) may have been used, but *Lambis truncata sebae* is the most common.

transformed *lambis* shape, and this shape is then turned into a human-bovine hybrid. The resulting skeuomorphic vessel is thus symbolically rich and meaningful, embodying a new being ideally placed to partake in ritual action. The engagement with aquatic and other beings is both distanced and deep in its representation in the tombs.

8.7 Conclusion

The material from the late third millennium BCE tombs at Ur reflects a range of relations between humans and aquatic beings, both in life and in death. Fish were less present than molluscs, perhaps partly because their physical remains are underrepresented, but the few extant remains indicate that they were considered appropriate grave offerings, most likely as food. Their representation in banquet scenes supports this role, and they appear to be part of an ideology that emphasises abundance and a range of available goods. Their depictions are fairly generic and may thus refer to fish - and fishing - in a general sense, rather than to the importance of specific species. The few examples of fishing equipment provide tantalising clues to some of the techniques and tools used, but rather than identify specific fishermen, they also seem to refer to the generic idea of fishing and the important role that rulers had in ensuring a supply of fish for food and other purposes (including medicinal and offerings to deities), and by extension, maintenance of water sources such as rivers and canals.

The roles of molluscs are even more complex and multifaceted. They were primarily procured through long distance trade with the Persian Gulf and further afield, and transformed into a variety of objects, sometimes maintaining their original shape, but most frequently changed to the point of being unrecognisable as marine shells. They represent complex meshworks of interaction and *chaîne opératoires*. Bivalve shells were closely linked with pigments in hues of especially black and green and were a key node in a meshwork of (personal) identity, death and transformation. Their shape was retained and extended to other 'luxury' materials, likely reflecting individual status and preference.

For cylinder seals, inlays and most of the jewellery, the finished product was so far removed from the actual animal or its shell as to have been completely unrecognisable, and what was left was perhaps a somewhat mysterious and appealing lustre. The comparative softness of the shells would have made them a popular material for craftspeople. Whether or not their origin as shell would have been generally known is not clear, but it is likely that their status as a non-local material would have been acknowledged. The process of transformation from raw material to finished (luxury) product may have been understood as a positive process (Verderame 2022: 164). In (urban) Sumerian thought, there was a division between the city and non-city landscapes, with the former perceived as 'civilised', and raw material coming from outside and made into fine objects by artisans, who themselves were primarily a product of the new urban spaces. The change from material to anthropogenic entity may thus also have been associated with this overall ontology.

The meshworks of human-fish and human-mollusc relations thus find expression in complex and transformed ways in the mortuary practices of mid to late third millennium BCE Ur, hinting at the role of aquatic beings not only in death, but also everyday life, symbolling, trade networks and personal identity.

Acknowledgements We are very grateful to Lorenzo Verderame for sharing his paper with us and to David S. Reese for sharing his personal notes on the presence of shells and fish in the Ur tombs. We also thank David S. Reese and Mari Yamasaki who aided us in the identification of some bivalve shell objects.

Bibliography

Alhaique, F., Romano, L. & D'Agostino, F. 2021. Between Sacred and Profane: Human-animal Relationships at Abu Tbeirah (Southern Iraq) in the Third Millennium BC. In: L. Recht & C. Tsouparopoulou, eds. *Fierce Lions, Angry Mice and Fat-tailed Sheep: Animal Encounters in the Ancient Near East.* Cambridge: McDonald Institute for Archaeological Research, pp. 63–75.

Aruz, J, ed. 2003. *Art of the First Cities: The Third Millennium B.C. from the Mediterranean to the Indus.* London: Yale University Press.

Aynard, J.M. 1966. Coquillages mésopotamiens. *Syria,* 43: 21–37. https://doi.org/10.3406/syria.1966.5873.

Baadsgaard, A. 2008. Trends, Traditions, and Transformations: Fashions in Dress in Early Dynastic Mesopotamia (PhD dissertation, Pennsylvania: University of Pennsylvania).

Baadsgaard, A. 2016. All the Queen's Clothes: Identifying Female Royalty at Early Dynastic Ur. *Near Eastern Archaeology,* 79: 148–55. https://doi.org/10.5615/neareastarch.79.3.0148.

Baadsgaard, A., Monge, J., Cox, S. & Zettler, R.L. 2011. Human Sacrifice and Intentional Corpse Preservation in the Royal Cemetery of Ur. *Antiquity,* 85: 27–42.

Baadsgaard, A., Monge, J. & Zettler, R.L. 2012. Bludgeoned, Burned, and Beautified: Reevaluating Mortuary Practices in the Royal Cemetery of Ur. In: A. Porter & G.M. Schwartz, eds. *Sacred Killing: The Archaeology of Sacrifice in the Ancient Near East.* Winona Lake: Eisenbrauns, pp: 125–58.

Bimson, M. 1980. Cosmetic Pigments from the "Royal Cemetery" at Ur. *Iraq,* 42: 75–77. https://doi.org/10.2307/4200116.

Borelli, N. 2021. An Ur III Account of Fish from Umma. *N.A.B.U.,* 2021: 13–14.

Boyd, B. 2017. Archaeology and Human–Animal Relations: Thinking through Anthropocentrism. *Annual Review of Anthropology,* 46(1): 299–316.

Costello, S.K. 2018. Rematerializing the Early Dynastic Banquet Seal. In: G. Jamison, M. Ameri, S.J. Scott & S.K. Costello, eds. *Seals and Sealing in the Ancient World: Case Studies from the Near East, Egypt, the Aegean, and South Asia.* Cambridge: Cambridge University Press, pp. 68-80. https://doi.org/10.1017/9781108160186.007.

Delougaz, P., Hill, H.D. & Lloyd, S. 1967. *Private Houses and Graves in the Diyala Region*. Oriental Institute Publication LXXXVIII. Chicago: University of Chicago Press.

Dickson, D. B. 2006. Public Transcripts Expressed in Theatres of Cruelty: The Royal Graves at Ur in Mesopotamia. *Cambridge Archaeological Journal,* 16: 123–44.

Dossin, G. 1968. L'inscription de Mesanepada. In: A. Parrot, *Mission archéologique de Mari IV: le "trésor" d'Ur*. Paris: P. Geuthner, pp. 53–59.

Edens, C. 1992. Dynamics of Trade in the Ancient Mesopotamian "World System". *American Anthropologist,* 94: 118–39.

Ellison, R., Renfrew, J., Brothwell, D. & Seeley, N. 1978. Some Food Offerings from Ur, Excavated by Sir Leonard Woolley, and Previously Unpublished. *Journal of Archaeological Science,* 5: 167–77. https://doi.org/10.1016/0305-4403(78)90030-4.

Englund, R.K. 1990. *Organisation und Verwaltung der Ur III-Fischerei*. Berliner Beiträge zum Vorderen Orient 10. Berlin: Reimer.

Evans, J.M. 2012. *The Lives of Sumerian Sculpture: An Archaeology of the Early Dynastic Temple*. Cambridge: Cambridge University Press.

Foster, B.R. 2005. *Before the Muses: An Anthology of Akkadian Literature*. 3rd ed. Bethesda, Maryland: CDL Press.

Frame, G., Jeffers, J. & Pittman, H. eds. 2021. *Ur in the Twenty-first Century CE: Proceedings of the 62nd Rencontre Assyriologique Internationale at Philadelphia, July 11-15, 2016*. University Park, Pennsylvania: Eisenbrauns.

Gansell, A.R. 2007. Identity and Adornment in the Third-millennium BC Mesopotamian "Royal Cemetery" at Ur. *Cambridge Archaeological Journal,* 17: 29–46.

Gensheimer, T.R. 1984. The Role of Shell in Mesopotamia: Evidence for Trade Exchange with Oman and the Indus Valley. *Paléorient,* 10: 65–73. https://doi.org/10.3406/paleo.1984.4350.

Hall, H.R. & Woolley, L. 1927. *Ur Excavations I: Al-'Ubaid*. Oxford: Oxford University Press.

Jawad, L.A. 2006. Fishing Gear and Methods of the Lower Mesopotamian Plain with Reference to Fishing Management. *Marina Mesopotamica,* 1: 1–37.

Kenoyer, J.M. 2008. Indus and Mesopotamian Trade Networks: New Insights from Shell and Carnelian Artefacts. In: E. Olijdam and R.H. Spoor, eds. *Intercultural Relations between South and Southwest Asia: Studies in Commemoration of E.C.L. During Caspers (1934–1996)*. Oxford: Archaeopress, pp. 19–28.

Kenoyer, J.M., Price, T.D. & Burton, J.H. 2013. A New Approach to Tracking Connections between the Indus Valley and Mesopotamia: Initial Results of Strontium Isotope Analyses from Harappa and Ur. *Journal of Archaeological Science,* 40: 2286–2297. https://doi.org/10.1016/j.jas.2012.12.040.

Laursen, S. & Steinkeller, P. 2017. *Babylonia, the Gulf Region, and the Indus: Archaeological and Textual Evidence for Contact in the Third and Early Second Millennia B.C.* Winona Lake (IN): Eisenbrauns.

Legrain, L. 1936. *Ur Excavations III: Archaic Seal-impressions*. London: British Museum.

Legrain, L. 1951. *Ur Excavations X: Seal Cylinders*. London: British Museum.

Lernau, O., Shapiro, J., Paz, S. & Greenberg, R. 2021. Fishing, Fish Consumption, Urbanism and Migrants at Tel Bet Yerah, 3200–2700 BC. *Antiquity,* 95: 885–899. https://doi.org/10.15184/aqy.2021.70.

Mackay, E. 1929. *A Sumerian Palace and the "A" Cemetery at Kish, Mesopotamia, Part II*. Chicago: Field Museum Press.

Moorey, P.R.S. 1994. *Ancient Mesopotamian Materials and Industries: The Archaeological Evidence*. Oxford: Clarendon Press.

Nissen, H.J. 1966. *Zur Datierung des Königsfriedhofes von Ur*. Bonn: Rudolf Habelt.

Oppenheim, A.L. 1954. The Seafaring Merchants of Ur. *Journal of the American Oriental Society,* 74: 6–17. https://doi.org/10.2307/595475.

Oppenheim, A.L. 1963. Mesopotamian Conchology. *Orientalia,* 32: 407–12.

Pilaar Birch, S.E. 2018. *Multispecies Archaeology*. Abingdon: Routledge.

Pittman, H. 1998. Cylinder Seals. In: R.L. Zettler, L. Horne, D.P. Hansen & H. Pittman, eds. *Treasures from the Royal Tombs of Ur*. Philadelphia: University of Pennsylvania, Museum of Archaeology and Anthropology, pp. 75–84.

Pollock, S. 1985. Chronology of the Royal Cemetery of Ur. *Iraq,* 47: 129–58. https://doi.org/10.2307/4200236.

Potts, D.T. 2003. The Gulf: Dilmun and Magan. In J. Aruz, ed. *Art of the First Cities: The Third Millennium B.C. from the Mediterranean to the Indus*. London: Yale University Press, pp. 307–308.

Potts, D.T. 2012. Fish and Fishing. In: D.T. Potts, ed. *A Companion to the Archaeology of the Ancient Near East*. Blackwell Companions to the Ancient World. Chichester: Wiley-Blackwell, pp. 220–35. https://doi.org/10.1002/9781444360790.ch12.

Reade, J. 2001. Assyrian King-lists, the Royal Tombs of Ur, and Indus origins. *Journal of Near Eastern Studies,* 60: 1–29.

Recht, L. 2011. Sacrifice in the Bronze Age Aegean and Near East: A Poststructuralist Approach (PhD dissertation, Trinity College Dublin, Dublin).

Recht, L. 2022. *The Spirited Horse: Equid-human Relations in the Bronze Age Near East*. London: Bloomsbury.

Recht, L. & Tsouparopoulou, C. eds. 2021. *Fierce Lions, Angry Mice and Fat-tailed Sheep: Animal Encounters in the Ancient Near East*. Cambridge: McDonald Institute for Archaeological Research.

Safar, F., Mustafa, M.A. & Lloyd, S. 1981. *Eridu*. Baghdad: Ministry of Culture and Information.

Sahrhage, D. 1999. *Fischfang und Fischkult im alten Mesopotamien*. Frankfurt am Mein: Peter Lang.

Schwartz, G.M., Curvers, H.H., Dunham, S. & Stuart, B. 2003. A Third-millennium B.C. Elite Tomb and Other New Evidence from Tell Umm el-Marra, Syria. *American Journal of Archaeology,* 107(3): 325–61.

Sykes, N. 2015. *Beastly Questions: Animal Answers to Archaeological Issues*. London: Bloomsbury Academic.

Tsouparopoulou, C. 2024. The Early Dynastic 'Maison des Fruits' at Tell K in Tello (Ĝirsu). *Bulletin of the American Schools of Overseas Research,* 391: 191–225. https://doi.org/10.1086/730298.

van Buren, E.D. 1948. Fish-offerings in Ancient Mesopotamia. *Iraq,* 10: 101–21.

Verderame, L. 2020. The Sea in Sumerian Literature. *Water History,* 12: 75–91. https://doi.org/10.1007/s12685-020-00244-6.

Verderame, L. 2022. Engendered Cosmic Regions in Ancient Mesopotamian Mythologies. In K. de Graef, A. Garcia-Ventura, A. Goddeeris & B. Alpert Nakhai, eds. *The Mummy under the Bed: Essays on Gender and Methodology in the Ancient Near East*. Münster: Zaphon, pp. 157–72.

Vidale, M. 2011. PG 1237, Royal Cemetery of Ur: Patterns in Death. *Cambridge Archaeological Journal,* 21: 427–51.

Vogel, H. 2014. Der Königsfriedhof von Ur und das Problem der so genannten Gefolgschaftsbestattungen. In: T. Link & H. Peter-Röcher, eds. *Gewalt und Gesellschaft: Dimensionen der Gewalt in ur- und frühgeschichtlicher Zeit*. Bonn: Rudolf Habelt, pp. 169–85.

Vogel, H. 2015. Beim-Stirnketten-Träger, Musikerinnen und dog-collar Trägerinnen als Bestattungen und MitBestattungen im Royal Cemetery in Ur. In: R. Dittmann & G.J. Selz, eds. *It's a Long Way to a Historiography of the Early Dynastic Period(s)*. Münster: Ugarit-Verlag, pp. 461–507.

Vogel, H. 2022. The Queen's Beauty: Leadership as an Aesthetic and Embodied Practice in Ancient Mesopotamia. In: U. Matić, ed. *Beautiful Bodies: Gender and Corporeal Aesthetics in the Past*. Oxford: Oxbow Books, pp. 23–48. https://doi.org/10.2307/j.ctv25wxc9n.

Winter, I. 1999. Reading Ritual in the Archaeological Record: Deposition Pattern and Function of Two Artifact Types from the Royal Cemetery of Ur. In H.J. Nissen, ed. *Fluchtpunkt Uruk archäologische Einheit aus methodischer Vielfalt*. Rahden: M. Leidorf, pp. 229–56.

Winter, I. 2000. The Eyes Have It: Votive Statuary, Gilgamesh's Axe, and Cathected Viewing in the Ancient Near East. In: R. Nelson, ed. *Visuality before and beyond the Renaissance*. Cambridge: Cambridge University Press, pp. 22–44.

Woolley, L. 1934. *Ur Excavations II: The Royal Cemetery*. Philadelphia: University of Pennsylvania Press.

Woolley, L. 1939. *Ur Excavations V: The Ziggurat and Its Surroundings*. Philadelphia: University of Pennsylvania Press.

Woolley, L. 1955. *Ur Excavations IV: The Early Periods, a Report on the Sites and Objects Prior in Date to the Third Dynasty of Ur Discovered in the Course of the Excavation*. London and Philadelphia: British Museum and the University Museum Philadelphia.

Woolley, L. 1962. *Ur Excavations IX: The Neo-Babylonian and Persian Periods*. London: British Museum.

Woolley, L. 1965. *Ur Excavations VIII: The Kassite Period and the Period of the Assyrian Kings*. Philadelphia: University of Pennsylvania Press.

Woolley, L. 1974. *Ur Excavations VI: The Buildings of the Third Dynasty*. London and Philadelphia: The Trustees of the Two Museums.

Woolley, L. & Mallowan, M.E.L. 1976. *Ur Excavations VII: The Old Babylonian Period*. Philadelphia: University of Pennsylvania Press.

Clamour from the Mediterranean to the Persian Gulf. Revised Dating and Distribution Patterns of Engraved *Tridacna squamosa*

9

Valery J. Schlegel

Abstract

Engraved *Tridacna* were used as decorated containers from the Mediterranean region to the Persian Gulf in the first millennium BCE. R. Stucky's work of 1974 is to date the most comprehensive catalogue of engraved *Tridacna*. In the 50 years since then, new finds have been published in various articles and are compiled in this chapter, along with a never adequately published and almost complete specimen housed in Leiden, and an example from a Nimrud tomb which shifts the former dating of engraved *Tridacna*. Furthermore, the chapter analyses the circulation of engraved *Tridacna* between their probable provenience(s) and the locations of their find spots, based on the published pieces and the new evidence for dating. This includes mapping the finds of over 140 specimens via revised dating and stylistic groups to draw conclusions on their provenience and distribution patterns.

9.1 Introduction

In the first millennium BCE, engraved *Tridacna squamosa* were used as decorated containers as far west as Vulci and as far east as Susa. Stucky's dissertation (1974) is to date the most comprehensive catalogue of engraved *Tridacna*. New finds have been published since in various articles and are compiled for the first time here in Table 9.1. The chapter also examines a never adequately published specimen in Leiden. Furthermore, the location of workshops processing the shells into artistic cosmetic vessels and the end costumers are as uncertain as the patterns of distribution are undiscussed. Therefore, this chapter analyses the patterns of distribution of engraved *T. squamosa* between their probable provenience(s) and the locations of their find spots based on additionally published pieces and new evidence for a revision of the former dating. The chosen approach to the research question is mapping the finds of now over 140 specimens via their (revised) dating and stylistic groups to draw conclusions on their provenience and circulation patterns.

9.2 Stylistic Groups of Engraved *Tridacna*

The large *Tridacna* clam species originates in the shallow waters of the Red Sea, the Persian Gulf and the Indian Ocean. After fishing for the shells or collecting the clams from the beach, the undulating scales of the outer shells were removed with a hammer, then ground down, polished and finally skilfully carved inside and out. The carvings were coloured with pigments to stand out from the whitish shell surface; various fragments still have traces of brown, black, red or often blue-green pigments (cf. Brandl 2001). The decoration has been divided into five groups by Stucky (1974: 19–21; 2007: 219–20) (see Fig. 9.1).

The first three groups have in common that the sculptured umbo (the "beak" of the bivalve) resembles a human face. This face has drilled pupils and a beauty mark on the right cheek, a diadem, a headdress out of crossed lines and "haircurls" resembling hanging leaves or rounded feathers. Typically, the nose and eyebrows (that is the nasal and frontal bone) are incised in merging lines, but an exception in deep carving is known from Vulci (Rathje 1986: 393–96). Sometimes ears and triangled earrings are also carved.

Group I examples are decorated with a central figure, its back protruding from the umbo-head. The arms are stretched out along the edges of the shell and a belt girdles its square patterned garment. Four wings emerge from the centre of the body. The figure has been called a siren or a "démon" (Stucky 2007: 219). It has been suggested that it is the goddess Ištar or Astarte (Berges 2006). Here, it will simply be called 'central figure'.

Group II examples follow the first but replace the back and legs of the central figure with an ornamental, central

V. J. Schlegel (✉)
Institut für Vorderasiatische Archäologie, Freie Universität Berlin, Berlin, Germany
e-mail: valery.schlegel@fu-berlin.de

C. Tsouparopoulou, L. Recht (eds.), *Human and Aquatic Beings: Interactions in and beyond the Eastern Mediterranean (3rd–1st Millennia BCE)*, Themes in Contemporary Archaeology, https://doi.org/10.1007/978-3-031-73643-8_9

Table 9.1 Catalogue of *Tridacna* shells

No.	Region	Location	Find spot	Dating	Stylistic Group	Decoration	Measurements (L × H × W in cm)	Reference Inv. No.	Reference
1	Iraq	Kish	Mound H at Ingharra		II	Wing with small triangles, lotus blossoms, backside of land animal (deer?); interior: triangle frieze, lotus'	4.2×3.7×0.6	Chicago, Field Museum FM 236219	Reese (1988a: 35–41)
2	Iran	Susa				Wing feathers		Paris, Louvre Sb 9204	Amiet (1976: 185–86)
3	Iran	Susa				Lotus blossoms, palm/tree of life		Teheran museum	Amiet (1976: 185–86)
4	–				II	Wing; interior: lotus blossom, triangle-frieze, palmette	7×8	Athens, Kanellopoulos Museum 1905	Reese and Sease (1993: 110)
5	–				II	Carved umbo, wing	8.3 (W)	Los Angeles, CMAA 84.301	Reese and Sease (1993: 110–11)
6	–					Carved umbo with triangles on front	3×2.6×2.9	Tokyo, Mikazuki gallery A-M 78	Reese and Sease (1993: 111)
7	–				IV	Wingtips, leg of central bird, sphinx with Egyptian double-crown, tree of palmettes and lotus blossoms; interior: lotus garland	4.2×6.4	Tokyo, Mikazuki gallery A-M 89	Reese (2002: 456); Reese and Sease (1993: 111–12)
8	Iran (probably)	–			II	Carved umbo, wing, head with crown, raised hand, lotus blossom	8.5×6.5	Tokyo, Mikazuki gallery A-M 90	Reese and Sease (1993: 112)
9	Iraq	Babylon/ Nippur			II	Wing, palmettes, lotus blossoms; interior: triangle and rhomboid friezes	7.6×9.5×2.1	Philadelphia UPM 320	Reese and Sease (1993: 113)
10	–				II	Wing; interior: sphinx with spread wings, lotus blossom	3.6×5.5×0.8	Private collection NY	Reese and Sease (1993: 113)
11	Israel	Tel Miqne-Ekron	Lower City, area III SE, phase IB	Iron IIC—630–604 BCE	IV	Umbo as falcon, polished	5.1/4.8×4.1×2.5	Jerusalem, IAA 99–2566 504	Brandl (2001: 49–53); Reese (2002: 456)
12	Israel	Tel Miqne-Ekron	Upper City, area I NE, topsoil debris	(630–580 BCE)	IV	Left leg and tailfeathers of central bird; traces of red pigments	4.8×4.8×0.5–1.2	Jerusalem, IAA 99–2567 6538	Brandl (2001: 54); Reese (2002: 456)
13	Jordan	Busayra/ Buseirah	B7.5.9, phase 5 between city wall and acropolis	725–300/ 200 BCE	IV	Umbo as bird, wing, tailfeathers, left leg of bird ending in lotus blossoms; interior: triangle-frieze; traces of black pigment inside and out	Ca. 10.5×8.2×7.7	Amman museum, Reg. 1120	Brandl (2001: 54. 60); Brandl (1984a: 19, 32); Reese and Sease (2004: 27); Reese (2002: 454–56)
14	Jordan	Amman	Citadel		IV/V	Carved umbo, bird?			Brandl (1984a: fig. 2b); Reese and Sease (2004: 27); Reese (2002: 456); Routledge (1997: 37–38)

15	Jordan	Amman	Citadel, Great Temple	10/9th—early 6th c. BCE	V	Polished, inlaid eyes, yellow faience, traces of red pigment under left eye, yellow pigment inside	**20.8×14**		Brandl (2001: 57); Brandl (1984a: fig. 7); Koutsoukou (1997); Reese and Sease (2004: 27); Reese (2002: 457); Routledge (1997: 37–38)
16	–				V	Originally described as bat, according to Brandl bee, according to Reese leatherback turtle (*Dermochelys coriacea*)		Jerusalem, Bible Lands Museum BLMJ 551	Brandl (2001: 57); Mienis (1994); Reese and Sease (2004: 29–30); Reese (2002: 457)
17	Syria	Tell Sekin			I	Carved umbo, central figure, wings, lotus	7.3×2.8×5.7	Los Angeles, CMAA 011-L0018	Reese and Sease (2004: 31–34); Reese (2002: 457)
18–24	Libya	Cyrene	Sanctuary of Demeter,	7th c. BCE—3rd c. CE		Seven fragments (not belonging to one specimen) with wingfeathers, umbo, lotus, palmettes (tree?), sphinx	4.7–10.7	Shahat Museum: Inv. 69–219, 71–657, 71–789, 74–750, 77–207, 78–300, 73–1136/ (467–473)	Reese and Sease (2004: 41 n. 43); Reese (1988a: 38 n. 26); Stucky (1974: 54); Warden (1990: 61) Personal communication to Reese from Donald White (excavation director)
25	Israel	Jerusalem	Shiloh exc. G2063, locus 703, gravel layer 7	2nd–1st c. BCE		'Small fragment, completely polished on both sides with incised figures'			Mienis (1992: 127); Reese (2002: 457)
26	Israel	Jerusalem	Shiloh exc. G4421, locus 780, fill. 7	2nd–1st c. BCE		'Heavily carved and polished fragment'			Mienis (1992: 127); Reese (2002: 457)
27	Israel	Shechem	Stratum VI (only assumed by Geva)			Tree of life; other side: sphinx, lotus	6×4.5×1.8	Leiden, Rijksmuseum van Oudheden B 1929 / 1.141	Geva (1980: 41–47); Reese (2002: 457) (contradictory regarding interior/exterior)
28	Lebanon	Byblos			V	Lion head with open mouth	6.85×6.2×5.15	Los Angeles, CMAA 021-L0004	Reese (2009: 190–93); Reese (2002: 457)
29	Iraq	Nippur	Area H	(7th c. CE)		Wing		Philadelphia UPM B9473	To be published by Reese (personal communication)
30	Iraq	Djokha		(650–580 BCE)		Carved umbo-head (human)	1.4×3.7	Paris, Louvre AO 6691	Reese (2009: 190–93); Stucky (2007: 382)
31	Turkey	Knidos	Sanctuary of Apollo of Emecik		II	Ornamental medallion, lotus blossoms, wingtips; inside: triangle-frieze, sphinxes; brown pigment residues (*Tridacna maxima*)	8.3×2.5×5.8	Museum Marmaris 10.47.2000	Berges (2006: 187); Reese (2009: 190–93)
32–33	Turkey	Knidos	Sanctuary of Apollo of Emecik			Wing; inside: sphinx, triangle-frieze; fragment possibly belonging to another fragment from Emecik with wing	6.3×1.5×3.4 and 6×1.8×6.1	Museum Marmaris 10.46.2000 and 11.38.99	Berges (2006: 187); Reese (2009: 190–93)

(continued)

Table 9.1 (continued)

No.	Region	Location	Find spot	Dating	Stylistic Group	Decoration	Measurements (L × H × W in cm)	Reference Inv. No.	Reference
34	Iraq	Nimrud	NW-palace, tomb I	730–694 BCE		Complete valve, eroded, carved umbo	**20×11**	Baghdad, Inv. Nr. 1988.81	Hussein (2016: 9. 69)
35	Turkey	Milet	Sanctuary of Aphrodite Zeytin-Tepe			13 fragments, one fragment with wing, one with sphinx, one with lotus' carvings			Furtwängler (2011: 49 n. 12); to be published by V. v. Graeve (personal communication)
36	Turkey	Ephesos	Sanctuary of Artemis		II	Central medallion (?), lotus blossom and bud, wing-tips, rear end of griffin (?); interior: triangle-frieze; sphinx, palm-tree/tree of life	4.45×0.3×4.8	Selcuk, Efes Müzesi ART90/K16	Bammer (1991/2: 44–48) Furtwängler (2011: 49 n. 12); Klebinder-Gauss and Pülz (2008: 203, 207)
37	Iraq	Uruk	Bit res, on pavement - HD-Qc 14–4	Post 625 BCE	II	Central medallion, wing?; interior: sphinx-head and torso, lotus bud	4.9×1.9×0.46	W 18062 / no. 1757	Van Ess and Pedde (1992: 246–47); Boehmer (1985: 134, 136)
38	Iraq	Uruk	N XIX, survey	(680–620 BCE)		Wing-tips of a sphinx (?), lotus buds, interior: wing-tips, frieze?	4.2×4.2×0.5	W 24502 / No. 1758 / IM 100204	Van Ess and Pedde (1992: 246–47); Boehmer (1991: 146); Boehmer (1985: 134–35)
39	Iraq	Uruk				Sphinx-head with Egyptian crown, lotus/palmettes; interior: lotus	ca. 5×6×1	Berlin, Pergamon-museum	Lindemeyer and Martin (1993: 295, Nr. 1991 pl. 127)
40	Turkey	Allegedly from grave in vicinity of Diyarbakir		(700–600 BCE)	II	Carved umbo, wing, ornamental medallion, sphinxes, lotuses & palmettes; interior: sphinxes, tree of life, palmettes; green-brown traces of pigments	**27.8×16.7**	Leiden, Rijksmuseum van Oudheden B 1998 / 12.3	Halbertsma (2001: 11); Stucky (2007: 223, 382)
41	–				IV	Umbo as Horus, inlaid garnets with gold as eyes, body of a bird with suns in claws on exterior; interior: winged disc below head, four rams and lotus blossoms on triangled frieze	ca. 10 (H)	Private collection (on loan in the Smith College Museum of art in 2004)	Caubet (2014: 163); Ertzner (2004: 93–95)

A find from Nimrud with Hittite hieroglyphs that Stucky (1974: 87) mentions is not a *Tridacna* but a fragment of a *Lambis truncata sebae* (according to Reese, personal communication; it was originally identified as *Xancus gravis,* new name *Turbinella pyrum,* see Rathje 1986: 394). A fragment of "incised *Tridacna*" from Milet, found in the Archaic sanctuary of Athena is not a *Tridacna* as the publication of Held suggests (2000: 168; Reese, personal communication). The "Tridacna-shell discs" Brandl published in 1984 are actually *Lambis* (Brandl 1984b: 15–41; Reese and Sease 1993: 128; Reese, personal communication). Two finds in the Ashmolean Museum from Al-Mina and Crete are not *Tridacna* either (Reese and Sease 1993: 128. For further earlier incorrect reports see Reese 1988a: 39 n. 26, 41 n. 67; Reese 2016: 12 n. 240, 413). There are some additional, unpublished engraved *Tridacna squamosa* in private collections, which could not be included in the table, such as a complete one with a bird of prey and black eyes, another one with the umbo as a female head and a bearded man in *knielaufschema* with horns of bearded, human-headed bull (?) and one with an umbo as a bird, probably an owl (various personal communications with Reese in 2020). For more detailed descriptions of decoration, see given literature. Noted dating is based on stratigraphy. Dating in brackets is only based on stylistic clues by Stucky. Bold measures indicate that the *Tridacna* specimen is (almost) complete

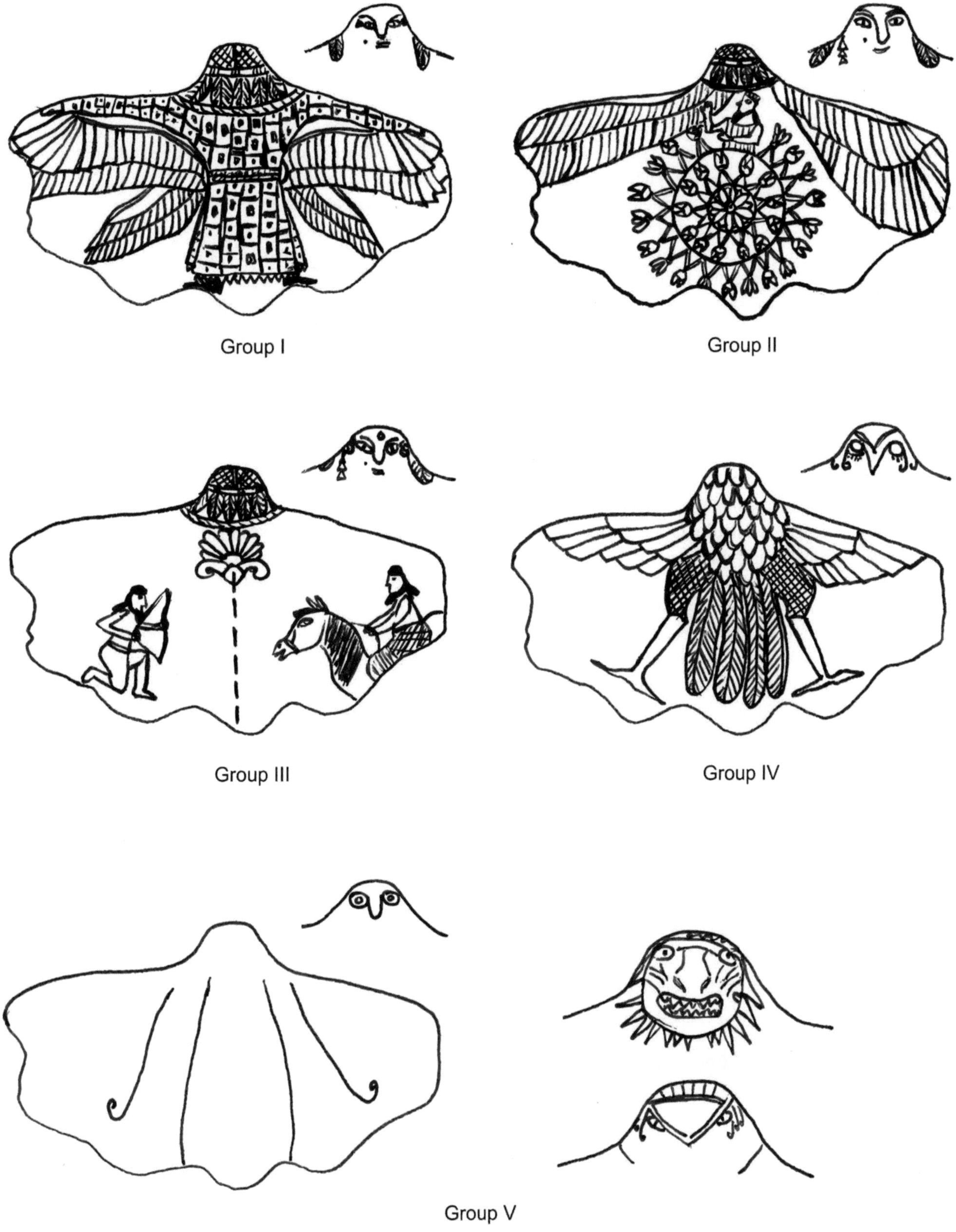

Fig. 9.1 Renderings of stylistic groups I-V of engraved *Tridacna*. The examples of group V are the bee/turtle specimen (Table 9.1, No. 16), a lion-umbo from Byblos (Table 9.1, No. 28) and an umbo from the Amman citadel (Table 9.1, No. 14). (Drawing by the author)

medallion consisting of triangles and alternating lotus blossoms and buds. Below the umbo's head and between the lotuses of the medallion is a bearded bust with a raised right hand while the left hand lays on the chest in a gesture of adoration. The figure has a crown of leaves and a squared patterned garment. This stylistic group is the most numerous (see Stucky 1974 and Table 9.1) and the specimen from Leiden presented below is an excellent example.

Group III examples are the rarest. They lack the central figure and place a field of interconnected lotuses and palmettes in the centre. Left and right, archers or horsemen move to or away from the centre in a symmetrical order. A specimen from Sippar belongs to the second group according to Stucky but is actually a combination of the second and the third group, since it has the central medallion with the bearded bust and two horsemen riding towards it from the left and right (Stucky 1974: 30; 2007: 383; Caubet 2014: 165–66).

Group IV was called a 'special group' by Stucky (1974: 19–21). It is special in that the umbo is not a human head but that of a bird of prey. On the exterior the bird-head receives wings, a short body with legs and tail-feathers.

Group V examples have the least amount of ornaments. The exterior is polished and hardly or not at all carved. The umbo is carved. Stucky only includes the types with bird-heads (Stucky 2007: 220), but I propose to include specimens where the shape of the umbo varies. An example with a lion head is known from Byblos (Reese 2002: 457; 2009: 190–93); Caubet (2014: 163) speaks of an owl head and Brandl describes an example as a bee, while other researchers agree it is a leatherback turtle (Brandl 2001: 57; 2002: 457; Reese and Sease 2004: 29–30).

The space between the main motifs was meticulously filled with floral and plant motifs, griffins, birds and sphinxes. All of the groups share the motifs of interior decoration (only in a frieze along the wavy edge since the middle part was filled with oil/cosmetics and not visible): typically triangle friezes, palmettes, lotus blossoms and sphinxes, sometimes also winged solar discs or rhomboid and pomegranate patterns. In a few examples, there are unique motifs, such as musicians in a *Tridacna* from Ashur (Stucky 1974: 22–23 No. 1; Reese and Sease 2004: 41; Caubet 2014: 164–65).

9.3 An Unpublished Engraved *Tridacna* from Leiden

One of the rare examples of an almost complete engraved *Tridacna* is the specimen in the Rijksmuseum van Oudheden in Leiden, Netherlands (Inv. Nr. B1998/12.3), which has never been adequately published.[1] It was bought from a private collector in 1998, who had acquired it in the 1990s, said to be from a grave around Diyarbakir, Turkey, but there is no evidence for its original findspot.

The condition is good: there are only three larger fractures and some smaller holes in the shell, as well as abrasions on the edges. Brown and black pigments filling the engravings are still visible. Winged sphinxes, lotus blossoms, and buds grace the interior of the shell next to three stylistic palm trees. The undecorated middle is delimited by a triangled frieze. Its umbo is carved as a female head with a prominent nose and makeup triangles next to the eyes. Two feathered hair curls are visible on the shoulders. The exterior of the shell is discoloured and, in some parts, the decoration has been destroyed. However, it is still possible to discern the stylistic group as number two, although the ornamental, central medallion is much damaged. Beneath the wings of the central figure rest winged sphinxes, lotus blossoms, and buds on the exterior of the shell. The remarkable condition makes it very likely that the specimen was found in a grave (see Figs. 9.2 and 9.3).

9.4 Previous Approaches to Provenience and Dating and the Need for Reevaluation

Researchers agree that the shells, forming natural containers, were used for cosmetics, since traces of a cosmetic paste such as kohl have been found in some examples (Stucky 2007: 220; Caubet 2014: 164). Nonetheless, other contents such as liquids, jewellery or even food have been suggested (Reese 1988a: 37). The discussion about the location(s) of the centre(s) of production is ongoing. The arguments are two-fold: stylistic origins and clusters of polished/unworked examples.

Previous research has often tied the stylistic and symbolic elements to the question of production centres: Warden (1990: 61) compared the engravings from Cyrene to the ones from Samos, Rathje (1986: 393) found the inner decoration of the Vulci example to correlate with shells from Nineveh, Naucratis and Lindos, and Berges (2006: 186) assumed the same workshop for the examples from Knidos, Rhodes and Vulci due to the great similarity in carving. Already Stucky (1974: 58–89) had deduced the provenience of engraved *Tridacna* from stylistic origins such as the winged figure from Assyria and Syro-Phoenicia. The lack of perspective and the *horror vacui* (here meaning the filling of all visible space of a filled container with plant motifs) were compared to the textiles of Assyrian reliefs (Stucky

[1] It featured in an unpublished BA thesis by Jansen 2013: 20–25. A photograph has been published by Akkermans et al. 2009: 92 and in Halbertsma 2001: 11. It has been catalogued but not discussed by Stucky 2007: 223 and was presented in a Dutch newspaper article by Vermeulen, Joost, 2009. Zwevend op een schelp. Parool: 9.

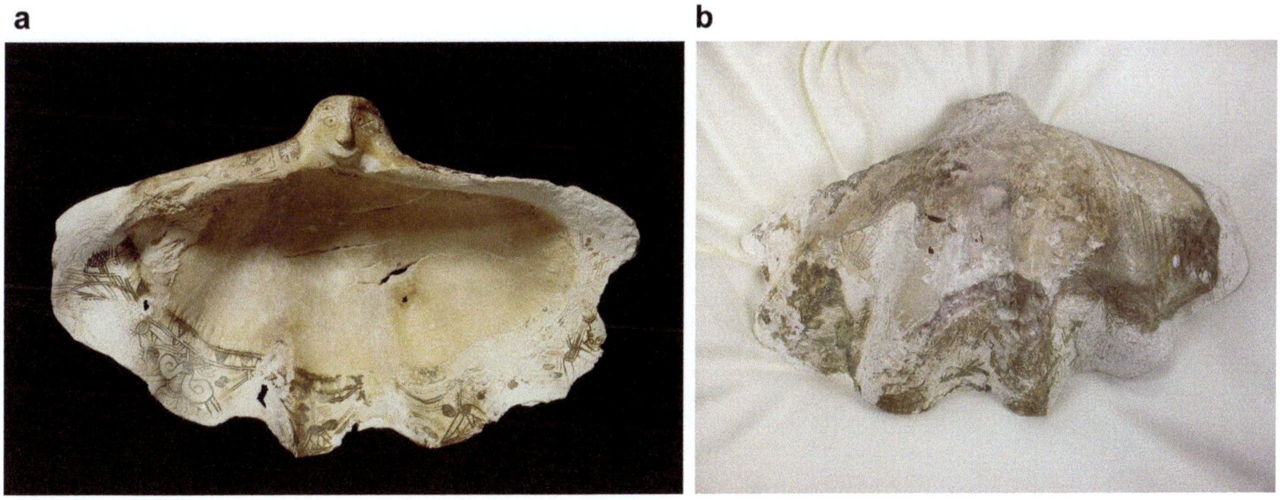

Fig. 9.2 Photographs of a complete engraved *Tridacna*, stylistic group II, eighth-sixth century BCE, from Turkey, H: 16.7, W: 27.8 cm: (**a**) front, (**b**) back. Rijksmuseum van Oudheden, Leiden, Inv. No. B 1998 / 12.3. (©RMO, Leiden)

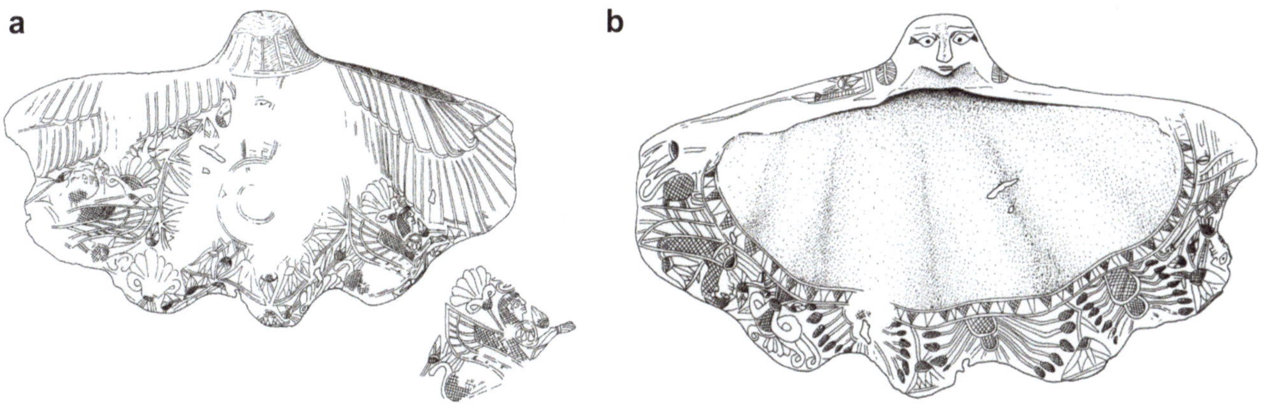

Fig. 9.3 Drawings of a complete engraved *Tridacna*, stylistic group II, eighth-sixth century BCE, from Turkey, H: 16.7, W: 27.8 cm: (**a**) back, (**b**) front. Rijksmuseum van Oudheden, Leiden, Inv. No. B 1998 / 12.3. (©RMO, Leiden)

1974: 79; Rathje 1986: 393), while single elements were said to be features of certain Near Eastern regions: the cross-hatching is typical for Phoenicia, the linear style of the dress for Nimrud ivories, the missing of overlapping or sculptural elements for Syrian ivories and bronze bowls (Stucky 1974: 86; Rathje 1986: 393–94). Akurgal called it an "interplay of styles fashionable in the East" (quoted from Rathje 1986: 393). Some researchers see the hybrid decoration of Egyptian and Assyrian motifs with Near Eastern patterns as an indication of Phoenician manufacture (Rathje 1986: 394; Berges 2006: 186–87).

Another argument for establishing provenience have been accumulations of unworked *Tridacna*, which have been found in Jerusalem, Kadesh Barnea, Busayra, Tawilan, and Petra (Reese 1988a: 40–41; 1988b: 264; 2002: 454, 457; Reese personal communication). This is why Stucky revised his original theory about workshops in the hinterland of Syria to workshops in Palestine and Transjordan (Stucky 1974: 79; 2007: 223). Using the ports of Gaza or Ashkelon instead of the Phoenician ports of Tyre, Sidon or Byblos would also explain the absence of *Tridacna* in Cyprus and Etruria (Stucky 2007: 223, with the exception of the Vulci specimen, see below).

These conclusions, paired with the find spots of engraved *Tridacna* (see Fig. 9.4), would prompt the assumption of a radial distribution from the zone between the east and west of the Dead Sea and the Phoenician coast. From here, the cosmetic shells would be distributed to the Mesopotamian palaces and graves in the east and the Greek sanctuaries in the northwest as well as southwest to Cyrene. However, as Warden (1990: 61) rightly acknowledged, the idea of these objects to be simply passed "along linear trade routes" does not hold up due to the multitude of new finds, which includes not only engraved but also polished and unworked examples. Additionally, the idea that cumulative finds of unworked *Tridacna* are exclusive evidence for workshops around the

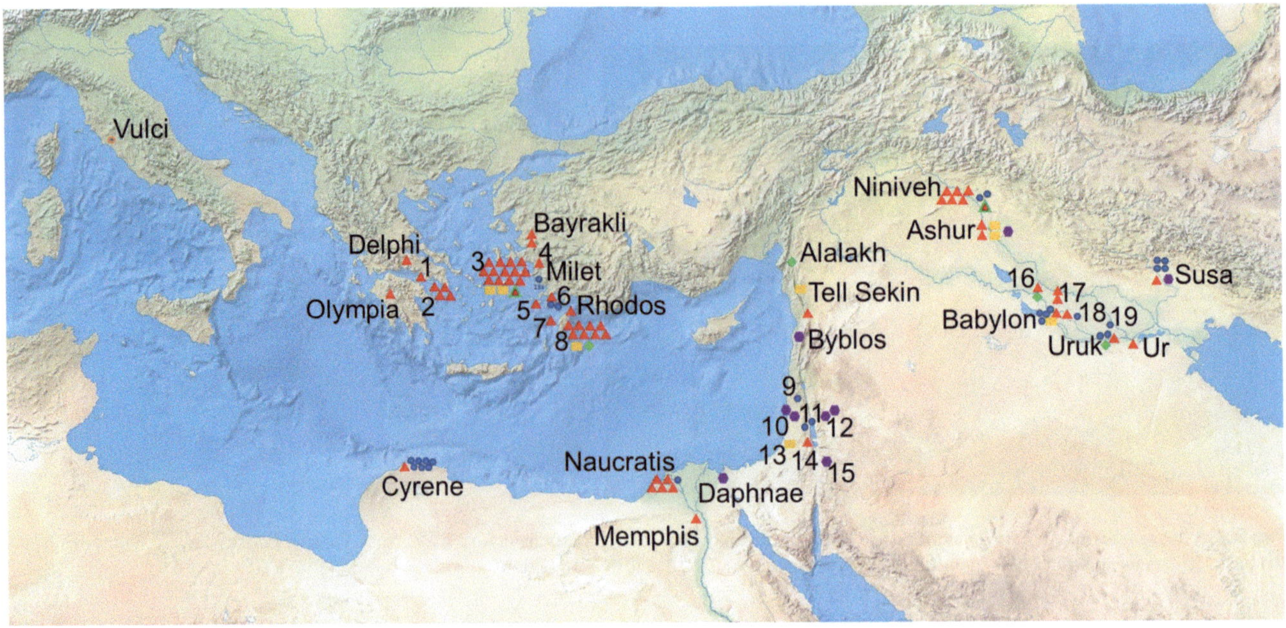

Fig. 9.4 Find spots of engraved *Tridacna*: 1. Perachora, 2. Aegina, 3. Samos, 4. Ephesos, 5. Kos, 6. Knidos, 7. Kamiros, 8. Lindos, 9. Shechem, 10. Tel Miqne-Ekron, 11. Jerusalem, 12. Amman, 13. Tell el-Far'ah (South Beth Pelet), 14. Tell Arad, 15. Busayra, 16. Sippar, 17. Kish, 18. Nippur, 19. Djokha (Umman). Geometric forms denote stylistic groups: blue circle = unknown stylistic group; yellow rectangle = Group I; red triangle = Group II (red triangle with green border: either Group II or III); green rhomb = Group III; purple hexagon = Group IV/V (Special Group). (Map by the author. Base map https://commons.wikimedia.org/wiki/File:Mediterranean_Basin_and_Near_East_before_1000_AD_topographic_map.jpg (created 6 April 2015, accessed 30 November 2022))

Dead Sea has been disproved by the dedication of unworked *Tridacna* in sanctuaries, as for example at Cyrene, Samos and Vroulia (Stucky 1974: 56; Reese 1988a: 37, 40–41; 1988b: 264; Warden 1990: 60; Reese personal communication; the example from the sanctuary of Apollo of Emecik is a *Tridacna maxima*, Berges 2006: 186–87). Moreover, the stylistic arguments for provenience (Phoenician manufacture) and the presumed location of workshops due to accumulations of unworked *Tridacna* (around the Dead Sea) seem contradictory.

Not only the provenience but also the dating of engraved *Tridacna* has been broadly based on the close stylistic similarities between the majority of *Tridacna*. This implied a relatively short production time of maybe 50–100 years, since different craftsmen and/or workshops seem to be recognisable, but with no significant chronological development (Stucky 1974: 84; Rathje 1986: 393). Furtwängler (2011: 48) surmised two generations of craftsmen, while Stucky (2007: 220) postulated a short production period of 650/30–600/580 BCE (also Brandl 2001: 58–59). Since the shells were durable, they could have survived as heirlooms long after the height of their production. However, a find from Nimrud's Northwest Palace, published in 2016 with a catalogue of the grave goods from the Nimrud Queens' Tombs by Hussein, challenges this time frame.

Earlier finds of *Tridacna* from the Northwest Palace in Nimrud could not be dated precisely since the palace was in use from 864 BCE until its destruction around 612 BCE (e.g. Stucky 1974: 90). But another engraved *Tridacna* was found in the sarcophagus of Tomb I. The photograph of the complete shell is blurred so that the decoration is not visible except for the umbo, which was carved as a human head (Hussein 2016: 69, Pl. 22a, Inv. No. ND 1988.81). Therefore, the *Tridacna* belongs to one of the first three stylistic groups. The tomb itself was built in the reigns of Ashurnasirpal II and Shalmaneser III between 883–824 BCE, based on the brick inscriptions (Hussein 2016: 5). Most of the grave goods cannot be dated specifically but a fibula with a *Pazuzu*-head (Hussein 2016: 60, Inv. No. ND 1988.19[2]) and a cylinder seal with a crescent most likely date to the end of the eighth century/early seventh century BCE (Herbordt 1992: 100–101; Theuer 2000: 326–27, 337–38, 348, 365–66; Hussein 2016: 7, 64, Inv. No. ND 1988.46b). The same is true for the enclosed alabaster vessels of Egyptian manufacture and their marble reproductions in use at the Neo-Assyrian court in the late eighth/early seventh century

[2] Fibulae with *Pazuzu* were used as protection against the female demon *Lamashtu*, who brought illness and death. There are examples in the British Museum (see Curtis 1994: 49–62, fig. 1–4) and they are known from incantations (Pedde 2003: 89–90; Curtis 1994: 49–62, fig. 1–4 and Pedde 2000: 79, 316–20, Pl. 73 date them to the late eighth/early seventh century BCE).

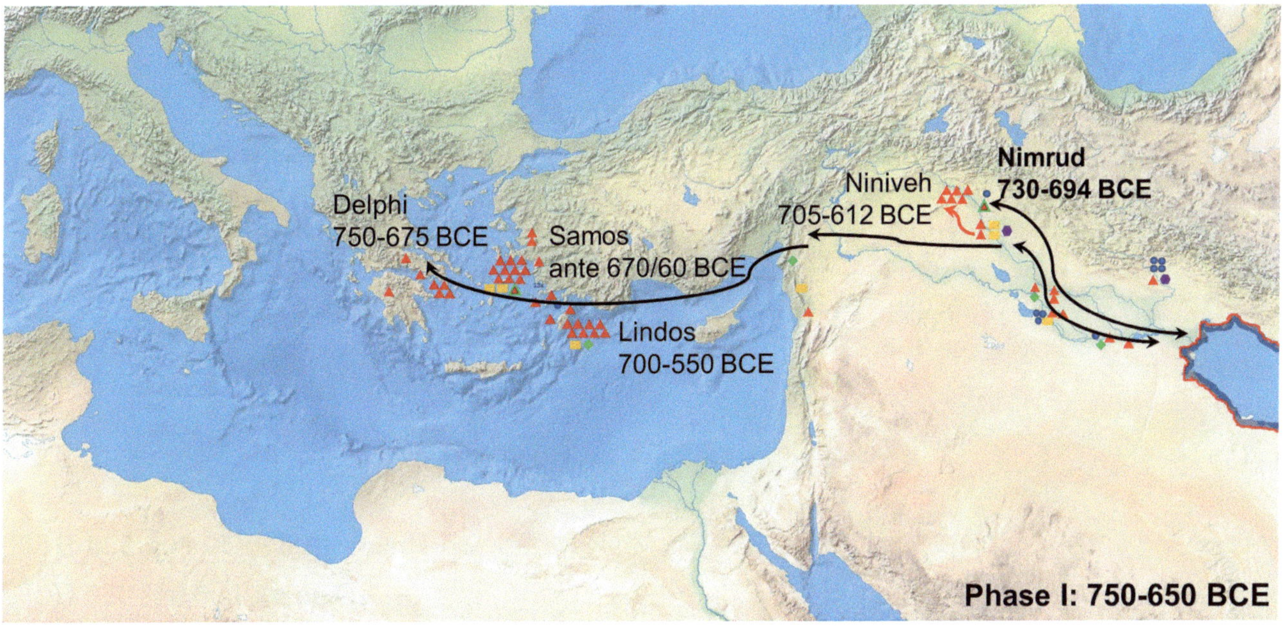

Fig. 9.5 Distribution of engraved *Tridacna* via revised dating. Phase I: 750–650 BCE. Map by the author. (Base map https://commons.wikimedia.org/wiki/File:Mediterranean_Basin_and_Near_East_before_1000_AD_topographic_map.jpg (created 6 April 2015, accessed 30 November 2022))

BCE (Hussein 2016: 65–69, Inv. No. ND 1988.53–59. ND 1988.65–67. ND 1988.72–73[3]). The royal household of the queen moved to Nineveh in 694 BCE, at which point the burials in the Northwest Palace of Nimrud ceased.[4] Therefore, the primary burial of Tomb I with the engraved *Tridacna* dates between 730–694 BCE. Such an early date does not match Stucky's (2007: 220)[5] narrow production time window. Thus, not only the provenience but also the dating of engraved *Tridacna* need to be re-evaluated.

The new finds that contradict former research theories on provenience and dating as well as the long overdue re-examination of engraved *Tridacna* — since Stucky's work of 1974 various new finds have been published but the corpus has not been compiled anew or revised — encouraged the approach taken in this chapter.

9.5 A Different Approach to Distribution Patterns by Revised Dating

In a not previously applied approach, first, all new finds of engraved *Tridacna* of the last 50 years have been listed in Table 9.1 (including publication references). As a next step, all specimens of which find spots are currently known have been mapped according to the stylistic groups introduced above (see Figs. 9.1 and 9.4). Finally, all known dates of *Tridacna* on the base of stratigraphy have been added and grouped (see Figs. 9.5 and 9.6). The result is a distribution pattern that differs from the radial distribution of *Tridacna* suggested in the last 30 to 70 years of research. Instead, three phases of distribution and two regions of production emerge.

Stucky's narrow dating of 650/30–600/580 BCE was simply derived from the closest overlapping dating of finds, while later examples were explained as heirlooms. This was based on the stylistic assumptions of a short production time. Now, the earliest example comes from Nimrud around 730–694 BCE. Only a few miles to the northwest, the examples from Nineveh, which could date as early as 705 BCE and as late as the city's destruction in 612 BCE, connect Mesopotamia to the Greek sanctuaries with specimens from the middle of the eighth to the middle of the seventh century BCE (see Fig. 9.5). These *Tridacna* probably derived from the Persian Gulf and were traded along the river and sea routes in both directions. They were most likely manufactured by Phoenician craftsmen because

[3] See for objects of comparison: von Bissing 1940: 153–59; Onasch 2014: 60–72, Pl. 8, no. 182–1.

[4] At that time, the correspondence of the queen's household moved to Nineveh and Ashur, the queen of Sennacherib received a new palace wing in Nineveh (built between 703 and 694/3 BCE) and in 673/2 BCE, a queen was buried in Ashur instead of Nimrud or Nineveh, Svärd 2015: 45, 187, 222–23. For a more detailed analysis on Tomb I of Nimrud, see Schlegel Forthcoming.

[5] While Stucky (1974: 87) agrees on an early production due to the find of a shell from Nimrud of the eighth century BCE, this shell is actually not *Tridacna* but *Xancus* (now called *Turbinella*, Rathje 1986: 394) which is why Stucky dismissed it in his 2007 publication.

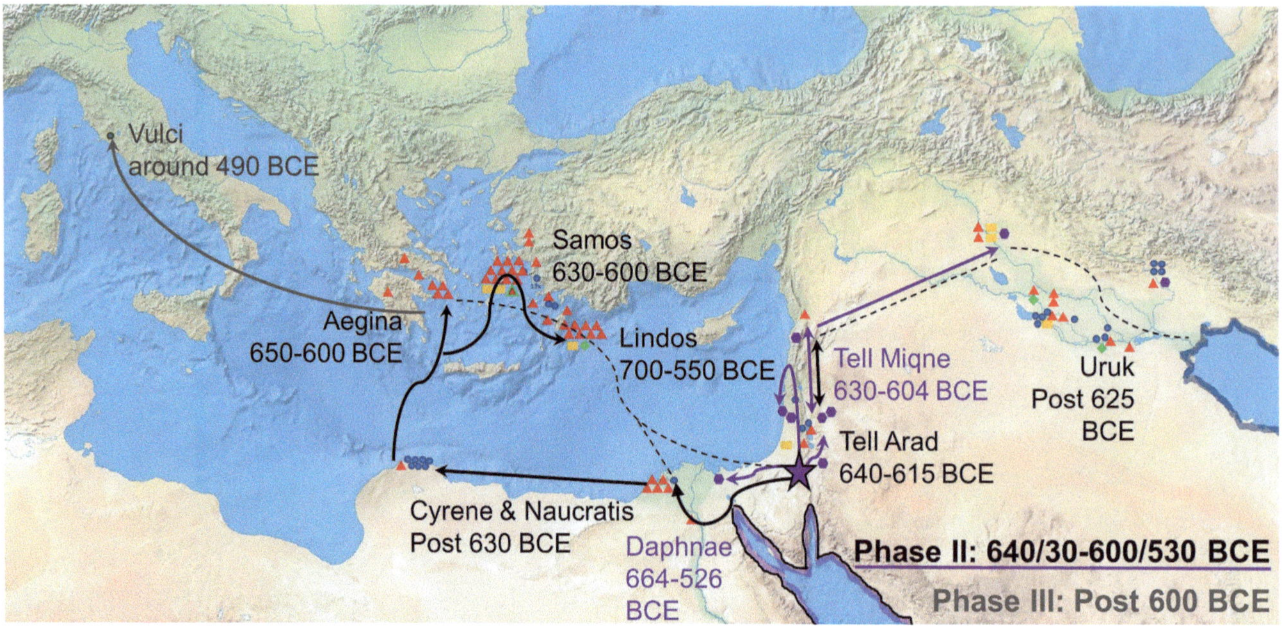

Fig. 9.6 Distribution of engraved *Tridacna* via revised dating. The purple arrows, text and hexagon relate to the distribution of stylistic group IV/V; their production centre is likely in the vicinity of the purple star due to its small distribution area. The black arrows and text refer to the distribution of the other stylistic groups. Dotted lines mark additional possible routes. Phase II: between 640/30 and 600–530 BCE; Phase III: Post 600 BCE. Map by the author. (Base map https://commons.wikimedia.org/wiki/File:Mediterranean_Basin_and_Near_East_before_1000_AD_topographic_map.jpg (created 6 April 2015, accessed 30 November 2022))

of the hybrid decoration of Egyptian and Assyrian motifs with Near Eastern patterns — but not necessarily in Phoenicia. Migrated or deported Phoenician craftsmen in the Tigris and Khabur regions can be considered as well, since the worked, as well as some unworked (see Stucky 1974: 56; Reese 1988a: 37, 40–41; 1988b: 264) shells had to be acquired from the Persian Gulf and were found in multiple spots between the Gulf and the Mediterranean coast. Thus, it might well have been that not only the shells but also the craftsmen travelled.

Conceivably in a smooth transition, a second phase began in the second half of the seventh century BCE. The *Tridacna* were now collected from the Red Sea, manufactured along the Dead Sea, due to the accumulation of unworked *Tridacna* around its shore (Reese 1988a: 40–41; 1988b: 264; 2002: 454, 457), and traded along the coasts and sea routes to the north and west.[6] *Tridacna* were still dedicated in Greek sanctuaries (where stylistic group II was the most popular) and some may even have reached Babylonia, but most *Tridacna* in the east are not dated. The patterns also suggest that Cyprus was not a necessary transit stop, since no specimens have been found there so far. An innovation of this phase was the stylistic groups IV and V with umbos carved not as human heads but as birds of prey and other animals, which were only traded in the Near East, especially around the Dead Sea (see Table 9.1 and Figs. 9.1 and 9.6). The divergence of motifs and quality could imply new craftsmen, who adapted former styles but also experimented. At the current state of research, this phase could have lasted until anywhere between 600–530 BCE (see Fig. 9.6). In a third and last phase, production had stopped, but some engraved *Tridacna* survived as heirlooms and were even reworked to fit the taste of the new time and location as proposed by the altered example from Vulci.[7]

9.6 Conclusion and Future Research

In the 50 years since the last extensive collection of engraved *Tridacna*, various new finds have been published in isolation. This paper compiles them (see Table 9.1) and presents an

[6] It is so far not clear what the reasons behind this change were. A movement of craftsmen, shift in trade routes or a lack of resources can be assumed.

[7] The westernmost find from an unidentified tomb in Vulci diverges not only geographically but also in its decoration: The exterior was polished, but traces of incisions on the edges reveal that it originally must have been decorated in a style similar to all other engraved *Tridacna*, since its interior with lotus blossoms and sphinxes resembles Near Eastern finds. The most remarkable difference is the deeply carved and detailed head. Rathje (1986: 393) sees in its "broad cheeks, the strong jaw, broad nose and fleshy lip" a Neo-Assyrian workshop of not yet specified location. Furtwängler (2011: 49–51) understands it as a transformation conducted by an eastern Greek craftsman who eradicated the Near Eastern motifs on the exterior but left the sphinxes of the interior because they are a well-known motif in all Mediterranean regions. The latter seems much more likely iconographically.

almost complete specimen in Leiden, which has not previously been adequately published.

Previous research inferred the provenience of engraved *Tridacna* via stylistic considerations to Phoenicia and via cumulative finds of polished *Tridacna* to workshops in Palestine/Transjordan. Not only did they formerly contradict each other but new evidence such as accumulations of unworked *Tridacna* in sanctuaries undermines this argument. Furthermore, an engraved *Tridacna* found in Tomb I of the Northwest Palace at Nimrud can now be dated to 730–694 BCE, pushing back the old dating of 650/30–600/580 BCE by several decades. These new finds encouraged the reevaluation of the distribution of engraved *Tridacna* via adjusted dating presented in this paper.

The chosen approach entailed mapping the known find spots of over 140 specimens via their stylistic groups and revised dating to analyse their provenience and distribution patterns (see Figs. 9.4, 9.5, and 9.6). This mapping illustrates that *Tridacna* from the Persian Gulf were engraved in three stylistic groups by Phoenician craftsmen in the Fertile Crescent between 750–650 BCE and traded in all of the eastern Mediterranean. Around the middle of the seventh century BCE, the production centre(s) moved to the Dead Sea and a fourth and fifth stylistic group were created, examples of which were only distributed in the Near East. These groups IV and V and at least stylistic group II remained in circulation until around 600–530 BCE. In the following decades, although production had ceased, some of the durable engraved *Tridacna* were still passed on and even reworked. These results await verification by a thorough revision of stylistic analysis of known and newly found engraved *Tridacna*. As this would go beyond the scope of this paper, future research shall pursue it.

Further prospects remain, such as for example the modes of distribution of engraved *Tridacna*. The objects could not only be traded but also change location as part of loot, gifts or personal objects (heirlooms) accompanying migration. Also, there are a number of objects which share features with *Tridacna* and should be added to the maps in future research. These objects include metal vessels resembling *Tridacna* (one even made out of gold with a bird's head now in Durham, Caubet 2014: 163), carved limestone/alabaster palettes from Jordan (Amman Citadel, Umm al-Biyara, Karak, Ghrareh, Tawilan, Petra, Tell Jezreel, cf. Rathje 1986: 393; Sease and Reese 1994; Reese 2002: 457; 2011; Stucky 2007: 221; one in Beirut, Rathje 1986: 393), engraved *Lambis* shell discs (wrongly described by Brandl as *Tridacna* in 1984b) and engraved ostrich egg shells with similar distribution patterns as *Tridacna* (Hodos et al. 2020). The approach introduced in this chapter could serve to map the patterns of distribution of these genres as well and to compare them with engraved *Tridacna*. This would reciprocally enhance the knowledge of their distribution modes, patterns and dating.

Acknowledgments I thank Lucas Petit and David Kertai from the Rijksmuseum van Oudheden, Leiden for allowing me to publish the specimen from Leiden and for providing photographs and renderings of the piece. My heartfelt gratitude goes to David Reese for his unhesitant help, review and expert comments, access to material and information for this article. Last but not least, I thank Laerke Recht and Christina Tsouparopoulou for the organisation of the session at the EAA 2021, the enabling of the publication and their kind review as well as the anonymous reviewer for their exceedingly positive feedback.

Bibliography

Akkermans, P., Bremer-den Hartog, R, Bomhof, P.J. & Kik, H. eds 2009. *100 Topstukken van het Rijksmuseum van Oudheden*. Leiden: Royal Museum of Antiquities.

Amiet, P. 1976. Tridacnes trouvés à Suse. *Revue d'assyriologie et d'archéologie orientale,* 70(2): 185–86.

Bammer, A. 1991/2. Multikulturelle Aspekte der frühen Kunst im Artemision von Ephesos. *Jahreshefte des Österreichischen Archäologischen Institutes in Wien,* 61/ Beiblatt: 17–54.

Berges, D. 2006. *Knidos, Beiträge zur Geschichte der archaischen Stadt*. Berlin: Phillip von Zabern.

Boehmer, R.M. 1985. Uruk–Warka XXXVII. Survey des Stadtgebietes von Uruk. VI. Kleinfunde. *Baghdader Mitteilungen,* 16: 119–48.

Boehmer, R.M. 1991. Kleinfunde aus Metall, Stein, Muschel u.a. In: U. Finkbeiner, ed. *Uruk. Kampagne 35-37 1982-1984. Die archäologische Oberflächenuntersuchung (Survey), Ausgrabungen in Uruk-Warka. Endberichte 4*. Mainz am Rhein: Phillip von Zabern, pp. 141–47.

Brandl, B. 1984a. The Restoration of an Engraved Tridacna Shell. *The Israel Museum Journal,* III: 76–79.

Brandl, B. 1984b. The Engraved Tridacna-shell Discs. *Anatolian Studies,* 34: 15–41.

Brandl, B. 2001. Two Engraved Tridacna Shells from Tel Miqne-Ekron. *Bulletin of the American Schools of Oriental Research,* 323: 49–62.

Caubet, A. 2014. Tridacna Shell. In: J. Aruz, S.B. Graff & Y. Rakic, eds. *Assyria to Iberia. At the Dawn of the Classical Age*. London: Yale University Press, pp. 163–66.

Curtis, J. 1994. Assyrian Fibulae with Figural Decoration. In: M. Dietrich & O. Loretz, eds. *Beschreiben und Deuten in der Archäologie des Alten Orients. Festschrift für Ruth Mayer-Opificius*. Altertumskunde des Vorderen Orients 4. Münster: Ugarit-Verlag, pp. 49–62.

Ertzner, A. 2004. Cosmetic Dish in the Shape of a Horus. In: C. Houser, ed. *From Myth to Life: Images of Women from the Classical World from the Celia and Walter Gilbert Collection, Smith College Museum of Art, Northampton, Massachusetts, March 12-October 10*. Northampton: Smith College Museum of Art, pp. 93–95.

Furtwängler, A.E. 2011. Tridacna – warum auch nicht mal griechisch? In: O. Pilz & M. Vonderstein, eds. *KERAUNIA. Beiträge zu Mythos, Kult und Heiligtum in der Antike*. Beiträge zur Altertumskunde 298. Berlin: De Gruyter, pp. 47–51.

Geva, S. 1980. A Fragment of a Tridacna Shell from Shechem. *Zeitschrift des Deutschen Palästina-Vereins,* 96(1): 41–47.

Halbertsma, R.B. 2001. Treasures around the Temple. The New Installation at the National Museum of Antiquities in Leiden. *Minerva,* 12(4): 8–13.

Held, W. 2000. *Das Heiligtum der Athena in Milet*. Milesische Forschungen 2. Mainz am Rhein: Phillip von Zabern.

Herbordt, S. 1992. *Neuassyrische Glyptik des 8.-7. Jh. v. Chr. unter besonderer Berücksichtigung der Siegelungen auf Tafeln und Tonverschlüssen*. State Archives of Assyria Studies I. Helsinki: Neo-Assyrian Text Corpus Project.

Hodos, T., Cartwright, C.R., Montgomery, J., Nowell, G., Crowder, K., Fletcher, A.C. & Gönster, Y. 2020. The Origins of Decorated Ostrich Eggs in the Ancient Mediterranean and Middle East. *Antiquity,* 94: 381–400.

Hussein, M.M. 2016. *Nimrud: The Queens' Tombs*. Baghdad/Chicago: Oriental Institute Miscellaneous Publications.

Jansen, L.J. 2013. Tridacna Revised (BA thesis, Leiden University). https://hdl.handle.net/1887/20074

Klebinder-Gauss, G. & Pülz, A.M. 2008. 'Fremdes' in der materiellen Kultur im Artemision von Ephesos. In: U. Muss, ed. *Die Archäologie der ephesischen Artemis. Gestalt und Ritual eines Heiligtums*. Vienna: Phoibos, pp. 201–207.

Koutsoukou, A. 1997. Miscellaneous Finds. In: A. Koutsoukou, K.W. Russell, M. Najjar & A. Momani, eds. *The Great Temple of Amman: The Excavations*. Amman: American Center of Oriental Research, pp. 135–56.

Lindemeyer, E. & Martin, L. 1993. *Uruk. Kleinfunde III. Kleinfunde im Vorderasiatischen Museum zu Berlin: Steingefässe und Asphalt, Farbreste, Fritte, Glas, Holz, Knochen/Elfenbein, Muschel/ Perlmutt/Schnecke*. Mainz am Rhein: Phillip von Zabern.

Mienis, H.K. 1992. Molluscs. In: A. de Groot & D.T. Ariel, eds. *Excavations at the City of David 1978-1985. Directed by Yigal Shiloh III: Stratigraphical, Environmental, and Other Reports*. QEDEM 33. Jerusalem: The Hebrew University of Jerusalem, pp. 122–30.

Mienis, H.K. 1994. Early Use of Giant Clams in the Middle East. *Clamlines,* 13: 18–19.

Onasch, H.U. 2014. *Ägyptische Alabastergefäße aus Assur*. Wissenschaftliche Veröffentlichungen der Deutschen Orient-Gesellschaft 128. Wiesbaden: Harrassowitz.

Pedde, F. 2000. *Vorderasiatische Fibeln. Von der Levante bis Iran*. Abhandlungen der Deutschen Orient-Gesellschaft 24. Saarbrücken: Harrassowitz.

Pedde, F. 2003. Fibeln in Gräbern. *Altorientalische Forschungen* 30(1): 85–92.

Rathje, A. 1986. A Tridacna Squamosa Shell. In: J. Swaddling, ed. *Italian Iron Age Artefacts in the British Museum. Papers of the Sixth British Museum Classical Colloquium*. London: Trustees of the British Museum, pp. 393–96.

Reese, D.S. 1988a. A New Engraved Tridacna Shell from Kish. *Journal of Near Eastern Studies,* 47(1): 35–41.

Reese, D.S. 1988b. Recent Invertebrates as Votive Gifts. In: B. Rothenberg, ed. *The Egyptian Mining Temple at Timna, Researches in the Arabah 1959–1984 I*. London: Institute for Archaeo-Metallurgical Studies, Institute of Archaeology, University College London, pp. 260–66.

Reese, D.S. 2002. Fossil and Recent Marine Invertebrates. In: P. Bienkowski, ed. *Busayra. Excavations by Chrystal-M. Bennett 1971–1980*. British Academy Monographs in Archaeology 13. Oxford: Oxford University Press, pp. 441–69.

Reese, D.S. 2009. On Incised Scapulae and Tridacna. *Eretz-Israel: Archaeological, Historical and Geographic Studies, 29*, 188–93.

Reese, D.S. 2011. Recent Shells and a Fossil Sea Urchin from Umm al-Biyara. In: P. Bienkowski, ed. *Umm al-Biyara. Excavations by Crystal-M. Bennett in Petra 1960–1965*. Levant Supplementary Series 10. Oxford: Oxbow Books, pp. 110–13.

Reese, D.S. 2016. The Invertebrates. In: E. Hallager & B.P. Hallager, eds. *The Greek-Swedish Excavations at the Agia Aikaterini Square Kastelli, Khania 1970–1987, 2001, 2005 and 2008, The Late Minoan IIIA:1 and II Settlements V:1*. Stockholm: Skrifter utgivna av Svenska Institutet I Athen, pp. 396–417.

Reese, D.S. & Sease, C. 1993. Some Previously Unpublished Engraved Tridacna Shells. *Journal of Near Eastern Studies,* 52(2): 109–28.

Reese, D.S. & Sease, C. 2004. Additional Unpublished Engraved Tridacna and Anadara Shells. *Journal of Near Eastern Studies,* 63(1): 29–41.

Routledge, B. 1997. Mesopotamian "Influence" in Iron Age Jordan: Issues of Power, Identity and Value. *Bulletin of the Canadian Society for Mesopotamian Studies,* 32: 33–41.

Schlegel, V. Forthcoming. Pazuzu & Bes: How the Woman of Tomb I at Nimrud Sheds New Light on the Conflicts between Assyria and Egypt. In: W. Spickermann, L. Recht & H. Galter, eds. *Clash of Civilizations. Proceedings of the 15th Symposium of the Melammu Project Held in Graz September 21–24 2022*. Vienna: Verlag der Österreichischen Akademie der Wissenschaften.

Sease, C. & Reese, D.S. 1994. An Unpublished Stone Palette. *Journal of Near Eastern Studies,* 53(4): 285–94.

Stucky, R. 1974. *The Engraved Tridacna Shells*. Dédalo 10, 19. Sao Paulo: Museu de Arqueologia e Etnologia, Universidade de Sao Paulo.

Stucky, R. 2007. Les Tridacnes à décor gravé. In: *La Méditerranée des phéniciens de Tyr à Carthage. Exhibiton cataloge Paris 6 November 2007–20 Avril 2008*. Paris: Institut du Monde Arabe, pp. 218–23, 382–83.

Svärd, S. 2015. *Women and Power in Neo-Assyrian Palaces*. State Archives of Assyria Studies XXIII. Helsinki: Neo-Assyrian Text Corpus Project.

Theuer, G. 2000. *Der Mondgott in den Religionen Syrien-Palästinas. Unter besonderer Berücksichtigung von KTU 1.24*. Orbis Biblicus et Orientalis 173. Freiburg: Universitätsverlag.

Van Ess, M. & Pedde, F. 1992. *Uruk. Kleinfunde II: Metalle, Asphalt, Farbreste, Fritte, Fayence, Glas, Holz, Knochen, Elfenbein, Leder, Muschel, Perlmutt, Schnecke, Schilf, Textilien, Ausgrabungen in Uruk-Warka Endberichte 7*. Mainz am Rhein: Phillip von Zabern.

von Bissing, F.W. (1940). Ägyptische und ägyptisierende Alabastergefässe aus den Deutschen Ausgrabungen in Assur. *Zeitschrift für Assyriologie und vorderasiatische Archäologie*, 46: 149–82.

Warden, P.G. 1990. The Small Finds. In: D. White, ed. *The Extramural Sanctuary of Demeter and Persephone at Cyrene, Libya. Final Reports IV*. University Museum Monograph 67. Philadelphia: University of Pennsylvania Press, pp. 1–86.

Correlations between Stones and Iconography: Fish-Cloaked Figures in Ancient Mesopotamia 10

Tiffany Ftaimi

Abstract

During the Neo-Assyrian and Neo-Babylonian periods, protective and apotropaic creatures were commonly depicted in ritual and adoration scenes in Mesopotamia. These creatures played an essential role in ritual practices, such as purification scenes in front of stylised trees, adoration scenes in front of deities, and the symbols of deities. Genii, hybrid creatures, and figures cloaked in animal skins can be identified as protective and apotropaic. Numerous representations of these creatures were depicted on cylinder and stamp seals from the Iron Age. This chapter focuses on fish-cloaked figures in Neo-Assyrian and Neo-Babylonian glyptic. This includes an examination of the types of stone on which fish-cloaked figures appear, and the aim is to clarify correlations between selected stones for seal production and the depiction of fish-cloaked figures for health and protective purposes.

10.1 Introduction

The composition of zoomorphic and anthropomorphic body parts increased with the beginning of the first half of the first millennium BCE. Hybrid creatures decorated large-scale wall reliefs in the Assyrian palaces and were favoured as subject matter in other artworks. The combination varies between applying the upper part or the lower part of humans and animals. The winged *lamassu* were placed as guardians in doorways of Assyrian palaces consisting of human heads and beasts' bodies. Protective and apotropaic figures attended in the embellishment of the internal space of royal residences owing defensive features. This chapter focuses on the representation and function of fish-cloaked figures, which were widespread in the visual representations of the Iron Age in Mesopotamia.

10.2 A Short Overview of Fish-Cloaked Figures in the First Millennium BCE

There are various descriptions of fish-cloaked figures in the literature. The figure has been interpreted as a priest (Klengel-Brandt 1968: 35), a fish-man, a fish-priest (Rittig 1977: 92), a fish genie (Klengel-Brandt 2014: 72), a protective deity, an anthropomorphic genie, an apotropaic figure, or an *apkallū.* In the Babylonian tradition, there were seven *apkallū,* 'wise men', who lived before the flood. Neo-Assyrian and Neo-Babylonian ritual texts mention their names as well as the cities they most probably came from (Black and Green 1992: 163). Further, in the *Gilgameš* epic, the Seven Sages are credited with the construction of the wall of the city of Uruk (Black and Green 1992: 163). *Apkallū* was a typical term for other apotropaic figures such as *Gilgameš*–like, fish–like and bird–like men (Reiner 1961: 5). Apotropaic figures were a favourite subject matter in palaces as reliefs, in seal iconography, and as small figurines that were used during the Neo-Assyrian period as a shield against evil forces. The purpose of buried *apkallū* clay figurines was to fight demons that cause diseases. To bring about this positive impact, individuals had to perform extensive rituals precisely and according to particular regulations to be effective (Kolbe 1981: 24–25). Indeed, the function of *apkallū* differs partly when they manifest in other visual media such as on reliefs, where they indicated in an impressive form that the royal family's protection from negative influences was ensured by the depiction of apotropaic figures (Kolbe 1981: 29).

The fish-cloaked figure was sometimes associated with Oannes, as found in the book of Berossus, along with references to the name U'anna in some cuneiform texts (Beaulieu 2021: 150). Oannes was a mythical hybrid creature in the shape of a fish-man who imparted knowledge to

T. Ftaimi (✉)
Centre for Ancient Studies, Institute for Prehistory, Protohistory and Near Eastern Archaeology, Heidelberg University, Heidelberg, Germany

C. Tsouparopoulou, L. Recht (eds.), *Human and Aquatic Beings: Interactions in and beyond the Eastern Mediterranean (3rd–1st Millennia BCE)*, Themes in Contemporary Archaeology, https://doi.org/10.1007/978-3-031-73643-8_10

humanity (Streck 2003: 2). The similarities between Oannes' features and the illustrations of the fish-cloaked figures promoted the integration of both these mythological characters. Kolbe has argued for the assimilation between Oannes and the fish-cloaked figures based on distinctions of their functions and names between Greek and previous traditions (Kolbe 1981: 26).[1]

The fish-cloaked figure makes its first appearance in the glyptic of the Kassite period (Stiehler-Alegria Delgado 1996: 160; Porada 1948: No. 581), but became a common motif on Neo-Assyrian and Neo-Babylonian objects, particularly cylinder and stamp seals. They occur on wall paintings[2] and on reliefs from the Neo-Assyrian palaces from the ninth century BCE (Rittig 1977: 89).

On seals, fish-cloaked figures mostly hold a so-called *Banduddû* bucket in one hand and a pinecone in the other, representing a purification scene. These purification scenes were not only popular illustrations on seals, but also on reliefs, as for instance the winged figures on reliefs from the North-West Palace at Nimrud (Oates 1983: 29).

Fish-cloaked figures are represented as long-bearded, with a fish cape, and commonly with horns as a head covering. The fish cloak differs in length over time. In the ninth century BCE, it reached only the waistline (Collon 2001: 91), while from the eighth century BCE onwards, the cloaks were depicted as longer (Black and Green 1992: 82). The dress of the fish-cloaked figures varies slightly on other finds. On wall reliefs, they are depicted wearing a long open-slit garment that leaves one leg visible (Rittig 1977: 88), with a short skirt underneath. On seals, the figures are depicted wearing, in addition to the fish cloak, a closed long-slit garment, a short skirt or they could even be naked underneath.

Fish-cloaked figures occur especially on seals and seal impressions of the first millennium BCE (Fig. 10.1). Contrary to what might be expected, after an examination of the Neo-Assyrian and Neo-Babylonian glyptic listed in the catalogues of museum collections, private collections, and excavations reports, it can be seen that these figures are not often represented with the water god, although such occurrences do exist, as for example on a limestone seal in the British Museum (BM WA 133031), where the water god and the fish-cloaked figure are depicted in front of a worshipper.[3] Rather, fish-cloaked figures are mostly depicted in ritual scenes on either side of a stylised tree, on which the winged sun disk or the sun god is depicted (Fig. 10.2).The figures are escorted by worshippers, bull-men, protective deities, or more rarely, by other deities. The representation of a fish-cloaked figure in front of a stylised tree and a winged disk is also found on a limestone seal (Stiehler-Alegria Delgado 1996: Pl. XXIX, No. 266a), now in the Newell collection (Von der Osten 1934: 61), dated to the Kassite period; thus, the fish-cloaked figure was not restricted only to the iconographic repertoire of the Iron Age.

The illustration of this figure was not only shown on cylinder seals, but also on some examples of stamp seals on which the figure stands in front of an altar. The figure does not occur in fight or hunting scenes, but instead occurs frequently in representations of ritual practices, healing processes, and exorcism, confirming its function as protector. The role of the fish-cloaked figure as a healer and purifier is shown on a Neo-Assyrian seal[4] flanking a sick kneeling bearded man holding the tapes of the winged sun disk above him (Meissner 1934: fig. 14).

Since fish-cloaked figures often occur with the sun god or with his symbol on glyptic material, it would be useful to understand the relation between these characters. In incantations and prayers, the sun god is usually mentioned as the god of justice, meaning that he can reverse an erroneous judgement against the patient, acquit him, and condemn the offenders and sorcerers instead of the patient by redirecting the harm spell back to those who first set it in motion (Schwemer 2007: 239). This could explain the frequent depiction of fish-cloaked figures in the company of the sun god, or in front of a stylised tree with the winged disk as a symbol of the sun god.[5] The symbol and the torso of the sun god help avert the witchcraft of the sick person and the role of the fish-cloaked figure is to purify and safeguard. On a cylinder seal made of blue chalcedony now in the British Museum (BM WA 104499, Collon 2001: 98), this scene with the god is repeated: a god in a winged disk is depicted along with a figure and a dog symbolising the healing goddess Gula.

[1] In works about fish written by well-known natural scientists (G. Rondelet and C. Gessner) in the sixteenth century CE, mysterious mythical creatures were added as a 'sea bishop' and considered as a possible relative of some creatures illustrated in ancient Mesopotamia (Stiehler-Alegria 2006: 4). However, this suggestion seems unlikely given the chronological, spatial, and iconographical divergences between the sea bishop and the fish-cloaked figure.

[2] There is a wall painting in room XXVII from the Assyrian palace in Til Barsib (Tell Ahmar) on which a fish-genie or similar figure is depicted, with fish-like extremities instead of anthropomorphic hands and feet (Thureau-Dangin and Dunant 1936: 61, Pl. LIIIb). The wall painting dates to the reign of Tiglath-pileser III (754–727 BCE) (Hrouda 1965: 114).

[3] The aquatic features of both figures may be an essential aspect of the relationship between them as well as a reference to the cleansing aspect of water.

[4] The seal is made of limestone, Museum ID: IB. SA: 105, stored in the collection of the Pontifical Biblical Institute, dated to the eight/seventh century BCE. The production technology is remarkable because the seal was engraved in relief compared with other seals and might have been used as an amulet rather than a seal (Van Buren 1940: 82).

[5] In earlier publications, the figure above the winged disk was identified with Ashur. However, recent re-examination of texts, symbols, and related figures that occurred with the sun god indicate an association with him rather than Ashur, who has neither symbols nor an attribute creature. See: Kertai 2020: 387.

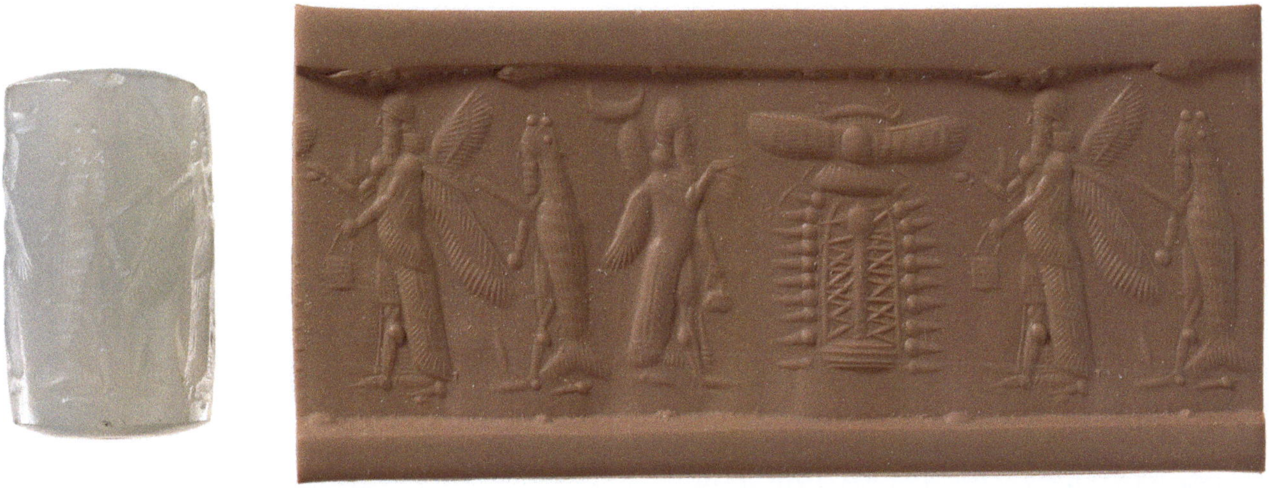

Fig. 10.1 Fish-cloaked figure with winged figures in front of a stylised tree. Neo-Assyrian cylinder seal and its modern impression. Chalcedony, found in Assur. VA 4231. Courtesy of Staatliche Museen zu Berlin—Vorderasiatisches Museum. (Photo: Olaf M. Teßmer. CC BY SA 4.0)

Fig. 10.2 Fish-cloaked figure with a *Banduddû*, next to a deity on a horse. Modern impression of Neo-Assyrian seal. Semi-opal, from the region of Assyria. VR 1981.110. (Courtesy of Foundation BIBLE+ORIENT, Fribourg Switzerland)

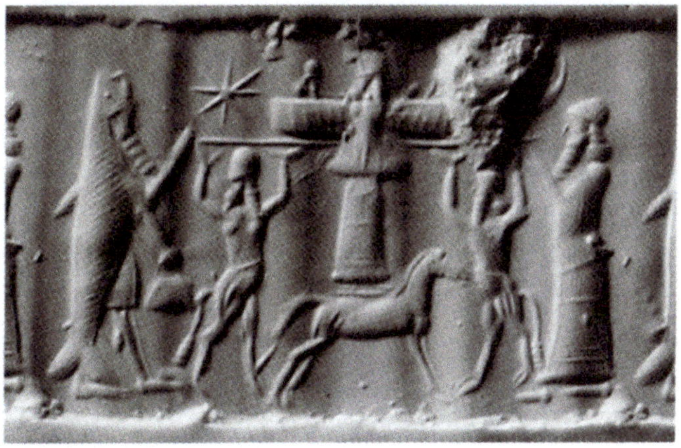

10.3 Stones and Seals

The importance of stones is revealed in several cuneiform clay tablets which record the role of various types of stones and their functions against daily problems. Precious stones held specific meanings in ancient Mesopotamia, and each type of stone was chosen for specific purposes. At the end of the ninth tablet of the epic of *Gilgameš* (*Gilgameš* Tablet IX 172–194), *Gilgameš* mentions a gemstone garden. Schuster-Brandis suggests (Schuster-Brandis 2008: 15) that he probably refers to an 'eternal' garden where gemstones grow like trees. Carnelian and lapis lazuli are mentioned:

> He went straight to look at the stone trees-
> The carnelian, it bears fruit,
> hung with a grape, nice to look at.
> The lapis lazuli has foliage,
> It bears fruit, also (it) pleasing to look at. (Schuster-Brandis 2008: 15)

Another text which records the destiny of stones is the myth of *Lugal-e*, in which Ninurta defeats Asakku, a hybrid creature who commanded a land with the support of several stones. This is followed in the text by an explanation and a determination of the stones' fate after the victory of Ninurta over Asakku and the stones. Soft stones which were cursed by Ninurta should be used in the crafts, and the blessed hard stones by Ninurta should be utilised in temples (Schuster-Brandis 2008: 16–17).

Stones were not only categorised and selected based on their destiny but also on their healing properties, as well as for the protection they seemingly offered against evil eyes and magic spells. Therefore, there were particular stones with particular effects. Some texts connect specific stones with protection and healing, as for example the stone identification book *Abnu šikinšu* ("the stone, its design…"). This text is a list of stones and their uses as remedies and prophylactics in rituals of incantation (Schuster-Brandis 2008: 11). It was believed that particular precious stones had disparate effects than other minerals. The functions varied depending on the type of stone: protective, curative, therapeutic, or benedictive. These functions were listed on cuneiform clay tablets with name, colour, and the various properties of each stone. Two citations from two different texts describe what a specific type of stone could offer.

From the stone identification book *Abnu šikinšu* (Text E) we can read:

> The stone whose design (is as follows): its blackness has red spots on it:
> *Šadânu ṣābitu* (magnetite) is its name.
> It is the stone of truth. May, he who puts it on, speak the truth. (Schuster-Brandis 2008: 39)

And from the text BAM 2, 194 (VAT 9587) VIII′ 9′-10′[6]:

> Who bears a seal made of *Šadânu*-stone - this person - he loses what he got.
> Who bears a seal made of *uqnû*-stone (lapis lazuli) - he gets the life force of God (?). This god rejoices in him. (Schuster-Brandis 2008: 50).

The type of stone used for a seal was carefully selected since the seal was personal property worn sometimes as a necklace, bracelet, or attached to one's dress with a pin. Seals were also used as a tool for validation rolled on contracts, letters and treaties (Otto 2010: 469–73). Therefore, the choice of stone was certainly not accidental or purely aesthetic, but also related to its perceived special properties and effects (Schuster-Brandis 2008: 48).

Clearly, semi-precious stones played an important role in rituals, for instance in fighting witchcraft. Certain seals, made of specific types of semi-precious stone, were symbolically used to exclude evil through the sealing (Schuster-Brandis 2008: 49): for instance, on a tablet from Ur which contains various rituals to protect against an evil wizard, a seal made of *Šadânu ṣābitu*-stone (magnetite) was used to seal the boat on which the figurines of witches were transferred (Schuster-Brandis 2008: 51). It has been suggested that the *ašpû*-stone, blue chalcedony (Fig. 10.3), associated with the moon god, also had the important property of being able to fend off and cure epilepsy (Stol 1993: 111; Schuster-Brandis 2008: 402).[7] This stone was used in many amulet-type jewellery pieces, sometimes tied around feet and hands, as were other stones believed to have the ability to cure epilepsy (AN.TA.ŠU.BA or *bennu*)[8] (Schuster-Brandis 2008: 130, 402). Instructions in typical prophylactical texts also refer to the production of necklaces (Schwemer 2007: 238) in order to prevent witchcraft or cure sickness.

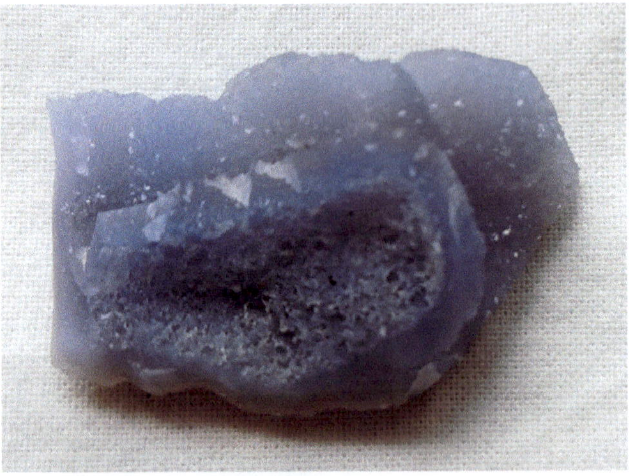

Fig. 10.3 Blue chalcedony stone. (Photo by Tiffany Ftaimi (06/2022))

10.4 Chalcedony Stones and Iconography of Fish-Cloaked Figures

The majority of seals from the first half of the first millennium BCE with depictions of fish-cloaked figures were made of semi-precious stones, especially chalcedony. The analysis of the seals presented here was based on glyptic from catalogues of museums and private collections.[9] A database of a glyptic assemblage exhibiting adoration and ritual scenes from the Neo-Assyrian and Neo-Babylonian periods was established by the author, including 1302 seals and seal impressions excavated in Anatolia, the Levant, and Mesopotamia. For this study, a total of 31 seals and eight seal impressions illustrating fish-cloaked figures were gathered and analysed iconographically. In the published catalogues, information about measurement, the material of seal stones, and further

[6]BAM 2 is an abbreviation for *Babylonisch-assyrische Medizin*, Bd. 2. and a part of the BabMed project. It contains texts from cuneiform tablets with medical recipes and treatments. The number 194 refers to the number of the text, VIII′ 9′-10′ refers to the eighth part of the cuneiform tablet and the lines numbers. Museum ID: VAT 9587. The text can be found here: https://www.geschkult.fu-berlin.de/e/babmed/Corpora/BAM-2/BAM-2_-194/index.html

[7]Stol (1993) mentions "jasper" rather than chalcedony; jasper is a variety of chalcedony and consists of it.

[8]Both names were confirmed as a designation for epilepsy. The first term is mentioned in the logographical form and means 'what has fallen from heaven', thus noting that the sickness did not stem from the earth. The second term indicates the association with the moon god, while the representative of the god is the demon *bennu*, whose name refers to a disease with epileptic symptoms (Hätinen 2021: 271–72).

[9]For instance, the British Museum, Pergamon Museum, Louvre Museum, Ashmolean Museum, the Danish National Museum, the collections of the Bibel + Orient of the University of Fribourg, the seal collection of the Walters Art Gallery, the Birmingham Collections, the Marcopoli Collection, the Collection of Chiha, the Kist Collection, the Collection of the Royal Ontario Museum, the Collection of Royal Museums of Art and History the Collection of Mr. Edward T. Nevell, the Collection of Mrs. William H. Moore, the Collection of Martin and Sarah Cherkasky, the Gorelick Collection and from the Nimrud Excavation, Assur and Tall Šēḫ Ḥamad.

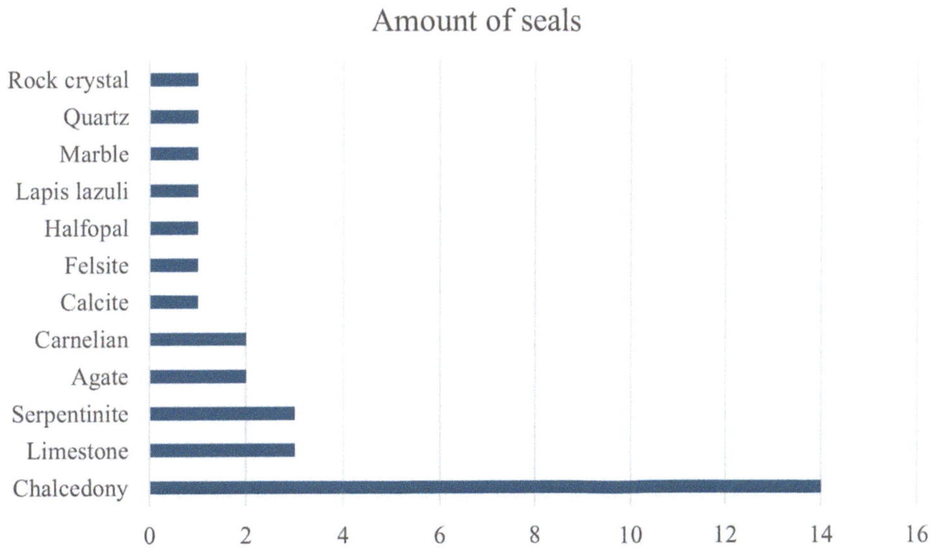

Fig. 10.4 Chart summarising the amount of 31 seals made of several stones with representations of fish-cloaked figures

details regarding the site or origin have been comprised. The methods applied in the identification of the stones were based on various methods: for instance, microscopical, XRF, and XRD analyses were used in the identification of the seal collection of the British Museum (Collon 2001: 18–19).

I performed a quantitative analysis with the software programme JMP for statistical purposes, adding the allocation of seal materials to my assemblage. The results have provided exhaustive information about 1302 seals and seal impressions which depict ritual practices from the aforementioned catalogues and further collections.[10] They indicate that 303 seals were made of stones identified as chalcedony. After calculating the percentage of this analysis, I obtained the following result: 38 per cent of the recorded seals were made of chalcedony, closely following the percentage of 36 per cent[11] of seals made of chalcedony in the Neo-Assyrian and Neo-Babylonian seal collection of the British Museum.

From the collected materials, 803 specimens are cylinder and stamp seals, and among these, 31 seals have scenes with fish-cloaked figures. Fourteen of these were made of chalcedony, six of which were dated to the Neo-Babylonian period and eight to the Neo-Assyrian period. Consequently, 45 per cent of the 31 seals, which is almost half of all collected seals representing this figure, were made of chalcedony (Fig. 10.4).

The proximity equability of seals made of chalcedony in the seal collection of the British Museum and the seal ensemble collected for the current study shows that chalcedony was a preferred stone used in seal production in these periods. The comparison between the 36 per cent, and 38 per cent, with the 45 per cent based on the statistic outputs in Fig. 10.4—which reflects the frequency of fish-cloaked figures illustrations on fourteen chalcedony seals—demonstrates that the results vary significantly from the first and second percentages. The collection of the British Museum includes seals with various subject matters and scenes, whereas my collection focused on seals with representations of deities and ritual practices. However, the amount of chalcedony used in the manufacturing process of seals in both collections is similar and underlines the popularity of this stone in these periods. Nevertheless, scenes portraying fish-cloaked figures were found on seals mainly produced of chalcedony.

This indicates that even though chalcedony was a popular stone for seals in general, fish-cloaked figures occur on this type of stone more often than on any other type, and more often than the overall average. The analysis thus suggests that fish-cloaked figures may not have been depicted randomly on semi-precious stones, but rather that especially chalcedony was a popular choice as a material efficient in combatting witchcraft and different diseases, such as epilepsy. Since the seals were also used as amulets, they not only protected the property of the owner, but also the owner themself (Salje 1997: 125). Moreover, it can be observed that fish-cloaked figures appear mostly in ritual and purification scenes and not in combat scenes, emphasising their apotropaic aspect.

Representations of fish-cloaked figures also occur on various other archaeological artefacts. This is for example the case on the so-called 'Hell Plaque', now in the Louvre, with

[10] The analysis was carried out in the framework of my Ph.D. thesis, entitled *Deities and Ritual Scenes on the Neo-Assyrian and Neo-Babylonian Seals and Seal Impressions*, at the Ruprecht-Karls-Universität Heidelberg. A detailed list of publications is included in the thesis.

[11] The analysis of all stone seals from both periods in the British Museum indicates that more than half the assemblage (52 per cent) was made of microquartz (Collon 2001: 20). Several types of stones belong in the microquartz category, including chalcedony. The types of stones of the 52 per cent are: 146 seals made of chalcedony, 44 of carnelian, seven of agate, four of sardonyx, and six of jasper (Collon 2001: 22–25). According to these numbers, 36 per cent of the seals from the Neo-Assyrian and Neo-Babylonian period at the British Museum collection were made of chalcedony.

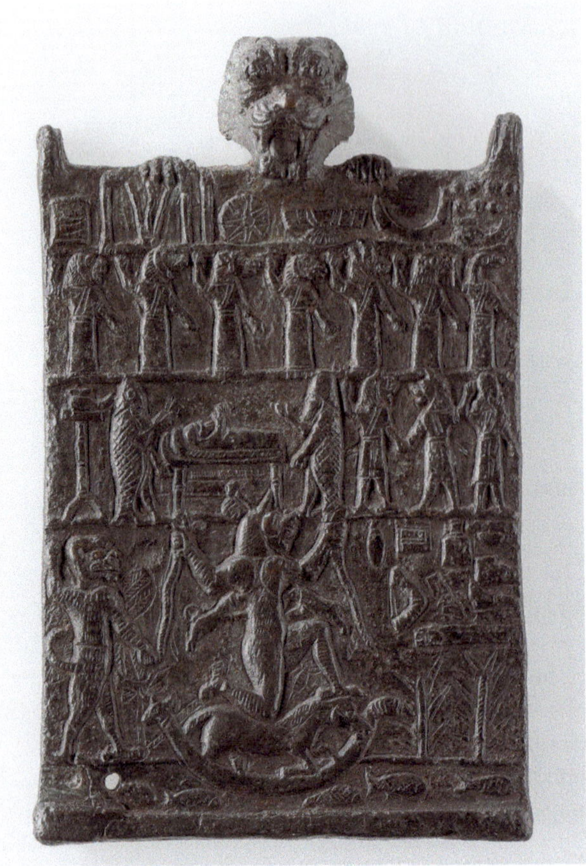

Fig. 10.5 Two fish-cloaked figures with Lamashtu and Pazuzu on a bronze plaque. Neo-Assyrian period. H: 13.8 cm, W: 8.8 cm. Musée du Louvre Paris AO 22205 ©2009 RMN-Grand Palais (Musée du Louvre). (Photo by Franck Raux. CC BY SA 4.0)

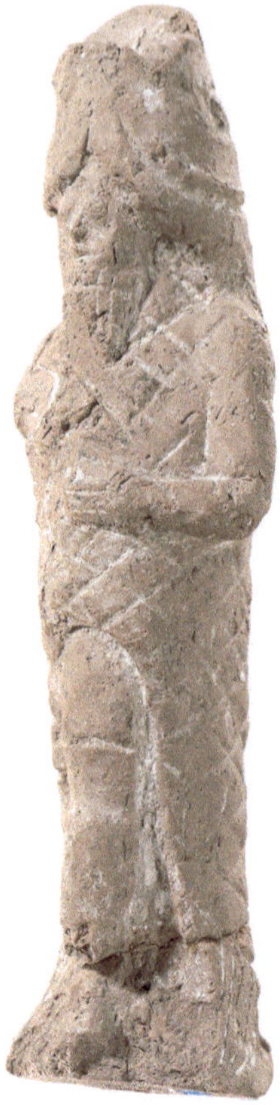

Fig. 10.6 *Apkallū* figurine: bearded fish-cloaked figure. Neo-Assyrian period. H: 14.8 cm, W: 4.8 cm, D: 4 cm. Clay. VA Ass 03727. (Courtesy of Staatliche Museen zu Berlin—Vorderasiatisches Museum. Photo by Olaf M. Teßmer. CC BY SA 4.0)

the female Lamashtu demon and male Pazuzu demon (Fig. 10.5). Pazuzu holds the whole plaque in the form of a modelled head and hands, and in the first and second registers, symbols of gods and seven figures are represented, asking Lamashtu to return to the Netherworld. In the third register, a sick person lying in a bed appears with raised hands, flanked by two fish-cloaked figures with *Banduddû* buckets, signifying a purification ritual to free the sick person's body from an evil demon (Meissner 1934: 14–15). This scene is repeated on a broken inscribed amulet from Assur (Meissner 1934: 131). Two fish-cloaked figures flank a sick person lying on a bed. The upper part of Lamashtu is recognisable and the lower broken part of the amulet is probably comparable with the bronze plaque with Pazuzu.

Other archaeological objects of interest are the so-called *apkallū* figurines (Fig. 10.6). These terracotta figurines have been found in groups of seven or 14 in capsules under floors at several sites,[12] including Nineveh, Ur and Nimrud (Rittig 1977: 86). Additionally, approximately 120 figurines from Assur were found in the 'House of the *Beschwöhrungspriester*', especially in foundation deposits (Nakamura 2019: 319). The practice of depositing apotropaic figurines in meaningful locations continued for 250 years, beginning in the ninth century BCE (Nakamura 2019: 319), and ending with the fall of the Assyrian empire (Green 1983: 87).

Cuneiform texts offer an opportunity to understand the function of these deposits under house floors as combatting and defending against magic and bewitchment (Rittig 1977: 151–62). A ritual text known today as KAR 298 (= VAT 8228) provides instructions for magical acts, which aim to expel evil from the house of a person and to protect the inhabitants from evil magic and aggressive witchcraft (Rittig 1977: 151). The text mentions a fish-cloaked figure and the deposition of figurines near the threshold of a sanctuary, behind a chair, and in houses for protective purposes (Rittig 1977: 164, lines 15–18), in order to prevent the entry of evil, particularly in residences of the deities Marduk and Ea (Wiggermann 1992: 65).

[12] Texts on cuneiform tablets also mention the utilisation of wooden figurines in apotropaic rituals, but such has not been found in the deposits (Green 1983: 91).

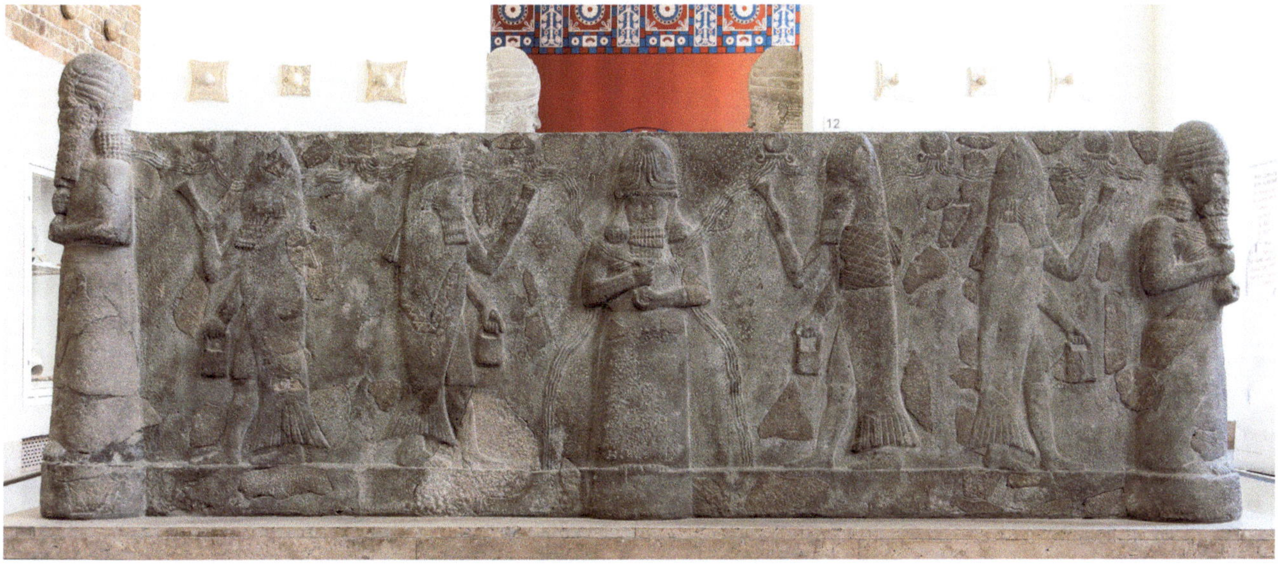

Fig. 10.7 Details of the holy water basalt basin from Assur Temple in Assur. Neo-Assyrian period. H: 117 cm, W: 312 cm. VA Ass 1835. (Courtesy of Staatliche Museen zu Berlin—Vorderasiatisches Museum. Photo by Olaf M. Teßmer. CC BY SA 4.0)

Another example of the association of fish-cloaked figures with purification is the appearance of these figures on a huge holy water basalt basin (Fig. 10.7), donated by Sennacherib and found in the Temple of Assur in Assur (Andrae 1977: 33). It was used for cultic purifications (Salje 1997: 133), and probably with the water sparkling from a well,[13] suggested by the possible original placement of the basin near the well (Gries 2017: I, 87). The engraved depictions on the side of the basin reflect its purpose: the figures and the sparkling water from the vessels were carefully chosen to illustrate the cleansing ritual, symbolising purification through the water and blessed by the presence of the aquatic apotropaic figures.

The fish-cloaked figure was also represented on reliefs in ritual acts and flanked a doorway of the Ninurta temple in Nimrud (Black and Green 1992: 83). The protective and additional purposes of this manifestation on reliefs were to add magical effectiveness to visitors entering the monuments (Rittig 1977: 93), especially near entrances of monumental buildings such as temples and palaces. The idea of presenting creatures from myths, the divine world and victories of rulers was to impress the dignity and solemnity of the place on visitors.

10.5 Chalcedony, Fish-Cloaked Figures and Epilepsy

The numerous fish-cloaked figurines and their representation on the Pazuzu plaque and seals demonstrate their role in purification scenes and as repellents of diseases, harms, and avoiders of epidemics, demons, and causes of evil. The fish-cloaked figure's role can also be determined by its appearance alongside further figures on a bell of copper (VA 2517) found in Assur (Braun-Holzinger 1984: 84). According to ritual texts, the bell had rung during healing procedures to fight against infections (Salje 1997: 133). Not only do the ritual texts mention the role of the bell, but the portrayed scene could also serve as an incantation scene, chiefly because of the presence of aquatic animals that relate to the water god Ea, who was the god of incantation and wisdom (Braun-Holzinger 1984: 85). The figure in the fish skin is considered a healer in addition to his protective and cleansing functions. The healing aspect of this figure is indicated on the Lamashtu bronze plaque (AO 22205) and on the seal from the Pontifical Biblical Institute (Museum ID. IB. SA: 105), where two fish-cloaked figures identical to those on the bell help in the treatment of a sick person.

The texts about the use of semi-precious stones as protective and therapeutic objects support the hypothesis that the selection of chalcedony and other precious and semi-precious stones was probably not a coincidence, although chalcedony was widely used in Babylonia (Collon 2001: 282–84). Chalcedony has often been called the stone of epilepsy *aban benni* in textual sources (Schuster-Brandis 2008: 402). Interestingly, the analysis of images of fish-cloaked figures on seals

[13]The well was found in the *Ostanbau*, which is located southeast of the Temple of Assur in Assur (Gries 2017: I, 140).

Fig. 10.8 Graph summarising the correlated elements related to fish-cloaked figures and epilepsy

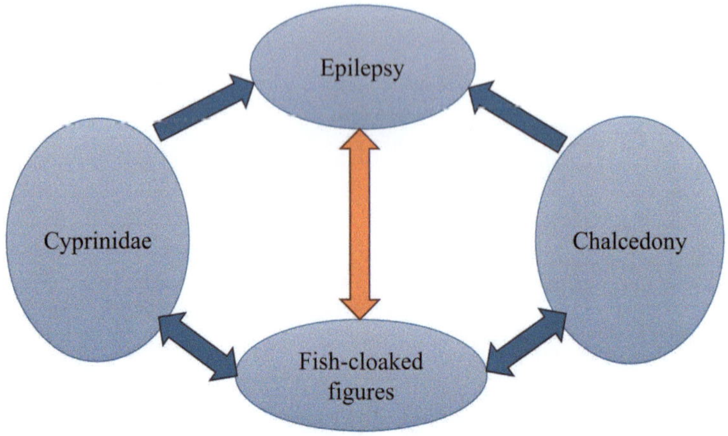

demonstrates that they occur more commonly on chalcedony than on other types of stone. This may reflect a correlation between the imagery and the choice of chalcedony as material for a seal, keeping in mind the function of the fish-cloaked figure to cure or to guard against the brain disease of epilepsy.

A relation between fish and epilepsy is known from cuneiform texts. The oil of fish was usually used as an ingredient in so-called "goat fumigation", as a treatment for epilepsy (Chalendar 2017: 463). Another part of a specific fish, the left eye of a large carp, could be wrapped in a bag and placed around the patient's neck to get rid of epilepsy and the ghostly hand (Chalendar 2017: 463–464). Fish played an essential role in the lives of ancient Mesopotamians since the two rivers crossed their lands and were a main source of nutrition. Worshippers offered fish to deities and used the skin of fish as a cloak and in specific medical treatments. Some texts[14] note that oil and parts of fish, especially the eye of a carp, were used to heal epilepsy (Chalendar 2017: 463–64). Further texts mention that fish-cloaked figures are not priests, but rather mythological figures with a clear indication of the carp fish *purād tāmit… ša ina nāri ibbanû* "carp of the sea … who were grown in the river" (Wiggermann 1992: 76). Carps are still a part of modern fisheries in Iraq and Syria (Jawad and Mutlak 2021: 907). In ancient Mesopotamia, the carp was probably consumed as meals besides the application of the fish's parts in medication and healing rituals.[15]

The textual evidence thus suggests the exploitation of fish belonging to the family Cyprinidae. Particular sorts of carp can grow up to 150 cm in length (Jawad and Mutlak 2021: 899) and some carp varieties have obvious scales. The depicted fish cloaks do not include teeth, in line with some types of carp. The marine biologist L.A. Jawad has discussed the large-sized fish cloaks represented on cylinder seals and interprets the fish covering the head and the body of the person based on ichthyological features as cyprinid fish sorts such as *Luciobarbus xanthopterus* and *Luciobarbus esocinus*, which are endemic in the Tigris-Euphrates riverine system (Jawad 2021: 95; note some incorrect dating of some of the finds). The length of the cloak on the representations differed over the course of time, perhaps reflecting the variety of the carp's grown length.

Considering these aspects, the connection between fish, epilepsy, and fish-cloaked figures is clear. The rendering of scales, size, and appearance of the depicted cloak on the heads of fish-cloaked figures compare favourably to those of a real carp. This may indicate that the represented fish cloak was based on knowledge of the carp's scales as observed through fishing and consumption. Moreover, van Buren argues that on the seal from the Pontifical Biblical Institute (Museum ID. IB. SA: 105), one figure of a set of two fish-cloaked figures is wearing a cloak of a carp, resembling presumably the freshwater of Mesopotamia in which the carp occurred (Van Buren 1940: 82 and Pl. IX, Nr. 82 a,b).

By way of conclusion, since the carp and the stone chalcedony were associated with epilepsy, and there appears to be a connection between the fish-cloaked figure and the carp, the manifestation of this figure on seals made of chalcedony is perhaps not a coincidence. Particularly, the healing and apotropaic role of the fish-cloaked figure can be observed on seals and other objects that were used to fight diseases and expel harm and demons. In other words, the shared features

[14] BAM II-1591: vi 34–37, BAM II- 166: 1–6, BAM II-197: 37–41, BAM III-311: 79′-80′, BAM V-476: r. 5′-6′.

[15] Based on excavations carried out in Eridu, Uruk, Lagash, and Girsu, many fish remains have been found in the temples of these sites (Ebeling 1971: 67). It has not always been easy to determine if among excavated fish remains, bones of carp (*Cyprinidae* sp.) were found. For Girsu, see Greenfield, this volume and Tsouparopoulou 2024. The reason for this difficulty in identifying the species might have been the lack of zoological examination and sometimes it could depend on the fact that the bones are poorly preserved. Fishes were offered on tables usually in front of the deities and often represented on Neo-Assyrian and Neo-Babylonian cylinder seals (Collon 2001: Pl. XI, Nr. 138, 149). It is worth mentioning that fishing and water were an essential part in Mesopotamian life and this was well documented in texts; for instance, 9600 carps were delivered in the fifth year of Lugalanda's reign in Lagash (Lambert 1971: 69).

between carp, chalcedony, and the fish-cloaked figure could establish a conjunction between these figures and epilepsy (Fig. 10.8).

10.6 Conclusions

Precious stones had a valuable meaning in the ideology and religion of the people of ancient Mesopotamia, but not only the material of the seal offered a protective function, the depicted scenes on them also provided an apotropaic effect for the owner, which is known from healing texts from the second millennium BCE (Salje 1997: 125). The appearance of the sun god, a fish-cloaked figure, and a dog (representing the healing goddess) on a blue chalcedony seal (BM WA 104499) in the British Museum nicely illustrates the connection between a chalcedony seal and its engraved motif. Having these three figures depicted on the same seal, considering the features of healing of each, could have allowed the seal to offer protection against demons, illness, and evil corresponding to the above-discussed function of chalcedony and of fish-cloaked figures. The depiction of a fish-cloaked figure on a seal made of chalcedony could thus guarantee a sort of double protection against epilepsy—through stone and iconography—for the patient or the owner of the seal.

This ideology and practice of associating precious and semi-precious stones with specific symbols or texts is also common today. Some specific stones are used as protective and healing products in different cultures and countries. For example, chalcedony is thought to protect against infection, agate is used to safeguard mothers and their unborn children during pregnancy, amber is used to decrease teething pain in babies, and turquoise is said to protect against the evil eye in many Middle Eastern societies. Today, a wide range of stones, with specific effects, in the form of raw and unpolished stones or as accessories is available in various stores and markets worldwide (Fig. 10.9). The impact of these precious stones depends on the beliefs and acceptance of the stones' purported benefits.

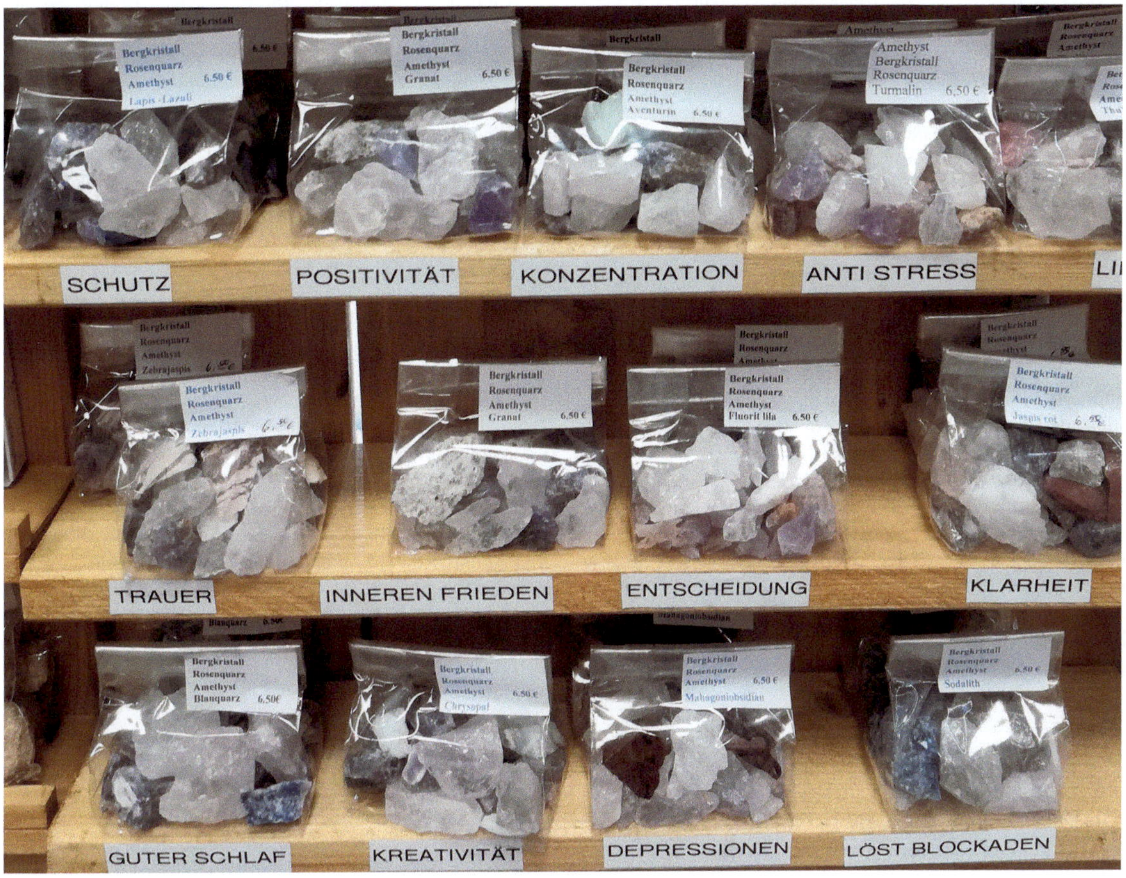

Fig. 10.9 Small bags with stones for healing and protection against various ailments. Heidelberger Steineladen. (Photo by Tiffany Ftaimi (06/2022))

Acknowledgments I am very grateful to Christina Tsouparopoulou, Lærke Recht and the anonymous peer reviewers for their remarks and suggestions on earlier drafts of this paper.

Bibliography

Andrae, W. 1977. *Das Wiedererstandene Assur*. München: C.H. Beck.

BabMed, BAM 2, 194 (VAT 9587) https://www.geschkult.fu-berlin.de/e/babmed/Corpora/BAM-2/BAM-2_-194/index.html [accessed 19 December 2022].

Beaulieu, P.A. 2021. Berossus and the Creation Story. *Journal of Ancient Near Eastern History,* 8 (1-2): 147–70.

Black, J. & Green, A. 1992. *Gods, Demons and Symbols of Ancient Mesopotamia*. London: Trustees of The British Museum.

Braun-Holzinger, E.A. 1984. *Figürliche Bronzen aus Mesopotamien*. München: Beck.

Chalendar, V. 2017. Quand l'animal soigne... Les utilisations thérapeutiques de l'animal dans le corpus médical assyro-babylonien (PhD dissertation, L'Université de recherche Paris Sciences et Lettres / Ecole Pratique des Hautes Etudes).

Collon, D. 2001. *Catalogue of the Western Asiatic Seals in the British Museum, Neo-Assyrian and Neo-Babylonian Periods.* Cylinder Seals V. London: Trustees of the British Museum.

Ebeling, E. 1971. Fisch als Opfer. *Reallexikon der Assyriologie und Vorderasiatischen Archäologie,* 3: 67.

Green, A. 1983. Neo-Assyrian Apotropaic Figures: Figurines, Rituals and Monumental Art, with Special Reference to the Figurines from the Excavations of the British School of Archaeology in Iraq at Nimrud. *Iraq, XLV*: 87–100.

Gries, H. 2017. *Der Assur Temple in Assur: Das assyrische Hauptheiligtum im Wandel der Zeit, I.* Wiesbaden: Harrassowitz.

Hätinen, A. 2021. *The Moon God Sîn in Neo-Assyrian and Neo-Babylonian Times*. Münster: Zaphon.

Hrouda, B. 1965. *Kulturgeschichte des Assyrischen Flachbildes*. Bonn: Habelt.

Jawad, L.A. 2021. Ichthyological Characteristics Available in the Fish Images Existed in the Art of Ancient Mesopotamia. In: L.A. Jawad, ed. *Tigris and Euphrates Rivers: Their Environment from Headwaters to Mouth*. Cham: Springer, pp. 85–126.

Jawad, L.A. & Mutlak, F. 2021. The Good and the Bad Realising the Grass Carp *Ctenopharyngodon idella* in the Freshwater System: Recommendations for Policymakers in Iraq. In: L.A. Jawad, ed. *Tigris and Euphrates Rivers: Their Environment from Headwaters to Mouth*. Cham: Springer, pp. 897–911.

Keel-Leu, H. & Teissier, B. 2004. *Die vorderasiatischen Rollsiegel der Sammlungen (Bibel+Orient) der Universität Freiburg Schweiz.* Fribourg, Göttingen: Academic Press Fribourg, Vandenhoeck & Ruprecht.

Kertai, D. 2020. "The "Sun(God) of All People" Šamaš's Presence in the Throneroom of Ashurnasirpal II at Kalḫu". In: N. Cholidis, E. Katzy & S. Kulemann-Ossen, eds. *Zwischen Ausgrabung und Ausstellung. Beiträge zur Archäologie Vorderasiens: Festschrift für Lutz Martin*. Marru Band 9. Münster. Zaphon, pp. 385–97.

Klengel-Brandt, E. 1968. Apotropäische Tonfiguren aus Assur. *Forschungen und Berichte,* 10: 19–37.

Klengel-Brandt, E. 2014. *Die neuassyrische Glyptik aus Assur.* Wiesbaden: Harrassowitz.

Kolbe, D. 1981. *Die reliefprogramme religiös-mythologischen Charakters in neuassyrischen Palästen*. Frankfurt am Main: Lang.

Lambert, M. 1971. Fischer(rei). *Reallexikon der Assyriologie und Vorderasiatischen Archäologie,* 3: 68–69.

Meissner, B. 1934. Siegelzylinder mit Krankheitsbeschwörungen. *Mitteilungen der Altorientalischen Gesellschaft,* 1/2/II: 14–26.

Nakamura, C. 2019. Ritual. In: A.C. Gunter, ed. *A Companion to Ancient Near Eastern Art.* Medford (MA): Wiley Blackwell, pp. 309–32.

Oates, D. 1983. The Excavations at Nimrud (Kalḫu) 1962. *Iraq, XXV*: 6–37.

Otto, A. 2010. Siegelpraxis. B. Archäologisch. *Reallexikon der Assyriologie und Vorderasiatischen Archäologie,* 12: 469–74.

Porada, E. 1948. *Corpus of Ancient Near Eastern Seals in North American Collections. The Collection of the Pierpont Morgan Library.* Washington: Pantheon Books.

Reiner, E. 1961. The Etiological Myth of the "Seven Sages". *Orientalia (Nova Series)*, 30(1): 1–11.

Rittig, D. 1977. *Assyrisch-babylonische Kleinplastik magischer Bedeutung vom 13.- 6. Jh. V. Chr.* Munich: Uni-Druck München.

Salje, B. 1997. Siegelverwendung im privaten Bereich (Schmuck) – Amulette- Grabbeigabe. In: E. Klengel Brandt, ed. *Mit sieben Siegeln versehen*. Mainz: von Zabern, pp. 125–37.

Schuster-Brandis, A. 2008. *Steine als Schutz- und Heilmittel, Untersuchung zu ihrer Verwendung in der Beschwörungskunst Mesopotamiens im 1. Jt. V. Chr.* Münster: Ugarit.

Schwemer, D. 2007. *Abwehrzauber und Behexung, Studien zum Schadenzauberglauben im alten Mesopotamien.* Wiesbaden: Harrassowitz.

Stiehler-Alegria Delgado, G. 1996. *Die Kassitische Glyptik.* Munich: Profil Verlag.

Stiehler-Alegria, G. 2006. Vom "Seebischof" zum Fisch-apkallu: Eine Zeitreise aquatischer Mischwesen. *Alter Orient aktuell,* 7: 4–8.

Stol, M. 1993. *Epilepsy in Babylonia.* Cuneiform Monographs II. Groningen: STYX Publications.

Streck, M.P. 2003. Oannes. *Reallexikon der Assyriologie und Vorderasiatischen Archäologie,* 10: 1–3.

Thureau-Dangin, F. & Dunant, M. 1936. *Til Barsib.* Paris: Geuthner.

Tsouparopoulou, C. 2024. The Early Dynastic "Maison des Fruits" at Tell K in Tello (Ĝirsu). *Bulletin of the American Society of Overseas Research,* 391: 191–225. https://doi.org/10.1086/730298

Van Buren, E.D. 1940. *The Cylinder Seals of the Pontifical Biblical Institute.* Rome: Pontificium Institutum Biblicum.

Von der Osten, H.H. 1934. *Ancient Oriental Seals in the Collection of Mr. Edward T. Newell.* Chicago: The University of Chicago Press.

Wiggermann, F. 1992. *Mesopotamian Protective Spirits: The Ritual Texts.* Groningen: STYX Publications.

Shells in Jewellery from Middle Kingdom Egypt 11

Maria Sofia Patrevita

Abstract

An interesting aspect of Egyptian maritime activities of the 12th Dynasty in the Red Sea is the impact on craft and jewellery. During this period, shell-shaped pendants made from a sheet of gold or electrum became very common in goldsmithing, such as the cowrie-shell girdles found in the tombs of the 12th Dynasty royal women at Lahun and Dashur. Used as ornaments from the period of the Badarian culture, cowrie-shells were fertility amulets linked to female sexuality and hathoric symbolism, but also a kind of insignia for members of maritime expeditions around the Red Sea. The metallurgical study of these reproductions of shells in precious metals is a means of researching the techniques, tools and ancient world of metallurgy. Experimental archaeology can help us research and understand the metallurgical processes behind this precious and symbolic artistic expression, connected to human-environment relations.

11.1 Introduction

Shells of the genus *Cypraea*, popularly known as cowries, are common finds in the archaeological record of the Near East, Egypt and Nile Valley, from the prehistoric period onward (Golani 2014; Wilkinson 1971). The taxonomic designation *Cypraea* derives from Cyprus, the mythological place of origin of the cult of Aphrodite, goddess of love. The name cowrie is of Indian root and in sixth century CE Sanskrit denoted "change" or "currency". The term is now used in reference to shells of the *Cypraea* family, including some 160 species worldwide. They generally inhabit shallow tropical or subtropical waters (Golani 2014). *Cypraea moneta* and *Cypraea annulus* (both from the Red Sea and the Indian Ocean) were used as "primitive coins" in northern Sudan, in West Africa and in Gojjam, northwestern Ethiopia. They were mainly used as amuletic jewellery (Johnson 1970; Smidt 2003).

From the Neolithic period, cowries with their rounded back removed appeared in Egypt and the Near East, used as necklaces of several beads together, bands, straps or bracelets. The shape was created by abrasion on a rough surface, such as a flat stone to facilitate the stringing of the shells as pendants or beads. During the Bronze Age, cowries became more common as grave goods, usually associated with burials of women and children. They were also popular in Predynastic Egypt and in the southern Levant of the Bronze and Iron Ages (Golani 2014). During the Middle Kingdom period, girdles of cowrie-shell beads appeared on the doll-like figures known as "concubines of the dead", wood figures of women and a steatite figure of a girl holding a cosmetic vase who has the cowries showing at her back (Wilkinson 1971). In addition to their association with women and children, cowries were occasionally found adorning domestic cats, suggesting a protective function of these shells and a connection between the cowrie and the Egyptian sacred animal (Golani 2014).

The important role of cowries and their extensive distribution attest to their immense popularity among many different ancient cultures. This particular shell appears in Neo-Assyrian written records, extending throughout the Fertile Crescent (Golani 2014). In these texts, cowrie shells are mentioned among precious items, including silver and gold, highlighting not only their economic value, but also their ritual and religious power. The use of cowries spread northwards into Europe, and ornaments made of Red Sea cowries reached southern Europe as early as the first century CE, becoming more common by the seventh century CE, at which time they are found as far north as Scandinavia (Golani 2014). Numerous shells have been found in the excavations of Pompeii and in the Vesuvius area. These include the exotic

M. S. Patrevita (✉)
Department of Humanistic Research and Innovation, University of Bari "Aldo Moro", Bari, Italy

Institute of Heritage Science, National Research Council of Italy, CNR-ISPC, Rome, Italy
e-mail: maria.patrevita@uniba.it

C. Tsouparopoulou, L. Recht (eds.), *Human and Aquatic Beings: Interactions in and beyond the Eastern Mediterranean (3rd–1st Millennia BCE)*, Themes in Contemporary Archaeology, https://doi.org/10.1007/978-3-031-73643-8_11

Cypraea, considered powerful amulets against female diseases. In addition, even at the end of the nineteenth century, women wore them around their necks to drive away infertility.

The frequent use of cowries as jewellery, especially in girdles, suggests that the shells were not just ordinary decorative beads, but were probably a strong aspect of Egyptian material culture and a powerful symbol of the role of women, mothers and warriors, with a function beyond ornamentation. The stylistic, technological and experimental studies of these shell imitations in precious metals are useful instruments to investigate techniques and tools involved in ancient gold working and goldsmithing.

11.2 Archaeological and Cultural Context

The reign of Sesostris II (Middle Kingdom, 1897–1878 BCE) seems to have been a peaceful period and the king probably started a scheme for land-reclamation and control of the flood waters of the Nile in the Fayum. He was buried in a pyramid of stone and brick at el-Lahun. Sesostris III came to the throne in 1878 BCE and made an expedition to Syria. The Second Cataract marked Egypt's southern boundary, and at el-Shellal, a canal was built to facilitate military operations against the Nubians. Under his reign in the area, eight forts were also built or rebuilt, including Semna, Kumma and Buhen, and a trading-post was set up at Kerma, south of the Third Cataract. Sesostris III was buried at Dahshur in the northern brick pyramid. The daughters of Sesostris II were Sit-Hathor-Int, buried beside Sesostris' pyramid at el-Lahun, Sit-Hathor-Yunet, possibly her elder sister, buried near the pyramid of Sesostris III at Dahshur, and Meret, wife of Sesostris III, who was buried beside his pyramid, not far from Sit-Hathor-Yunet (Wilkinson 1971).

The most famous hoards of jewellery, including cowrie-girdles, come from Dahshur and el-Lahun. In 1889–1890, Flinders Petrie excavated the pyramid of Sesostris II at el-Lahun, but did not complete his operation. In December 1913, he returned and in February 1914 his workmen found the entrance to a pit leading to an antechamber, sarcophagus-room and offering-room, all plundered in antiquity, except for a cache of jewellery remained hidden in the northern side of the antechamber. Among the most important pieces of jewellery were a diadem, hair beads, a pectoral with the name of Sesostris II, lion-amulets from bracelets, gold claw-pendants, a girdle with lioness-leopard head beads and a cowrie-shell girdle. All these objects showed signs of wear, suggesting that they had been used by the princess during her lifetime (Wilkinson 1971).

In 1894, de Morgan began to excavate the northern brick pyramid of Sesostris III at Dahshur. He found the tomb of Sith-Hathor-Yunet (her name means 'Daughter of Hathor of Dendera'), which consisted of a simple shaft with a chamber opening to the north. At the north side of the chamber, a huge uninscribed sarcophagus had been placed. In a cavity on the floor, de Morgan found a wooden box which was reduced to powder, but with a silver inlay on the lid that gave the name of its owner: "Sit-Hathor-Yunet". Among the jewellery of this treasure were a pectoral with the name of Sesostris II, two cylindrical pendant-amulets (*charm-cases*), claw-pendants from anklets, gold shell-pendants, gold lion amulets and a cowrie-shell girdle (Grajetzki 2014/2015). The next day, de Morgan discovered another hoard of jewellery, again hidden under the floor. The treasure, which he called "Second Treasure", belonged to the princess Meret, wife of Sesostris III. Among the main items were a pectoral with the name of Sesostris III, cylindrical pendant amulets (*charm-cases*), oyster-shell amulets, claw amulets from anklets, lion amulets from bracelets, lioness-leopard head beads and cowrie-shell beads from girdles (Wilkinson 1971; Aldred 1978; Melandri 2012).

11.3 Symbolic Meanings of Cowries and Their Imitations in Precious Metal in Middle Kingdom Egypt

In Egypt, first with the Badarian culture (Early Predynastic Period, 4400–4000 BCE), then during the following culture of Naqada (Predynastic Period, 4000–3000 BCE), an important category of Predynastic amulet consisted of natural objects such as claws and shells of various types, including cowries (Andrews 1994; Midant-Reynes 2014). But it was in the Middle Kingdom Egyptian jewellery that there was an increase in their use as amulets and their imitation in precious metals. This increase is probably linked to the Egyptian maritime activities of the 12th Dynasty (Middle Kingdom, 1991–1786 BCE) in the Red Sea (Manzo 2011).

During the 12th Dynasty, especially in its second half, the shell-shaped pendants became part of the repertoire of jewellery workshops (Manzo 2011). Similar shell-shaped pendants were recorded in the jewellery of the 12th Dynasty royal ladies, Sit-Hathor-Int, Sit-Hathor-Yunet and Meret. These types of shell imitations in precious metals, like gold, electrum and silver, introduced in this dynasty, continued in use until the end of the Middle Kingdom/Second Intermediate period and the early New Kingdom. The 12th Dynasty jewellery found in the tombs of the royal ladies were different and innovative from the traditional Egyptian ornamental types and most of them were worn during their lifetime. This suggests that, from the early 12th Dynasty, the shells, real or precious metal reproductions, became a part of the repertoire of the workshops which may have been attached to the court and probably made by the same goldsmith or group of master goldsmiths (Wilkinson 1971; Manzo 2011).

Cowrie-shells and their reproduction in precious metal were powerful amulets characterised by different symbolisms and apotropaic meaning. The lengthwise, serrated opening resembling a female vulva makes the cowrie an amulet of fertility linked to female sexuality and to protect against sterility. In addition, they may also resemble the shape of an eye linked to the "Eye of Horus" or the "Eye of Ra". The "Eye of Ra" may be regarded as the organ of creation, "the one who creates" and contains the tears which created both deities and humankind. The "Eye of Ra" is also identified with the lioness goddess Sekhmet, the violent manifestation of Hathor (Manzo 2011). If on the one hand, the shell embodies hathoric symbolism, the protection of the womb, the sacred feminine area and women's sexuality (thus, fertile, mother and generator of life), on the other hand, it was linked to the "Eye of Ra", recalling another aspect of feminine nature: the lioness goddess Sekhmet.

An interesting aspect of the cowrie girdles is the presence of amethyst and carnelian beads, two semiprecious stones with an intriguing meaning, connected to the hathoric and leonine symbolism of the cowrie. Carnelian, a translucent form of chalcedony, ranging in colour from red-brown or orange to a barely red-tinged transparency, is found in considerable quantities in the Eastern Desert and Nubia (Andrews 1994). Like jasper, it was associated with menstrual blood and its protective power, and widely used for the production of women's amulets such as the "Knot of Isis", in ancient Egyptian *tit* (Connor and Facchetti 2016). Amethyst is mentioned with relative frequency in Egyptological literature, generally in reference to jewellery (Hackley 2014). It is a translucent quartz with a glassy sheen, ranging in colour from deep violet to a barely violet-tinged transparency. Its chief source during the Middle Kingdom was Wadi el-Hudi, southeast of Aswan, although older working areas have been found northwest of Abu Simbel (Andrews 1994). Used as an ornament from the Predynastic period, amethyst is attested throughout the Early Dynastic Period and the Old Kingdom. During the Middle Kingdom, amethyst appears as beads, scarabs and amulets, such as cowrie, claw and lion, related to one another by their association with amethyst's place of origin, the southern desert, and by mythological connections, recorded in the "Myth of the Solar Eye of Ra", also known as the "Myth of the Distant or Wandering Goddess". Hathor was associated with foreign lands, the desert, the mining regions and the products of these regions. She was called "Mistress of the Desert, Mistress of Foreign Lands, Mistress of Turquoise, and Mistress of Amethyst" (Hackley 2014).

Another interesting aspect of the double symbology of the cowrie and the link Hathor-Sekhmet is that, among the grave goods of the 12th Dynasty royal ladies, precious metal reproductions of shells are often associated with other jewellery composed of crouching lions, feline heads, feline claws and beads of amethyst (Manzo 2011). Moreover, the 12th Dynasty witnessed the production of the female sphinx, a new type of royal representation characterised by a leonine body and female head with hathoric coiffure. The connection between the goddess, the queen, and the lioness is attested in Egyptian myth: Hathor was identified as the wild lioness, Sekhmet or Hator-Tefnut coming back to Egypt from Punt (Manzo 2011). This dual aspect of the Egyptian goddess Hathor/Sekhmet, mother and warrior, is highlighted and attested by the royal jewellery of the 12th Dynasty with the cowrie girdles and the feline jewellery, symbols of female identity. During the New Kingdom, women's girdles made of particular beads that resemble the shape of the cowrie, called *wallet-beads,* appear (Wilkinson 1971). Moreover, a popular form of scarab with a circular or elongated oval base and markings on its back which show that it is based on the cowrie shell is called a "cowroid scarab" (Andrews 1994).

In Nile Valley goldsmithing, among the Early Kerma cowrie shell ornaments, there were also imitations and reproductions made in precious metals and stone, gold and calcite (Markowitz and Doxey 2014). Surely, the most eloquent example of the great goldsmith mastery achieved by Nubian artisans is provided by the jewellery found in pyramid no. 6 in Meroe, belonging to one of the best known "queens of the south": Amanishakheto (Markowitz and Lacovara 1996). Among these splendid jewels, there are different types of so-called shield rings, massive in their gold structure and found in numbers of nine. The shield rings with the face of the fertility god Sebiumeker or the leonine deity Apedemak are accompanied by pendants in the shape of cowrie-shells, as amulets of fertility and female warrior power. Cowries are present also in assemblages from Kerma recently investigated in the Fourth Cataract. Moreover, these shells, fixed on leather bands and used as personal body adornments, were found in Gash Group tombs (early second millennium BCE) (Fattovich et al. 1994; Manzo 2012). An exquisite example of a gold cowrie necklace, imported from Egypt or made locally in Egyptian style, comes from grave 3 at Uronarti, one of the Middle Kingdom Egypt strategic forts, such as Askut, Buhen, Mirgissa, Semna and Kumma, linked to trade and gold mining operations (Markowitz and Doxey 2014; Lacovara & Markowitz 2019).

The use of the cowrie-shell as an amulet is still attested even in East Africa today, in particular in Ethiopia (Silverman 1999; Smidt 2003). In Amharic *zagul* and in Tigrinya *za egol,* in Tigray and in particular in the Aksum area, this precious shell is used as decoration on leather clothes (a sort of baby backpack), worn by women and known in the area of Adwa as *mahasal*. The protection is reinforced by ornaments, in particular necklaces, worn by women and children, considered *talsam* or talismans in Tigray (Silverman 1999; Smidt 2003).

11.4 The Cowrie-Shell Girdles of the 12th Dynasty Royal Ladies

During the Middle Kingdom (1991–1786 BCE), Egyptian goldsmithing reached very high technical-artistic levels, refining the knowledge already developed in previous periods, expanding the typologies and consolidating the symbolic language. Among the innovations in the production of ceremonial girdles of this period, there are the cowrie-shell girdles. The most beautiful cowrie girdles belonged to the daughters of Sesostris II: Sit-Hathor-Yunet, Meret and Sit-Hathor-Int.

In Sit-Hathor-Yunet's treasure, eight large gold cowrie shells, 32 acacia-seed beads of gold, cast in pairs side by side, 31 similar single beads of green felspar, and 31 of carnelian were found. Sit-Hathor-Yunet's cowries were made of tin sheet gold and contain copper-silver alloy pellets that jingle. The entire girdle is 84.3 cm long, and each cowrie is 4.7 cm in length (Cowrie Shell Girdle of Sithathoryunet, Middle Kingdom, The Metropolitan Museum of Art: https://www.metmuseum.org/art/collection/search/545533; Brunton 1920; Winlock 1934; Wilkinson 1971).

Sit-Hathor-Int's cowrie-girdle is strung with acacia-seed beads made of gold, green felspar and carnelian, very similar to that of Sit-Hathor-Yunet. Unfortunately, the published elements do not appear to be complete. There are only six cowrie shells, each a centimetre shorter than the shells of Sit-Hathor-Yunet's girdle, and only 120 acacia beads of gold, green felspar, lapis lazuli, and carnelian (Winlock 1934).

The cowrie-shell girdles of Queen Meret were probably two, but it is difficult to determine from the publications how the beads and the gold shells were found and whether they had been unstrung. One set consisted of eight large shells, each 5.7 cm long, in which there are pellets like those from el Lahun, but larger than Sit-Hathor-Yunet. With them, two smaller shells were found, each 5 cm long. The cowries' shape is identical on each side with holes for two threads and all shells made for a single girdle are identical (Winlock 1934; Wilkinson 1971). The other group consists of 20 small shells, each of which was only 1.6 cm long. Even if the assembly in one single girdle seems possible, the numbers of each sort make asymmetrical combinations, and therefore it seems probable that Mereret had two girdles: one of large cowries similar to those of the other two princesses, and another of small shells (Winlock 1934).

The separate halves appear to have been punched with a stamp and possibly finished with *repoussé* technique; even the notches of the shell pattern are chiselled. The two halves were joined with soldering (Winlock 1934; Wilkinson 1971). There are several hypotheses about the technique with which the shells were created. Brunton does not mention soldering, but Winlock says the el-Lahun beads were soldered. They are burnished and one bead seems to show signs of soldering. Brunton says that they were cast *cire-perdue,* but Winlock disagrees (Brunton 1920; Winlock 1934; Wilkinson 1971).

11.5 The Cowrie-Shell Girdle from Museo Civico Archeologico, Bologna (Italy)

The cowrie-shell girdle (Fig. 11.1), preserved in Museo Civico Archeologico, Bologna (Italy), is a splendid example of this particular type of cowrie jewellery. It consists of a series of carnelian and amethyst beads, alternated with large, embossed silver cowrie-shells (Bresciani 1975). According to the catalogue of the third collection, Giuseppe Nizzoli reports that the original ancient thread of the cowrie-girdle was not well enough preserved to hold the weight of the beads and was therefore replaced with a silk thread but maintaining the original sequence of the girdle elements (Bresciani 1975). However, the shells have two sets of piercings, suggesting that the girdle originally had two threads of beads. The cowrie-shell girdle from Museo Civico Archeologico of Bologna is very similar to another girdle with cowrie-shells dated to the Late Middle Kingdom (end of 12th–13th Dynasty). In 1912–1913, this piece of jewellery was found, along with other pieces, in a plundered chamber reached by a shaft in the portico of a rock-cut tomb (Tomb MMA 840) in the Asasif section of the Theban necropolis, Upper Egypt, now in the Metropolitan Museum of Art of New York (Girdle with cowrie shells, Late Middle Kingdom, The Metropolitan Museum of Art https://www.metmuseum.org/art/collection/search/544420). The cowrie-shells are made of electrum and alternate with beads of amethyst and carnelian; the entire girdle is 66 cm long, and each shell is 2.5 cm long. The provenance of the cowrie-shell girdle of Museo Civico Archeologico of Bologna is not known, but it can be attributed by stylistic comparison to the 12th Dynasty or to the period immediately following and is likely from the area of Thebes.

11.6 Goldsmithing in Ancient Egypt

The study of the paintings and reliefs, as well as the inscriptions that accompany metallurgical and goldsmith scenes, is essential to understanding the evolution of techniques and tools, and the innovations and developments of metallurgy over the centuries. In Pharaonic Egypt, as well as in other ancient civilisations and even today, there was a complex hierarchy among the craftspeople involved in the various stages of metallurgical processing. After the production of metal plates (smelting and casting), which were hammered to be made thinner by the so-called "blacksmiths",

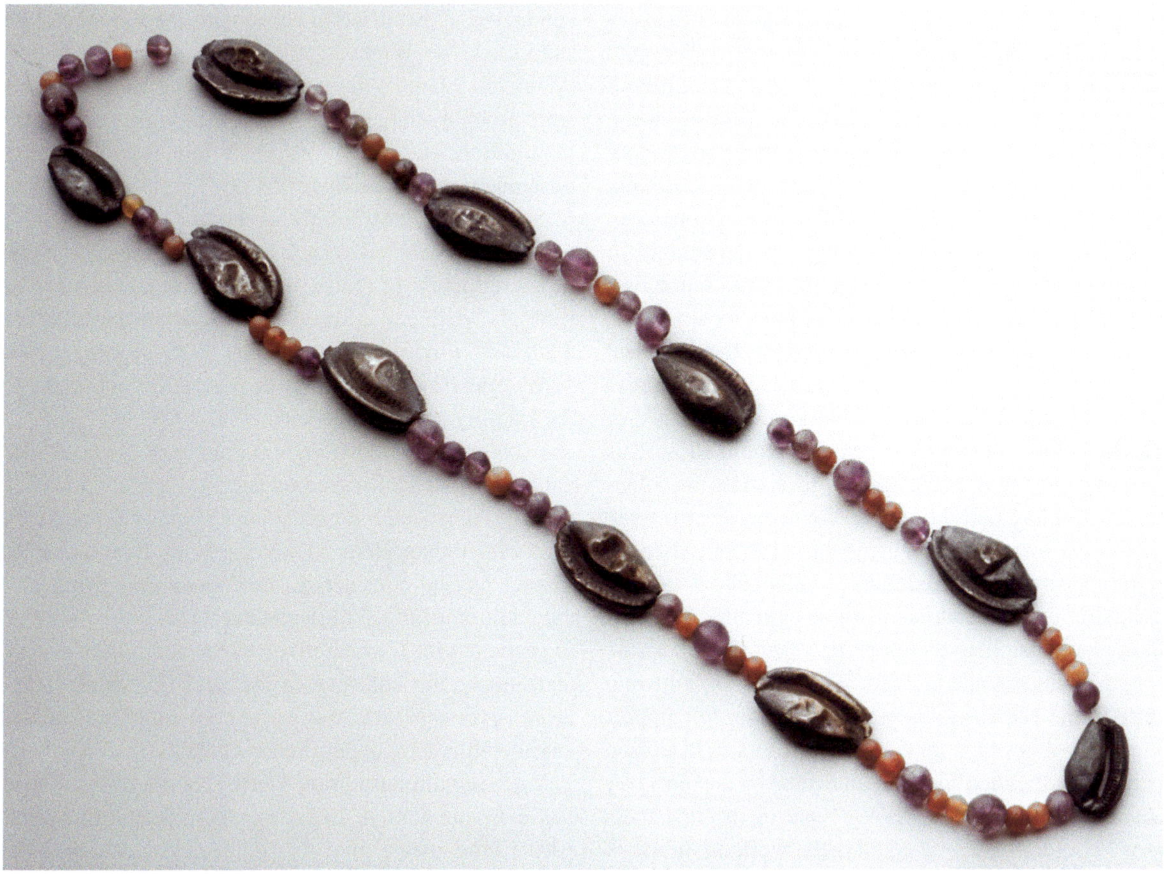

Fig. 11.1 The silver cowrie-shells girdle from Museo Civico Archeologico, Bologna (Italy) (Photo by Patrevita (2021))

or metallurgists, the "whitesmiths" (goldsmiths, silversmiths), the "finishers", received the material, and with it created the most precious objects such as articles for funerary use, vases and jewellery.

11.6.1 Repoussé

The ancient and elegant technique of *repoussé* is a combination of embossing and chiselling. The design is embossed from behind in order to obtain a relief image on the front, which is then finished with a chisel (Giardino 1998). The metal sheet was fastened to a soft base (such as wax, soft wood, or a leather pillow filled with sand or clay) so that the back of the metal object could be worked with suitable tools made of bone, hardwood, stone or metal (Scheel 1989). Taking a short detour to ancient Mediterranean goldsmithing, we can note a selection of bone tools from Mycenaean palatial workshops (fourteenth-thirteenth century BCE). With their short but used edges, these tools might have been used as punches to press a thin gold leaf onto a relief surface (Konstantinidi-Syvridi et al. 2014).

11.6.2 Welding

Welding is one of the essential techniques for the modern and ancient craftsperson. It is used to join two separate pieces of metal together, by sliding the soldering along the two edges or surfaces to keep them together, as if they were one piece. The first welding technique developed by the goldsmiths consisted of placing, along the edges of the pieces to be joined, fragments of low-melting gold. Each piece was placed, in the desired position, on the coals which, fed by the bellows, brought the heat to the necessary temperature, until the solder melted and flowed along the edges, joining the pieces. However, it was discovered that in order to make a cleaner welding in which the soldering did not overflow due to uncontrolled heat and ruin the delicate filigree or granulation decoration, it was necessary to add a small amount of very fine copper which was then brought in close contact with the soldering gold to melt together with it and the gold surface of the piece, creating a stronger and cleaner soldering.

Analyses conducted on Egyptian objects have demonstrated the use of welding from the 4th Dynasty, while in Mesopotamia it was practiced even earlier

(c. 4100–1750 BCE), suggesting technological influences that reached Old Kingdom Egypt (Scheel 1989). The gold found in nature was not pure, but contained quantities of silver or copper (Scheel 1989). From the 12th Dynasty, different soldering mixtures of gold, silver and copper began to be used. Electrum was also an important soldering, a natural alloy of gold and silver. Furthermore, it has been determined that when gold objects remain underground for many centuries or millennia, the metal undergoes mutations, especially the soldering or binder metal. Whether it is gold, silver, copper or electrum, it is dissolved by the chemical action of the moist soil, and a thin patina of pure gold appears on the entire surface of the object. On the surface there is thus fine gold and below, or within this, unaltered gold alloy. This is why in ancient works there are no differences in colour between the welding and the rest of the object (Ferro et al. 2008). With the exception of pure gold, all metals used by goldsmiths and silversmiths oxidise and do not allow welding when heated in contact with air because oxidation obstructs the melting process. It is therefore necessary to cover the surface with a metal soldering flux such as borax, allowing the sliding of the soldering. In Egypt, instead of this important compound of boron, a famous 'salt' was used, also used in the procedure of mummification, indicated by the term *ntry* (pure, divine): natron (Scheel 1989). One of the first tools necessary for soldering, melting and metal working, in use in the Old and Middle Kingdoms, is the blowpipe. This tool consisted of a barrel with a clay tip inserted into the end closest to the furnace, so as not to burn it. The Egyptian blowpipe is very similar to a tool still used today in modern jewellery production in the Middle East, North Africa and East Africa (Tigray, Ethiopia). Today, the Axum goldsmiths use a blowpipe, in Tigrinya called *menafia*. The principle of the *menafia* is practically the same as the ancient Egyptian blowpipe (Sernicola 2019).

11.7 Methodology

This study is mainly experimental. Since no laboratory analysis and no microanalytical studies on this specific type of jewellery has been performed, the study presented here is based on the experimentation of ancient tools and techniques, with the exception of the welding, which requires further in-depth studies and experiments. The illustrated experiment constitutes a starting point for the continuation of the research.

The first phase of the research consisted of the analytical study of the cowrie-shell girdle preserved in the Archaeological Museum of Bologna (Italy). The analytical study conducted allowed me to agree with the hypothesis already supported by Winlock: the welding of the two valves (Winlock 1934; Wilkinson 1971). The individual cowries were probably made using a matrix and finished with chisels. In addition, the shells of the Bologna girdle are of two different sizes, some larger (3 cm long) and others smaller (2.5 cm long). Interestingly, at the line of conjunction of the two valves (identical to each other, therefore it is an unrealistic reproduction of the shell), the welding applied in ancient times is visible. Unfortunately, a laboratory study (SEM *Scanning Electron Microscope*, EDS *Energy Dispersive X-ray Spectrometer*) cannot at present be undertaken given the fragility of the find (Ferro et al. 2008).

Most work on metallurgical joining of precious metals in antiquity have not focused on ancient Egyptian jewellery and most of the research for other ancient cultures has focused on gold rather than silver. Three methods for gold and silver joining can be considered: *hard soldering*, which involves the addition of metal to the soldering; *colloidal hard soldering* (granulation), which involves the addition of copper salts and an organic adhesive to the soldering; and *autogenous welding* (fusing), which is carried out without the addition of any metallic component (Schorsch 1995).

The second phase of the study consists of the archaeological experimentation. Experimental archaeology is undoubtedly a fundamental tool for archaeological study, where the goal is not only the practical result, but the set of knowledge that is obtained during the research and which constitutes support of theoretical and ethnoarchaeological studies (Ferro et al. 2008; Botwid and Eklöv Pettreson 2016).

11.8 Experimental Reproduction of a Cowrie-Shell in Precious Metal (Silver)

The archaeological experiment was divided into the following phases: bone tools (burins and chisels), *repoussé*, polishing and welding, and final polishing.

11.8.1 Bone Tools (Burins and Chisels)

Splinters and pieces of bone, obtained through the use of a hammer stone, were polished on a flat stone with sand and water. These bone tools used for embossing/chiselling the metal sheet were shaped based on what is needed for the production of the replica. Rounded tip burins were used for the embossing phase, while square tip chisels were used for the finishing of the shell shape and for the realisation of its characteristic notches.

11.8.2 Repoussé

At first, the thin silver sheet (constantly annealed to keep it malleable) was worked on a small wooden matrix that recalls the cowrie-shell shape, sketching its design. Keeping the metal warm, the embossing phase was started on a beeswax base (sometimes a leather base) using the bone burins and chisels (Fig. 11.2). Trying to follow the original shape of the cowrie girdle preserved in Museo Civico Archeologico of Bologna, the characteristic notches of the *Cypraea* shell were made. The best result seems to be the use of a soft base (wax or leather) that allows the metal to be modelled. The chisel used had a square and very thin tip. The characteristic notches have been chiselled on both sides, along the edges, imitating the long, narrow, slit-like opening, toothed at the edges.

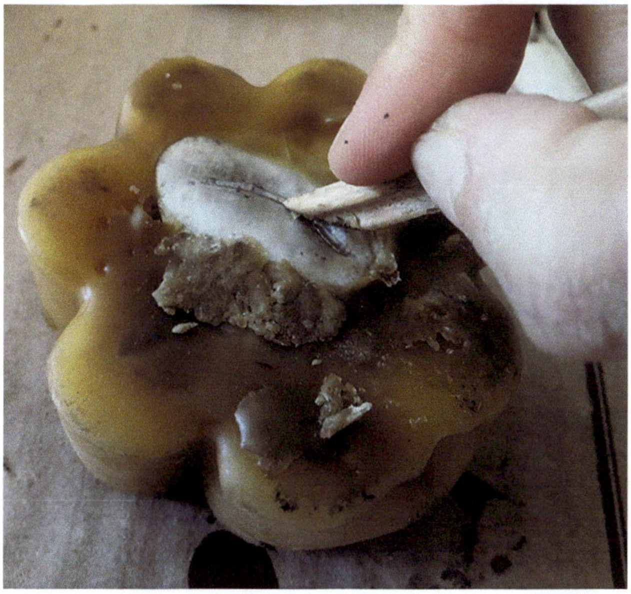

Fig. 11.2 *Repoussé* technique on a beeswax base with bone chisel (Photo by Patrevita (2021))

11.8.3 Polishing and Welding

Once both valves of the shell were finished, they were polished on a flat stone using water to smooth the perimeter along which the welding must flow and therefore to produce a better result (Fig. 11.3). For the welding, borax and silver soldering were used (the latter with a melting point lower than that of the silver sheet used for the shell, so that it melts first and allows the two pieces to be soldered together). Both halves were soaked in liquid borax and small soldering squares were placed along the perimeter to be welded. Using a small torch powered with beeswax and a copper blowpipe, the shell must initially be heated gently, drying the borax in order to block the pieces. Then the intensity of the flame was increased by directing the heat with the blow under the shell. In a short time, the metal object turned red and the soldering melted, sliding along the smooth perimeter and joining the two halves of the shell into a single piece (Fig. 11.4).

11.8.4 Final Polishing

The last phase of the experimental goldsmithing was the final polishing. The protruding edges along the welded perimeter had to be smoothed, first through the use of a smaller stone and then with a larger flat stone, water and sand. After this phase, the final polishing was done with a piece of leather, water and sand, thereby obtaining the characteristic colour of silver.

Fig. 11.3 Smoothing the perimeter of the valves of the shell on a flat stone with water (Photo by Patrevita (2021))

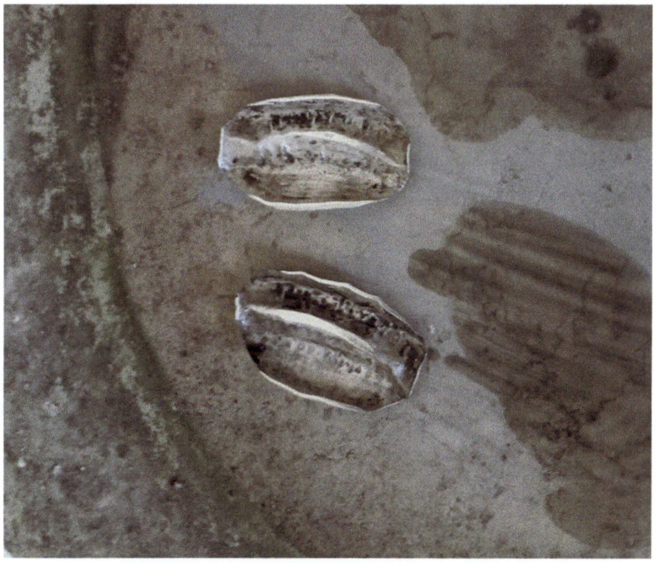

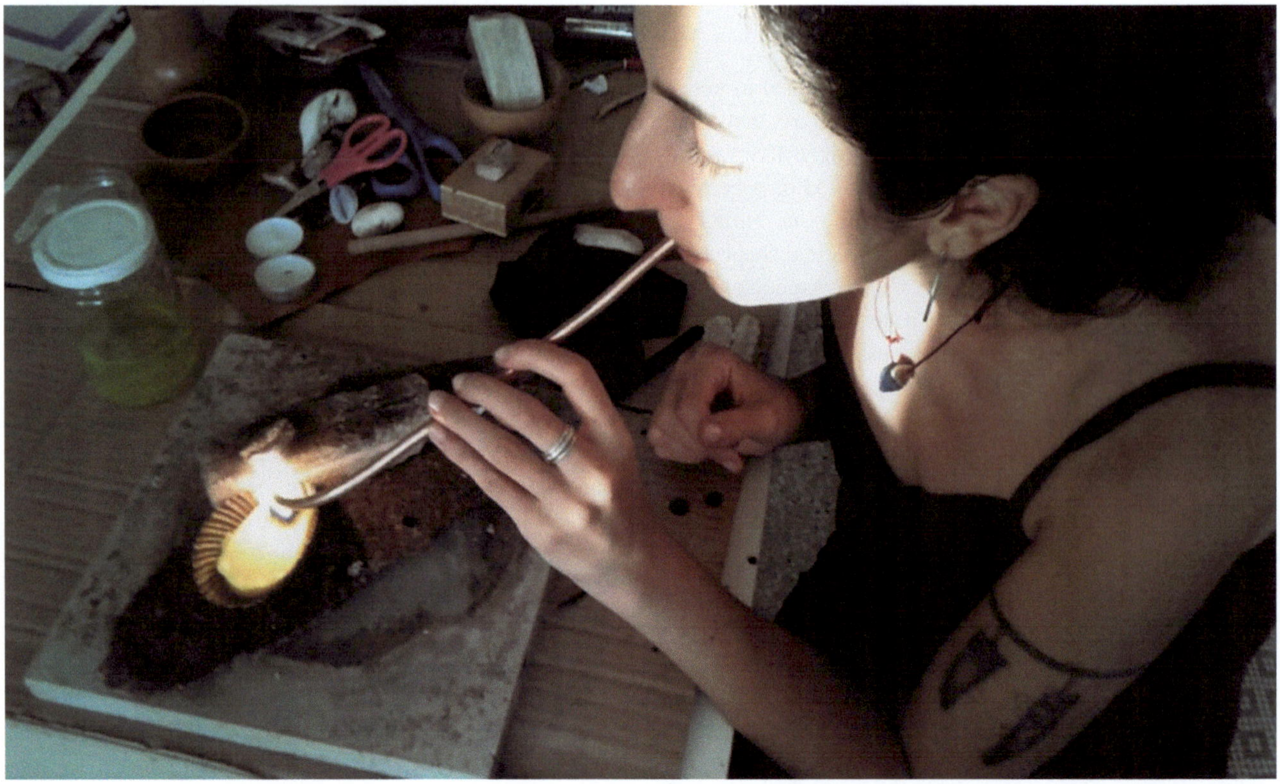

Fig. 11.4 Hard-soldering using blowpipe and small torch (Photo by Patrevita (2021))

11.9 Results and Conclusions

Humans have linked themselves to the environment that surrounds them through the use of certain natural amulets, thereby expressing identity, belonging to a specific area and spiritual vision and beliefs. The role and the apotropaic value that the cowrie-shell had in the Egyptian context and in the ancient Mediterranean, as well as their precious metals imitations, were an important and characteristic identity marker of Middle Kingdom Egypt jewellery. In conclusion, typological, technological, symbolic and experimental studies of ancient metallurgy and goldsmithing are fundamental in understanding the role of jewellery in ancient Egypt, especially in a context where jewellery making reached such technologically high levels. Experimental archaeology, goldsmithing study and ethnographic research constitute an interdisciplinary and dynamic tool to investigate the ancient metallurgical universe.

The experiment concerning the reproduction of a cowrie-shell in silver presented here was fully successful (Fig. 11.5). The aim of this preliminary study was the creation and reproduction of the *repoussé* technique, the tools used for its realisation and the use of the blowpipe for the welding. Future research will broaden the knowledge and experimentation of ancient Egyptian metallurgical techniques, also attested in other areas of the Mediterranean and the African continent. The next step will be the experimental realisation of different phases of the metallurgical and goldsmithing processes involved in ancient Nile Valley goldworking, trying to delineate the complete gold *chaîne opératoire.*

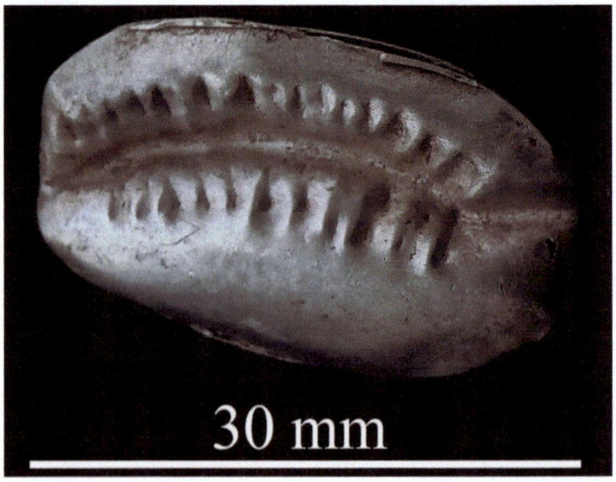

Fig. 11.5 Silver replica of a cowrie-shell (Photo by Patrevita (2021))

Acknowledgments With thanks to Dr. Daniela Picchi for guidance and permission to study the cowrie-shell girdle from Museo Civico Archeologico, Bologna (Italy).

Bibliography

Aldred, C. 1978. *I gioielli dei Faraoni. Le tecniche degli antichi gioiellieri egiziani.* London: Thames & Hudson.

Andrews, C. 1994. *Amulets of Ancient Egypt*. London: British Museum Press.

Black, A.J. 1973. *Storia dei Gioielli*. Novara: Istituto Geografico De Agostini.

Botwid, K. & Eklöv Pettreson, P.E. 2016. Use Traces on Crucibles and Tuyères? An Archeological Experiment in Ancient Metallurgy. In: P. Eklöv Pettersson, ed. *Prehistoric Pottery Across the Baltic*. BAR International Series 2785. Oxford: Archaeopress, pp. 21–28.

Bresciani, E. 1975. *La collezione egizia nel Museo Civico di Bologna*. Ravenna: Longo-Editore.

Brunton, G. 1920. *Lahun I. The Treasure*. London: British School of Archaeology in Egypt.

Chirikure, S. 2015. *Metals in Past Societies: A Global Perspective on Indigenous African Metallurgy*. London: Springer.

Chirikure S., Burrett, R. & Heimann, R.B. 2009. Beyond Furnaces and Slags: A Review Study of Bellows and Their Role in Indigenous African Metallurgical Processes. *Azania: Archaeological Research in Africa,* 44(2): 195–215. https://doi.org/10.1080/00671990903047108

Connor, S. & Facchetti, F. 2016. *Amuleti dell'Antico Egitto*. Modena: Franco Cosimo Panini.

Drici, F. 2016. The Chaîne Opératoire of Bronze Working in Ancient Sudan: An Attempt at Reconstituting the Manufacture of Kushite Weapons. *Dotawo: A Journal of Nubian Studies,* 3(1): 9–28. https://doi.org/10.5070/d63110039

Fattovich, R., A. Manzo and D. Usai 1994. Gash Delta Archaeological Project: 1991, 1992–93, 1993–94 Field Seasons. *Nyame Akuma, 42*, pp. 14–18.

Ferro, D., Formigli, E., Pacini, A. & Tossini, D. 2008. *La saldatura nell'oreficeria antica. Indagini archeometriche e archeologia sperimentale*. Rome: Edizioni Kappa.

Giardino, C. 1998. *I metalli nel mondo antico. Introduzione all'archeometallurgia*. Rome-Bari: Editori Laterza.

Golani, A. 2014. Cowrie Shells and Their Imitations as Ornamental Amulets in Egypt and the Near East. *Polish Archaeology in the Mediterranean,* 23(2)*:* 71–83.

Grajetzki, W. 2014/2015. The Tomb and the Treasures of Sithathoriunet. In: J.P. Phillips, ed. *Ancient Egypt. The History, People and Culture of the Nile Valley*. Manchester: Warners, pp. 18–24.

Hackley, L. 2014. Amethyst, Apotropaia, and the Eye of Re (MA Thesis, American University of Cairo). https://fount.aucegypt.edu/etds/1370

Johnson, M. 1970. The Cowrie Currencies of West Africa. Part I. *The Journal of African History, 11*, pp. 17–49.

Klemm, R. & Klemm, D. 2013. *Gold and Gold Mining in Ancient Egypt and Nubia. Geoarchaeology of the Ancient Gold Mining Sites in the Egyptian and Sudanese Eastern Deserts*. Berlin: Springer.

Konstantinidi-Syvridi E., Goumas A., Papadimitriou N., Philippa-Touchais A. 2014. Gold Working Techniques in Mycenaean Greece (17th/16th–12th Century BC): Some New Observations. In: H. Meller, R. Risch & E. Pernicka, eds. *Metalle der Macht. Fruhes Gold und Silber/Metals of Power-Early Gold and Silver, Halle (Saale)* (pp. 335–348), Salzland Druck GmbH & Co. KG.

Lacovara P. & Markowitz Y. J. 2019. *Nubian Gold. Ancient Jewelry from Sudan and Egypt*. American University in Cairo.

Manzo A. 2011. Punt in Egypt and beyond. *Egypt and the Levant*, 21: 71–85. https://doi.org/10.1553/AEundL21s71

Manzo, A. 2012. The Italian Archaeological Expedition to the Eastern Sudan of the Università degli Studi di Napoli "L'Orientale". An Overview of the 2010 and 2011 Field Seasons. *Newsletter di Archeologia CISA, 3*, pp. 313–335.

Markowitz, Y.J. & Doxey, D.M. 2014. *Jewels of Ancient Nubia*. Boston: MFA Publications.

Markowitz, Y.J. & Lacovara, P. 1996. The Ferlini Treasure in Archaeological Perspective. *Journal of the American Research Center in Egypt,* 33: 1–9. https://doi.org/10.2307/40000601

Maryon, H. 1971. *Metalwork & Enamelling*. New York: Dover Publications.

Melandri, I. 2012. A New Reconstruction of the Anklets of Princess Khnumit. *Vicino Oriente,* XVI: 41–53.

Midant-Reynes, B. 2014. Prehistoric Regional Cultures. *UCLA Encyclopedia of Egyptology*, 1(1): 1–13.

Rehder, J.T. 1994. Blowpipes versus Bellows in Ancient Metallurgy. *Journal of Field Archaeology,* 3: 345–50. https://doi.org/10.1179/009346994791547562

Scheel, B. 1989. *Egyptian Metalworking and Tools*. Aylesbury: Shire Publications.

Schorsch, D. 1995. The Gold and Silver Necklaces of Wah: A Technical Studies of an Unusual Metallurgical Joining Method. In: C. Brown, F. Macalister & M. Wright, eds. *Conservation in Ancient Egyptian Collection*. London: Archetype, pp. 127–35.

Sernicola, L. 2019. Archaeological Expedition at Aksum (Ethiopia) of the UniversitÁ degli Studi di Napoli "L'Orientale" and Ismeo – 2018 and 2019 Field Seasons: Seglamen. *Newsletter di Archeologia CISA, 10*, pp. 391–405.

Silverman, R. 1999. *Ethiopia. Traditions of Creativity*. Seattle: University of Washington Press.

Smidt, W.G.C. 2003. Cowrie(s). In: S. Uhlig, ed. 2003. *Encyclopaedia Aethiopica*, 1: A-C. Wiesbaden: Harrassowitz Verlag, pp. 294–95.

Troalen L., Guerra, M.F., Tate, J. & Manley, B. 2019. Technological Study of Gold Jewellery Pieces Dating from the Middle Kingdom to the New Kingdom in Egypt. *ArcheoScience,* 33: 111–19. https://doi.org/10.4000/archeosciences.2095

Winlock, H.E. 1934. *The Treasure of el-Lahun*. New York: The Metropolitan Museum of Art.

Wilkinson, A. 1971. *Ancient Egyptian Jewellery*. London: Taylor & Francis.

Shell Discs from First Millennium BCE Iran

12

Parisa Naseri, Abbas Motarjem, and David S. Reese

Abstract

Marine shells are one type of material culture found in archaeological excavations at many sites in the Iranian plateau. The shells come from open waters according to their biological origin and can be caught in certain coastal areas, but have also been found in distant locations. This indicates that they were traded over long distances, sometimes many hundreds of kilometres, and it seems that a special attitude towards these shells was related to aesthetics or religious/magical beliefs. Based on this, the main aim of this chapter is to explore the possible stimulus behind the transfer of shells from the area of origin to the place of consumption and deposition.

12.1 Introduction

This chapter examines the distribution and role of shell discs in the Iranian plateau, with a focus on the material from that area and from the first millennium BCE, although examples from earlier periods and adjacent areas are also discussed to provide further context.

Small discs made of marine shells are commonly found in archaeological excavations in the Iranian plateau and in southwest Asia more broadly. They appear as early as the Palaeolithic and continue at least into the Iron Age and are especially found as part of funerary assemblages. There is a wide variety of types, with different levels of decoration and drilling, and some with a central hole, also suggesting a range of functions and symbolic meanings. Although it is not always possible to identify the exact species, we know that the discs were made from marine specimens, and in some cases travelled long distances of several hundred kilometres from their source, presumably adding to their economic and social value. The discs have been variously interpreted as buttons, medallions, pendants, discs or bosses, spindles, shield decorations, musical instruments, part of horse harnesses or chariots, furniture, belts, or buckles.

In this study, in terms of methodology, the analysis has been performed in a descriptive and analytical way, directly and indirectly, and our statistical results were included in the research based on all the samples that could be studied by either of the two methods. The study of the direct method includes all the samples of shells discovered from archaeological deposits that we have been able to access directly; most of this group is in the National Museum of Iran in Tehran. The indirect method was used for those shells that have been previously published, or where information is available on the website of the museums where they are stored. After collecting all the examples, the analysis was based on a comparative method. The contexts in which they were discovered, such as temples and cemeteries, suggest the ritual importance of these objects but also their use as personal objects. The existence of an indirect interregional trade system has also been observed, which decreased in number as the distance from the origin increased.

Further, the examination of these finds reveals information about the source of shells and their final destination (trade routes), how shells were modified (technology), and how they were used (as exchange goods, funeral gifts, decorations, and so on) (Trubitt 2003; Bar-Yosef Mayer 2011: 186).

P. Naseri (✉) · A. Motarjem
Department of Archeology, Faculty of Art and Architecture, Bu-ali Sina University, Hamadan, Iran
e-mail: p.naseri@art.basu.ac.ir

D. S. Reese
Division of Anthropology, Peabody Museum of Natural History, Yale University, New Haven, CT, USA

C. Tsouparopoulou, L. Recht (eds.), *Human and Aquatic Beings: Interactions in and beyond the Eastern Mediterranean (3rd–1st Millennia BCE)*, Themes in Contemporary Archaeology, https://doi.org/10.1007/978-3-031-73643-8_12

12.2 The Shell Discs

12.2.1 Chronological and Geographical Distribution

The shell discs include a varied range of forms, with a wide distribution from the southern Levant to western Iran (Fig. 12.1). Most of the shell discs in Iran date to the second and first millennia BCE, although there are older examples such as those from Haji Firouz (dating to the Neolithic period).

Shells from the Persian Gulf and Oman were regularly imported into Iran. They initially appeared during the Epipalaeolithic and Neolithic periods. From the Palaeolithic period, shells from the Indian Ocean and the Persian Gulf were exchanged over vast distances across the ancient Near East, the eastern Mediterranean, and further north and west. In the Neolithic period, marine shells have been observed in some sites such as Ganj Dara, and a single shell disc has been found at Haji Firouz. In the Chalcolithic period, there are very few shell samples, but in the Bronze Age, we see a peak and intensity of the trade of shells, and disc-shaped shells are also part of the identified shells.

Shells were therefore particularly well-known during the Bronze Age when various species, such as *Lambis*, *Pleuroploca* and *Chicoreus*, were imported for local manufacture of items like rings, bangles, containers, and seals, and found in temples or as grave offerings. These shells are believed to be proof of the expansion of Mesopotamian trade with the cultures of the Persian Gulf. This is also believed to have significantly impacted the supply of shell species from the Gulf of Oman (*Lambis*, *Pleuroploca*) to the cities of the Indus Valley civilisation (Kenoyer 1983: 460; Moorey 1994: 130–37). They were also traded to locations in parts of the Zagros regions of Iran, including Shahr-i Sokhta, Shahdad, Konar Sandal and Tepe Yahya, Susa, Luristan, Kurdistan, Tepe Hissar, Marlik, Ziewyeh, and Hasanlu. It is also important to note the likely shell trade between different Bronze Age polities in the Persian Gulf (Mashkour et al. 2013: table 4).

During the Iron Age, a variety of shells were transported over long distances from the Persian Gulf and the Gulf of

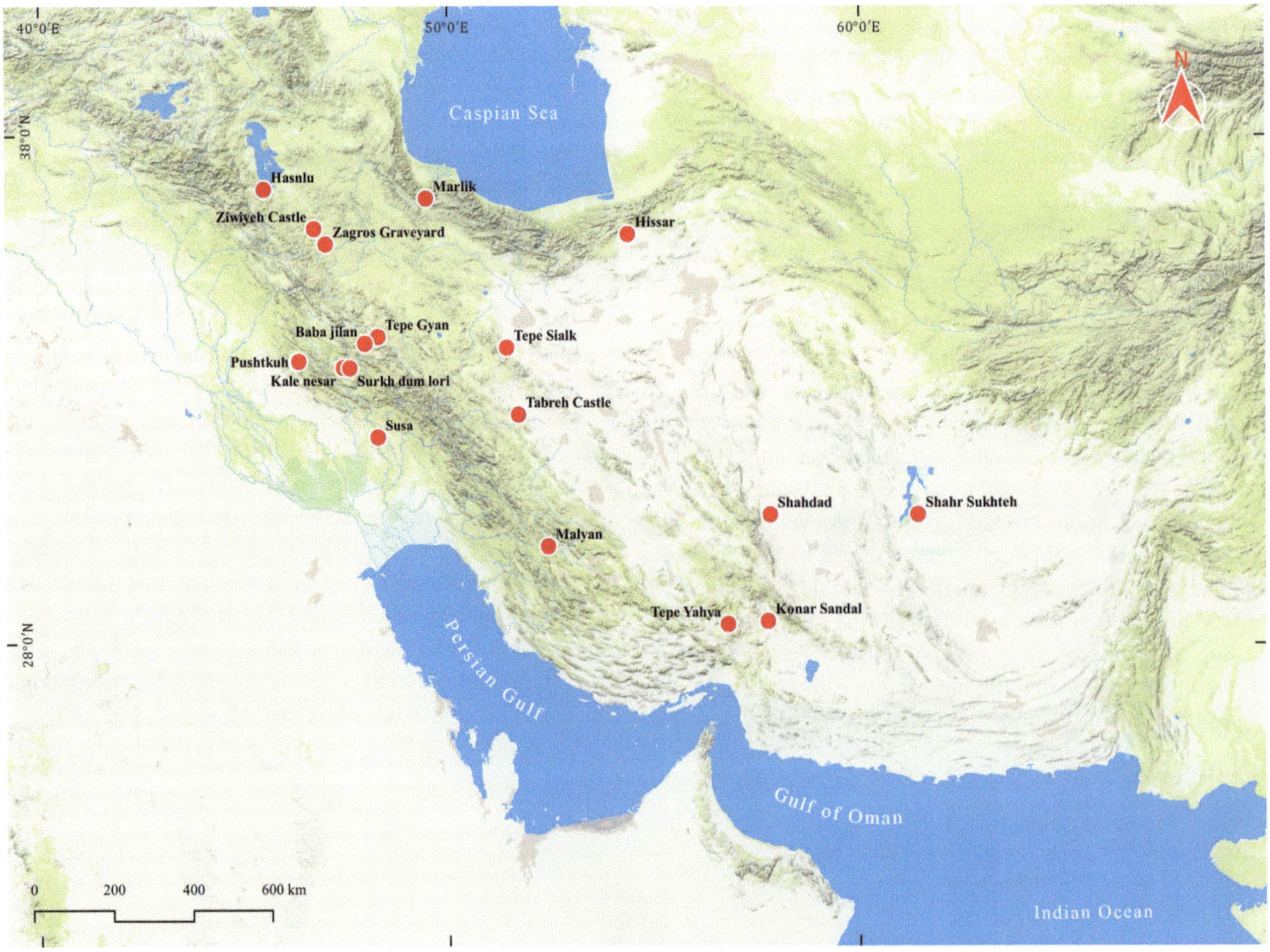

Fig. 12.1 Location of sites mentioned in the text. (Map by Alirezazadeh. M; QGIS 3.1416)

Oman. Shell trade thus continued, examples of which have been found in most of the archaeological sites of the Iranian plateau. For example, 7800 shells were excavated in Hasanlu in northwestern Iran, 98 per cent of which came from the Persian Gulf, 850 km away (Reese 1989: 80).

The wide distribution of shell discs includes western Iran. Many samples of disc-shaped shells and other types of worked and natural shells have been recovered from sites in the Iranian plateau, which covers a long period from the Palaeolithic to the Iron Age, demonstrating shell trade through land and sea routes. Shell discs have frequently been found in graves, typically described as buttons, for example in Posht Kouh, Tepe Sialk (Cemetery I), Hasanlu (Level IVB [Burnt Building IV]), Tepe Yahya (Level IVA), Workbud Cemetery, Cham Zhi Muweh Cemetery, Tepe Giyan. This interpretation is based on the view that they were decorations added to the clothing of the dead in the Iron Age.

In the western Iranian Iron Age Zagros cemetery of Chamahzi Mumah, shell discs were part of both male and female burials, and might also have decorated clothing (Haerinck and Overlaet 1999: 40; Amelirad et al. 2012: 49–50). The Iron III Luristan cemeteries of War Kabud (Haerinck and Overlaet 2004: 75–77, fig. 37) and Surkh-i Dum also include shell discs. There are worked and unworked shells at Surkh-i Dum (Schmidt et al. 1989: 368, 378–79, Pls. 225a–c, e). At least four shells with bronze nails (or studs) were found in the Iron Age Zagros 'cemetery' near Sanandaj (Amelirad et al. 2012: 49–50), and a similar object was found at Tepe Giyan (Herzfeld 1941: 140, Pl. 254 top right; Barnett 1963: Pl. XVIg). Another example is an undecorated shell disc with a central hole from the late ninth century BCE discovered in Burnt Building IVB at Hasanlu, now in the National Museum of Iran (Reese 1989: 86).

Susa also had shell discs, several of which are now in the Louvre (Biscione et al. 1981: 54, fig. 18). Examples from earlier periods can be found further to the east and south: undecorated shell discs with a central hole from Cemetery A at Tepe Sialk (1350–1000 BCE) (Biscione et al. 1981: 55, fig. 19) and from Yahya, apparently from the beginning of the second millennium BCE (Biscione et al. 1981: figs. 19, 33; the Yahya fragment [third millennium] is said to have been made from a bivalve rather than a gastropod), and even from the initial levels of the IVC period from the end of the fourth millennium to the beginning of the third millennium BCE Luristan in the Iron Age (Overlaet 2003; made from *Lambis*: Pl. 189). They sometimes include metal nails inserted through a central hole. Another example is from Baba Jilan (Luristan), where an iron rod was attached to the shell.

Various suggestions have been put forward concerning the uses of shell discs in Iran, including as part of musical instruments, shield ribbons, spindle whorls, horse harnesses, and possible attachments to furniture, chariots, or even battering rams; they are recorded as buttons in many Iranian sources. Simple examples of these shell discs have been found at Haji Firouz and Malian Fars, the cemeteries behind Luristan (Herzfeld 1941: 139–40; Curvers 1989: 368, Pl. 25 a-c; Reese 1995: 269–71).

Shell discs are also seen among Iran's neighbours. The origin of these shells is difficult to determine. The style, species of shell, and sometimes decoration are also very similar to examples from northern Arabia, Saruq al-Hadid, Sumhuram (Lischi 2020: figs. 2–5 and Weeks et al. 2019: figs. 2–6), with parallels from Nimrud, Tell Halaf (Mallowan 1966: 125, 452, Pl. 66), Kuyunjik (the Nineveh palace mound; Barnett 1963: 83, 85, Pl. XVIf), Jerusalem (Reese 1995: 269, fig. 4 a-b), Uruk, Ur, Kish, Sippar, Dibba, and other archaeological sites (Weeks et al. 2019: 214). It might be suggested that there was a single Iron Age tradition in the Persian Gulf and Iran.

12.2.2 Species Used for the Shell Discs

The largest of the Iranian shell discs must have been made from large marine shells. The discs underwent intensive processing during creation, which removed distinguishing morphological characteristics, making it challenging or even impossible to identify the species of shell utilised (Moorey 1994: 130). The exact species of shell that was used is not known, but the size and shape of the discs mean that we can limit the options to a few options. Two large gastropod species are the most likely candidates: *Lambis truncata sebae* and *Pleuroploca trapezium*. Southern Oman and the Indian Ocean are the present sources of both species (Gensheimer 1984: 67).

12.2.3 Manufacturing Techniques

Traditionally, stone tools and flint-tipped string drills were used to shape shells into discs, smooth them down, and drill holes in their centres. More recently, hammers, metal cylinders, braces, and bit drills have been used. The final polishing was done with stones and a sand-and-water abrasive; in modern times, imported grinding blocks are also used. The shell processing was probably divided into three or four phases: the cutting of the shell to obtain the desired shape, smoothing of the surfaces, perforating the centre with a large hole and the interior suture ledge with one or more small holes, and engraving the decorative motif on the exterior (Lischi 2020).

The interior suture ledge of the Iranian shell discs was perforated with three or four small holes, probably for attaching it to some other object. These holes were most likely drilled with small metal points, with a variation in the size of the points. Drills were used, and the drilling directions were synchronised to design a space with a central hole, with two or more channels coming together at the centre. Some holes exhibit an enlarged oval shape, suggesting that the drill was used in an oblique position to create an acute angle. The average size of the drill point is 0.2 to 0.5 cm (Weeks et al. 2019; Lischi 2020).

12.3 Shell Discs in Iran

In Iran, there are two main types of shell discs: one with simple decoration, and the other with miniature patterns. Simple decoration discs are perforated only in the centre of the disc. The shells with miniature patterns have various motifs such as ring-and-dot patterns, ibexes, pomegranate or poppy flower, parallel squares, stars, and so on. Below, we will discuss some of these decorated shells, all dated to the first millennium BCE.

1. Incised *Lambis* Shell Disc, now in the British Museum, BM 128870
   The disc measures (cm): Length: 10.35, Width: 9.15, Diameter bottom hole: 0.20, Diameter centre hole: 0.35. Its provenance is unknown, but is said to be from Tabarak/Esfahan, Iran. The disc is decorated in low relief in quadrants arranged around a central circle with a hole in the middle; a cross radiates out from the centre; each of the quadrants contains a stylised ibex; the hole at the top has a copper alloy nail (or stud) broken off in it; the eyes of the ibexes still contain traces of bitumen for missing inlays. Bibliography: Herzfeld 1941: 140, fig. 254 left; Barnett 1963: 83–84, Pl. XVIIa; Weeks et al. 2019: fig. 6 (as from Tabarak/Esfahan); personal analysis by Reese.
2. Two shell discs from Luristan, now in the National Museum of Iran (Fig. 12.2). The discs are probably from *Lambis* shells. One is simple (Fig. 12.2 right), and the other is engraved with ring-and-dot motifs (Fig. 12.2 left). Comparanda can be seen on shells from Marlik and Tepe Hissar which are worked with ring-and-dot motifs (see Schmidt 1937: 231–32, Pl. LXX:H 1973; Negahban 1996: 92). The purpose of these objects is unclear, but since many have nails in them (and all would have had them during use), it seems likely that they were attached to some other object. Previous suggestions have included being attached to leather, wickerwork, furniture, horse trappings, and shield bosses (Herzfeld 1941: 140; Barnett 1963: 83–84; Mallowan 1966: 125, 452; Brandl 1984: 40). This use seems more likely than their being used as clappers or castanets for dancing girls, as some writers have suggested.
3. Shell disc from northwestern Iran, now in the Metropolitan Museum of Art, MMA 1989.281.29 (Fig. 12.3).
   The disc's diameter is 6.4 cm. This dome-shaped object may have decorated a pommel, the knob on the end of a dagger or sword handle opposite the blade. It is composed of a shell disc attached to a copper alloy backing by a nail (or stud) going through the centre. The head of the nail is covered with gold foil. Around the edge of the disc runs a decorative pattern, formed of a gold foil frame arranged in a pattern of interlocking triangles, into which small inlays were set. Many small inlays have deteriorated and been lost entirely. Those which remain appear to have been made of carved shell. Inlaid into the disc's surface are four rams, shown in profile and arranged head-to-tail with

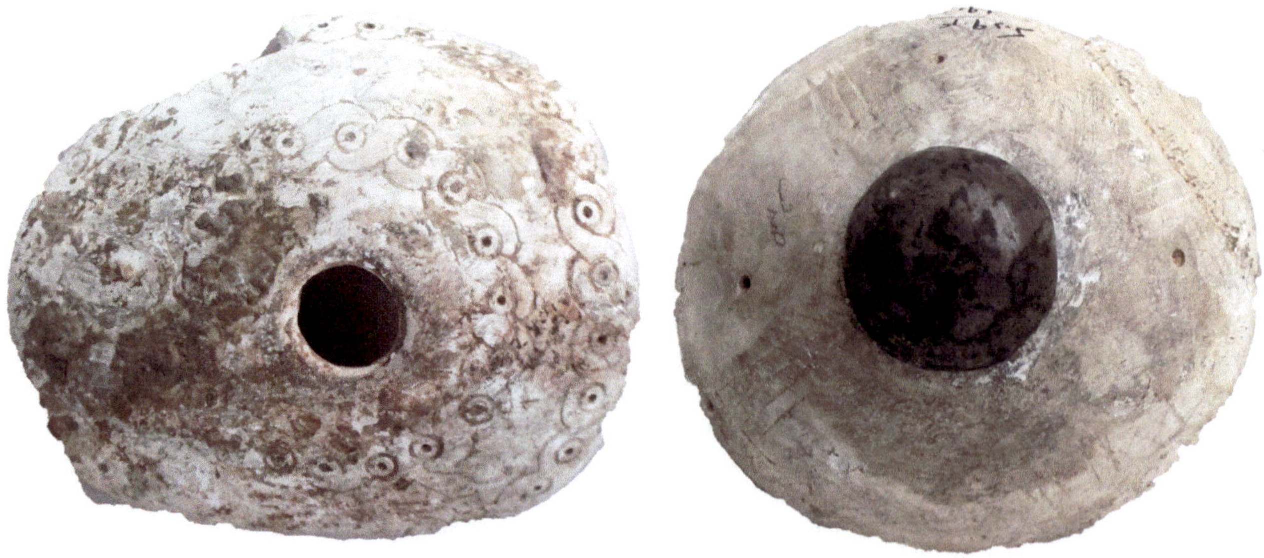

Fig. 12.2 Shell discs from Luristan, Iran. (National Museum of Iran Tehran)

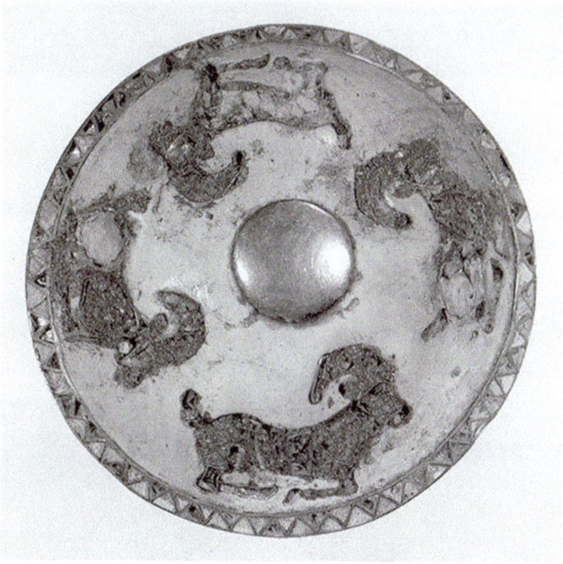

Fig. 12.3 Shell disc from northwest Iran (maybe Ziwieyh), now in the Metropolitan Museum of Art, New York City (MMA 1989.281.29). Shell, gold foil, stone inlay, Gift of Norbert Schimmel Trust, 1989, OA Public Domain. (©The Metropolitan Museum of Art)

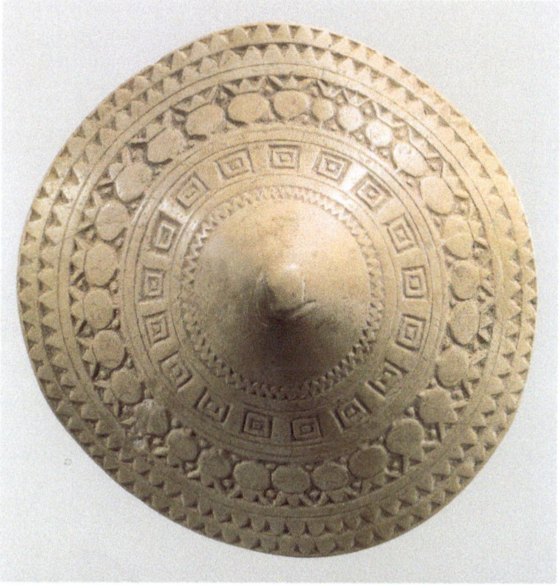

Fig. 12.4 Shell disc from Iran, now in the Metropolitan Museum of Art, New York (MMA 54.45), OA Public Domain. (©The Metropolitan Museum of Art)

heads toward the centre. A black residue is visible in many areas of the surface that were incised to form the bodies of the rams. This may be traces of bitumen (a tar-like substance often used as an adhesive). The outlines of the rams are formed from strips of gold foil that may initially have been set into bitumen. Their bodies were inlaid in a contrasting material, perhaps a soft stone, which has now deteriorated into a powdery substance. The same style was found in metal covers at Ziwieyh. The image of a goat is also engraved on a metal specimen. Bibliography: Muscarella 1992: 55, no. 23.

4. Worked shell disc, possibly made from *Lambis* or *Pleuropoca*, in the Metropolitan Museum of Art, MMA 54.45, of unknown provenance (Fig. 12.4). The exterior is decorated, from the interior outwards, with concentric decorative bands, a narrow zig-zag, concentric double squares, pomegranates with triangular calyxes, and two narrow bands of excised triangles. The interior has three drilled perforations toward the centre and a smooth, largely unworked inner surface. The interior small suture ledge has two drilled perforations. It was obtained in Iran in the late 1920s or 1930s and purchased by the Metropolitan Museum of Art in 1954. It is said to date to the eighth century BCE (Pope 1938: Pl. 57 B). Bibliography: Weeks et al. 2019: 4, fig. 4.
5. Undecorated shell discs, from Luristan (Fig. 12.5).
In some cases, only the centre of these shells has been pierced. This type of shell is cut and polished; sometimes, there is a metal nail in the centre. The following example is from Baba Jilan in Luristan, which has a metal handle.

12.4 Discussion: Uses and Symbolic Roles of Shell Discs

The shell discs discussed here have undergone unique processes to transform them into worked objects, and some travelled hundreds of kilometres from their source. A number of functions has been suggested, and they are variously referred to as buttons, medallions, pendants, discs, bosses, spindles, shield decorations, musical instruments, attachments to horse harnesses or chariots, furniture, belts, or buckles.

As can be seen from these examples and those discussed above, shell discs are often found in burials, where they are recognised as valuable or exotic goods or offerings (Allen and Payne 2017: 1). Shells can be used to reconstruct certain aspects of past social and economic conditions. Archaeological excavations have uncovered grave offerings in the cemeteries of the Iranian plateau since the Neolithic period. Many interpretations have been made about why shells are associated with the deceased. So far, many researchers have conducted research in this field. For example, Alekshin et al. (1983) studied the relationship between the deceased and grave offerings from the perspective of their significance for the construction of social identity.

Shell discs in the Near East, especially in the Early Iron Age, have been interpreted in terms of the complex economic and cultural contacts that underlie the connections and significance of these artefacts. Shell discs can be considered socially meaningful as a personal decoration that reflects part of the identity of the buried person and positively

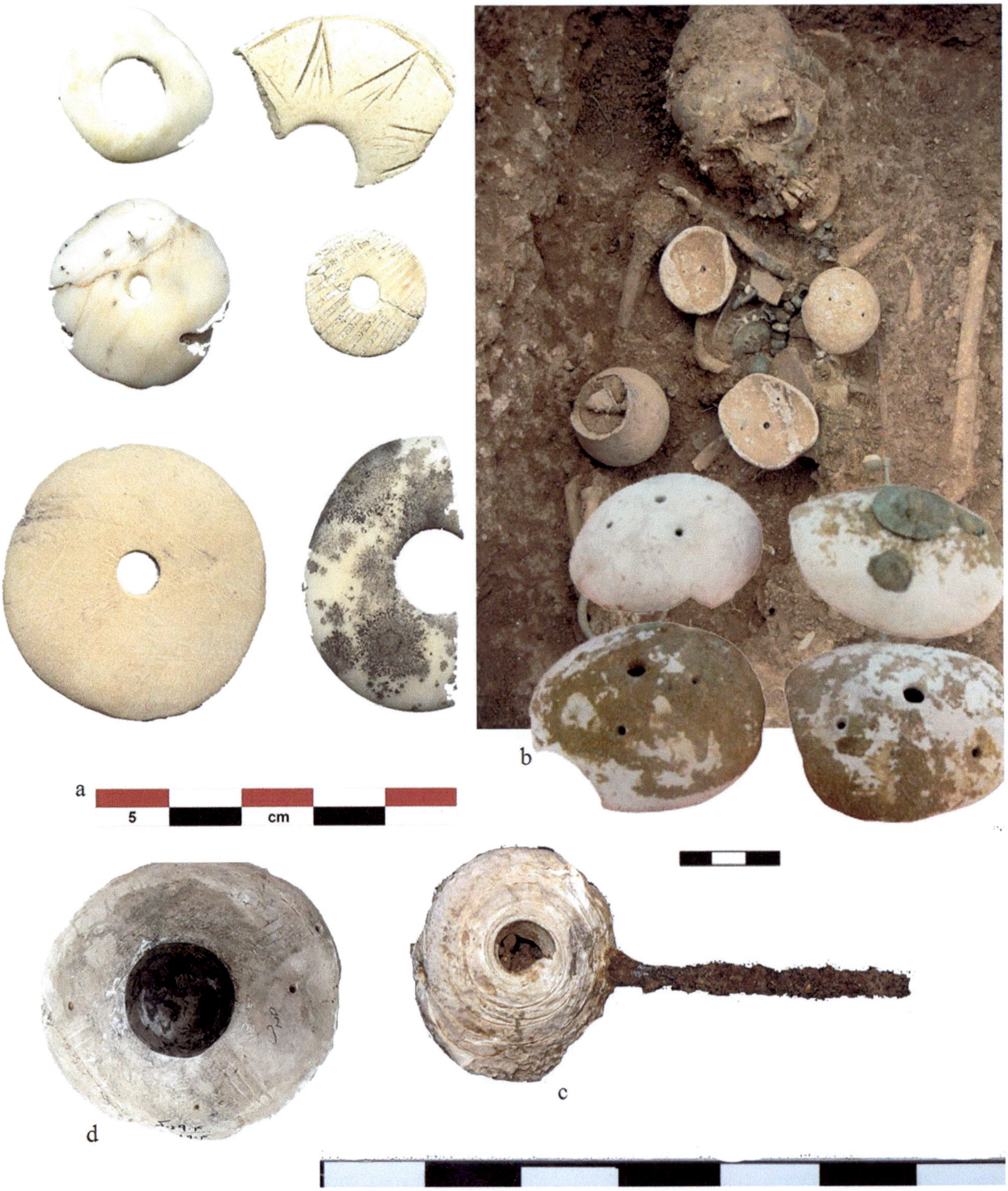

Fig. 12.5 (**a**) Shell discs from Susa, SB 24218 (©2013 Musée du Louvre/Antiquités orientales), (**b**) "Zagros Graveyard" Tomb A12. (From Amelirad et al. 2012: pl. 16), (**c**) Baba Jilan in Luristan. (From Hasanpur et al. 2016), (**d**) in National Museum of Iran (number 24)

influences the type of life after death. On the other hand, these shell discs may have been a symbol of the power and social and economic status of their owners.

Most were likely sewn on clothing. In some western regions of modern-day Iran, some older people call the button *syafe* and use the same name for marine shells. It is possible that in the past, they sewed the shell on their clothes as buttons. It has been said that everything incorporated in a culture's rituals, customs, folk beliefs, and traditions is a combination of belief and reality. Every fact and truth over time undergoes changes that, in later stages, can become a superstition. With this process, superstitions are formed and changed and take their place in society. Spells, supplications, and magical elements with extraordinary power that have existed in different periods, with the help of human imagination, have led to the emergence of belief in Iranian popular culture (Alipour and Sheikhzadeh 2009: 65). For many years, marine shells have been regarded as a symbol of divine blessing, counteracting the evil eye and promoting beauty, female fertility, and healing from disease. An analysis of shell artefacts in their archaeological and ethnographic contexts shows that the people of western Iran believed in the magical and supernatural powers of these artefacts. This is partly due to their specific characteristics: strength, durability, and exceptional lustre and colours.

Rare metals such as gold have enjoyed similar value and prestige among various cultures that had neither shared heritage nor cultural contact with each other. The number of ornaments being reused in ancient times is a complicated topic. As is still the case in some parts of the Middle East and Central Asia, stray finds were probably always prized as talismans. However, it is likely that the modern widespread reuse of old shells only occurred when necklaces were known to be found in ancient burials. At Hasanabad, massive shell discs with holes—called *nafeperi* or 'fairy's navels'—were occasionally sewn on garments or everyday items. A visiting ethnographer observed that these were "usually old and worn looking, and I was informed they were from ancient graves" (Watson 1979: 195). Indeed, these are the same discs discovered at the Early Iron Age Chamahzi Mumah cemetery.

The shell discs provided a symbolic language that kept the people of the ancient Iranian plateau in touch with the sea. The natural shape of the shell is symbolically linked to water and the sea. Moreover, their value increases with distance from the coast. Their use dates back to the Palaeolithic period in Iran and continues to this day. Raw shells have been turned into ornaments and beads, further increasing their value. As luxury goods for adornment and display, marine shells have been used as symbols of status and ritual use, examples of which can be seen in prehistoric Iranian cemeteries, temples, and palaces. Shell is a very durable natural material. According to archaeological evidence, various forms have been used in different ways from the Neolithic period onwards (Kenoyer 1985; Reese 1989).

12.5 Conclusions

Many shell discs in Iran were found in burials. However, the nature of the discs and the association between specific artefacts and other burial gifts suggest that they were probably used by people of high social status. Shell discs can be seen as both a material manifestation and a symbol of the social and economic status and life experiences of their owners and were closely associated with the biographies of the exotic raw materials from which they were made—inherently associated with the sea and possibly health and fertility, and subsequently nurtured through collection, carving, decoration, exchange, and use. As signs of status, insignia of rank, and/or perhaps talismans, they were worn by their wearers neither passively nor without wider social 'consent'. The importance of shell discs is related both to individuals and their broader social contexts in ancient Iran. The existence of these disc-shaped shells contributes to our understanding of water-based artefacts and their social roles, and their uncertain origins and uses shed light on extensive trade exchanges in prehistoric times.

Bibliography

Alekshin, V. A., Bartel, B., Dolitsky, A.B., Gilman, A., Kohl, P.L., Liversage, D. & Masset, C. 1983. Burial Customs as an Archaeological Source [and Comments]. *Current Anthropology,* 24(2): 137–49.

Alipour A.R. & Sheikhzadeh, M. 2009. Magical Objects in Iranian Popular Culture and Its Influence on Saqakhaneh School Painting. *Bagh of Nazar,* 6(11): 53–68.

Allen, M.J. & Payne, B. 2017. Molluscs and Archaeology: An Introduction. In: M.J. Allen, ed. *Mollusks in Archaeology: Methods, Approaches, and Applications.* Studying Scientific Archaeology 3. Oxford: Oxbow Books, pp. 1–4.

Amelirad, S., Overlaet, B. & Haerinck, E. 2012. The Iron Age "Zagros Graveyard" near Sanandaj (Iranian Kurdistan): Preliminary Report on the First Season. *Iranica Antiqua,* XLVII: 41–99.

Barnett, R.D. 1963. Hamath and Nimrud: Shell Fragments from Hamath and the Provenance of the Nimrud Ivories. *Iraq,* 25: 81–85.

Bar-Yosef Mayer, D.E. 2011. Nawamis, Shells, and Early Bronze Age Pastoralism. In: M.S. Chesson, ed. *Daily Life, Materiality, and Complexity in Early Urban Communities of the Southern Levant.* Winona Lake (IN): Eisenbrauns, pp. 185–95.

Biscione, R., Tosi, M. & Durante, S. 1981. *Conchiglie, Il commercio e la lavorazione delle conchiglie marine nel medio oriente dal IV al II millennio a.C. Roma Palazzo Brancaccio, 14 maggio-19 luglio 1981.* Rome: De Luca Editore.

Brandl, B. 1984. The Engraved Tridacna-shell Discs. *Anatolian Studies,* 34: 15–41.

Curvers, H.H. 1989. Iron Age Bone and Shell Objects. In: E.F. Schmidt, M.M. van Loon & H.H. Curvers. *The Holmes Expedition to Luristan.* Oriental Institute Publications 108. Chicago: University of Chicago Press, pp. 363–80.

Gensheimer, T.R. 1984. The Role of Shell in Mesopotamia: Evidence for Trade Exchange with Oman and the Indus Valley. *Paléorient,* 10: 65–73.

Haerinck, E. & Overlaet, B. 1999. *Chamahzi Mumah: An Iron Age III Graveyard.* Luristan Excavation Documents 2. Acta Iranica 35. Leuven: Peeters.

Haerinck, E. & Overlaet, B. 2004. *The Iron Age III Graveyard at War Kabud, Pusht-I Kuh, Luristan.* Leuven: Peeters.

Hasanpur, A., Hashemi, Z. & Overlaet, B. 2016. The Baba Jilan Graveyard near Nurabad, Pish-i Kuh Luristan: A Preliminary Report. *Iranica Antiqua,* 50: 171–212.

Herzfeld, E. 1941. *Iran in the Ancient East.* New York: Oxford University Press.

Kenoyer, J.M. 1983. Shell Working Industries of the Indus Civilization: An Archaeological and Ethnographic Perspective (PhD dissertation, University of California, California, Berkeley).

Kenoyer, J.M. 1985. Shell Working Industries of the Indus Civilization: A Summary. *Paléorient,* 10: 49–63.

Lischi, S. 2020. Decorated Shell Discs from Sumhuram, Oman. In: C. Coppini & F. Simi, eds. *Interactions and New Directions in Near Eastern Archaeology. Volume 3. Proceedings of the 5th "Broadening Horizons" Conference (Udine 5-8 June 2017).* Trieste: EUT Edizioni Università di Trieste, pp. 359–70.

Mallowan, M.E.L. 1966. *Nimrud and Its Remains* I-II. London: Collins.

Mashkour, M., Tengberg, M., Shirazi, Z. & Madjidzadeh, Y. 2013. Bio-archaeological Studies at Konar Sandal, Halil Rud Basin, Southeastern Iran. *Environmental Archaeology,* 18(3): 222–46.

Moorey, P.R.S. 1994. *Ancient Mesopotamian Material and Industries: The Archaeological Evidence.* Oxford: Clarendon.

Muscarella, O.W. 1992. Roundel. In: Ancient Art: Gifts from the Norbert Schimmel Collection no. 23. *The Metropolitan Museum of Art Bulletin* 49(4): 55.

Negahban, E.O. 1996. *Marlik: The Complete Excavation Report.* Vol. 87. Philadelphia: University of Pennsylvania Museum of Archaeology.

Overlaet, B. 2003. *The Early Iron Age in the Pusht-i Kuh, Luristan.* Vol. 4. Ghent: Gent University.

Pope, A.U. 1938. *A Survey of Persian Art: From Prehistoric Times to the Present.* Oxford: Oxford University Press.

Reese, D.S. 1989. Treasures from the Sea: Shells and Shell Ornaments from Hasanlu IVB. *Expedition,* 31(2–3): 80–86.

Reese, D.S. 1995. Marine Invertebrates and Other shells from Jerusalem (Sites A, C, and L). In: I. Eshel & K. Prag, eds. *Excavations by K.M. Kenyon in Jerusalem 1961–1967* IV. British Academy Monographs in Archaeology 6. The British School of Archaeology in Jerusalem. Oxford: Oxford University Press, pp. 265–78.

Schmidt, E.F. 1937. *Excavations at Tepe Hissar Damghan.* Philadelphia: University of Pennsylvania Press.

Schmidt, E.F., Van Loon, M.N. & Curvers, H.H. 1989. *The Holmes Expeditions to Luristan.* Chicago: Oriental Institute, University of Chicago.

Trubitt, M.B.D. 2003. The Production and Exchange of Marine Shell Prestige Goods. *Journal of Archaeological Research,* 11(3): 243–77.

Watson, P.J. 1979. Archaeological Ethnography in Western Iran, 1959–1960. *Ethnoarchaeology,* 2(2): 235–48.

Weeks, L., David-Cuny, H., Avanzini, A., Lischi, S., Genchi, F., Caputo, F., Al Ali, Y.Y. & Boraik, M. 2019. Worked and Decorated Shell Discs from Southern Arabia and the Wider Near East. *Arabian Archaeology and Epigraphy,* 30(2): 1–26. https://doi.org/10.1111/aae.12126

Concluding Remarks: Creatures of the Water 13

Tina L. Greenfield

From the river, oh Inana, let my man eat
Sumerian proverb

Abstract

This chapter aims to bring together some of the cross-cutting themes that transcend the volume. The volume exposes humans' enduring bond with aquatic life, from its essential role in sustaining early civilisations to its profound cultural significance. By integrating diverse scholarly perspectives, including archaeological, textual, and zooarchaeological analyses, this volume offers a holistic exploration of this intricate relationship through a number of case studies. The chapter uncovers some of the vivid tapestry of our interdependence with water creatures, as discussed in the contributions of the volume, but it also offers some new and fresh evidence from Mesopotamia to these shared themes.

13.1 Introduction

The idea that water brings food to the table is a concept as old as humanity. We are linked to the oceans, seas and rivers for sustenance, and the creatures that are found within provide means to celebrate life, death, power, and sacred actions. They provide the reasons and avenues for trade and interconnected worlds. The impact of aquatic beings on humans and their world is undeniable and can be seen throughout every facet of society. This book incorporates narratives by scholars that are rarely found together and represent a variety of approaches to understanding the relationships between humans and the creatures of the rivers, streams and seas. Water is the essence of life and it is therefore crucial to contextualise its fundamental role in nourishing life in the seas and related bodies of water.

To gain a more comprehensive understanding of the role of aquatic life and its relationship to humanity, contributions by the various scholars in this volume incorporate many research dimensions, including archaeological, textual, iconographical and zooarchaeological ones. They weave the contributions from the various archaeological sciences into humanistic, anthropological, and historical theoretical frameworks. While the chapters in this volume are arranged geographically, their research cross-cuts these traditional disciplinary and theoretical boundaries. As such, the introductory chapter provides a framework (see themes below) that enhances knowledge and understanding of the role of aquatic life for early societies.

> The fate of humans and water is inextricably linked. In the words of the Whanganui River Tribe's proverb, 'Ko au te awa, ko te awa ko au' – I am the river, the river is me. (UN Reports 2021)

Water, one of the four main elements in Greek philosophy, brings forth and sustains life in every aspect of the cultures discussed in this book. From the Sumerian water god Enki (Fig. 13.1) in the Near East to the Greek water gods of Poseidon and Aphrodite, one 'sees' (pun intended!) powerful narratives within creation myths. Thales, a sixth-century BCE Greek philosopher, proposed water as the fundamental substance for life and the universe (Bertrand 1947).

A holistic multidisciplinary framework wherein the cultural and physical environment are both considered provides a greater understanding of the complexities of cultural behaviour. These can include both a consideration of the physical landscape as well as the resource management strategies of acquiring creatures from a multitude of sources—in this case, oceans, seas, rivers, and other water sources. One must consider not only economy (e.g. exploitation strategies),

T. L. Greenfield (✉)
Department of Anthropology, The University of Winnipeg, Winnipeg, MB, Canada
e-mail: t.greenfield@uwinnipeg.ca

C. Tsouparopoulou, L. Recht (eds.), *Human and Aquatic Beings: Interactions in and beyond the Eastern Mediterranean (3rd–1st Millennia BCE)*, Themes in Contemporary Archaeology, https://doi.org/10.1007/978-3-031-73643-8_13

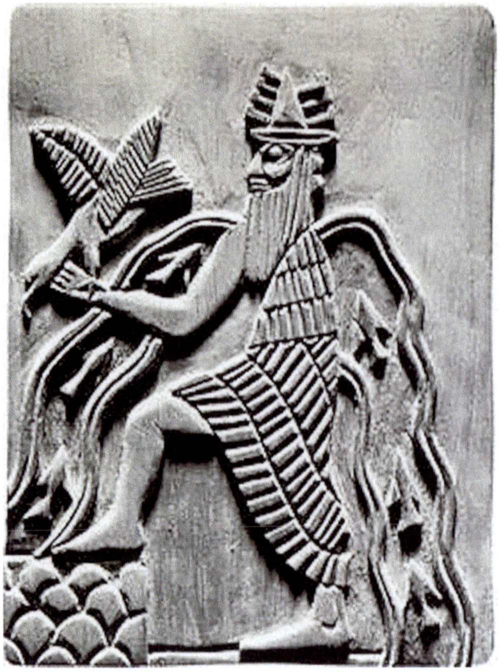

Fig. 13.1 Sumerian god Enki. Modern reproduction of a detail of the Adda seal. (Photo by Xalva. Wikimedia Commons)

but also the varying perceptions of cultures according to each unique society. These can vary not only according to its use and value, but also how the creatures found within their depths and along their shores are treated, hunted, farmed, and represented in their art and in local theology.

Around the world, creation myths often involve the emergence of all lifeforms from the rivers and seas that sustain people's livelihoods (Leeming and Leeming 1994, 2009). The studies in this volume illustrate the collective relationship between not only humans to water, but also the relationships held between the divine and the mortal. Mortals must worship and pray to the divine aquatic beings for sustenance, protection and healing—these are displayed over and over again in life and art (see Ftaimi, this volume, and here Fig. 13.2). Both the divine imagery and the words of prayer are recognition of divine power over life and death.

13.2 Some Thoughts on the Contributions

The introductory chapter in this book highlights continuities in research between different periods and regions across the ancient Eastern Mediterranean (Aegean and Cyprus) and Near Eastern worlds (Egypt and Southwest Asia). It showcases the breadth of the data from the various geographical regions that contribute to our understanding of the importance of water to human life in ancient times. Eight broad and

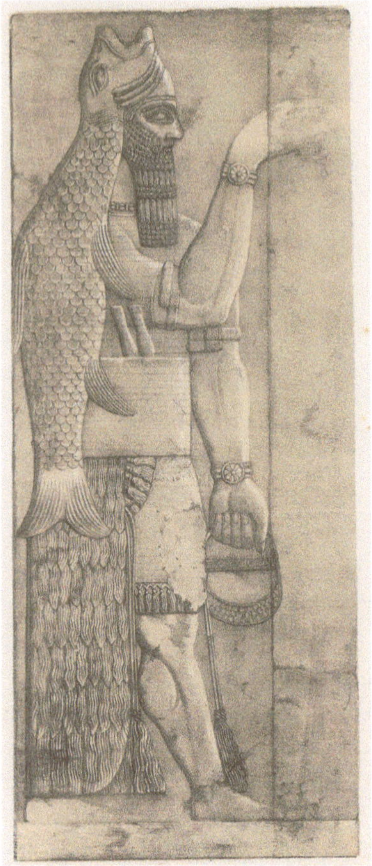

Fig. 13.2 Drawing of bas-relief of an Apkallu from the temple of Ninurta at Nimrud. (After Layard, 1853: plate 6)

interlocking themes are discussed in the introductory chapter: human-aquatic relations; underwater spaces; fish and mollusc remains and human diet; fishing and collecting; trade, shells and fish as raw material: transformations and materiality; iconography and symbolling of aquatic beings; and finally aquatic beings and identity.

Themes that bind humans to their bodies of water (introduced in Chap. 1 and supplemented below) cross over from the standard disciplines of iconography and texts to archaeological sciences (isotopes) and zooarchaeology. The strength of this book is the multi-faceted investigation of aquatic beings and humans. This results in a synergy of information across time and space. The integration of results achieved by historians, archaeologists, philologists, art historians, anthropologists, radiogenic scientists and many other scholars who study aquatic beings, are uniquely compiled in this collection of case studies. The chapters in this book elicit the manifold forms of evidence that tell the stories of how *water impacts* humans in every aspect of life. The supplemental and interlocking themes that also link the various studies, include:

13.2.1 Sustainability, Resilience, and Adaptability to Climate Change

From the smallest river beings to the behemoth sea creatures, humanity has played a role in their acquisition—whether on the edge, surface, or depths of the bodies of water that lap at their feet (Friesen and Arnold 1995; Monks 2001; Davis et al. 2008). As water ebbs and flows, humans have to adapt. With climate change no longer at our doorstep, but within our homes, villages, and lifestyles (Gibbard et al. 2022), the sustainability of aquatic beings must be addressed. Extinction is happening all around us. We are in the midst of the sixth major biological extinction event since the beginning of the world (Monks 2017; Cowie et al. 2022). It is clear from the many studies in this book how important it is for humans to understand the world around them. How important it is for humans to maintain healthy relationships with the species and environments upon which they depend. This is true not only in antiquity but extends to our modern lives. To ignore or damage the environment has long lasting negative consequences. This has been long recognised in Mesopotamian studies (Jacobsen and Adams 1958). If we cannot maintain ourselves in a stable environment, how are we ever to survive climate change? Moving forward, the challenges we face not only as scholars, but also as members of the global community, is to find the balance between *sustainability* and *resilience* (in both cultural and physical environments), in the face of climate and environmental change. We do this by modifying or adapting our acquisition of resources to provision our societies when deemed necessary.

13.2.2 Food as Prestige

Information on diet, motivation, acquisition and in particular *food as prestige* is woven into the research of several papers in this volume. The gathering of information on aquatic beings with a cross-cultural and geographically wide lense is invaluable to understanding human behaviour, as is clearly shown in Yamasaki, Schlegel, and Recht and Tsouparopoulou's research. Texts and images provide information that is not always recognised in the fauna from archaeological deposits (see Ftaimi this volume). In some cases, we can see how certain animal foods are high prestige as they are associated with the gods and goddesses. In other instances, we can see how human-like creatures of the deep are visualised (e.g. fishman in the image of Oannes, Fig. 13.3). These images and relationships are not visible from the zooarchaeological record. At the same time, the zooarchaeological record informs on data missing from textual and iconographic sources. Both play an essential role in informing on the nature of human exploitation of aquatic environments.

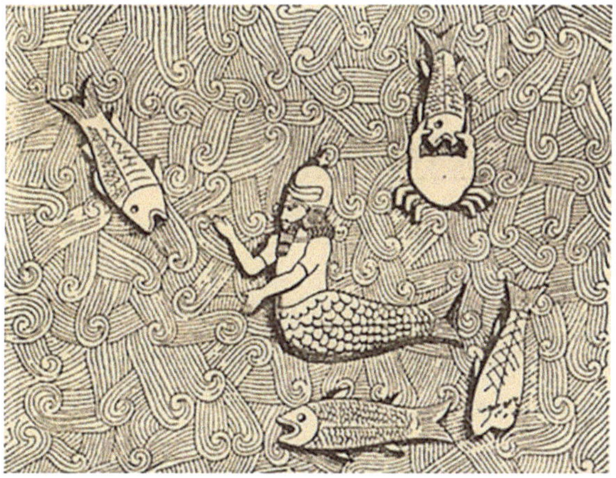

Fig. 13.3 Detail of relief from Khorsabad with Oannes (Kulullû). (Image by Lozman. Wikimedia Commons)

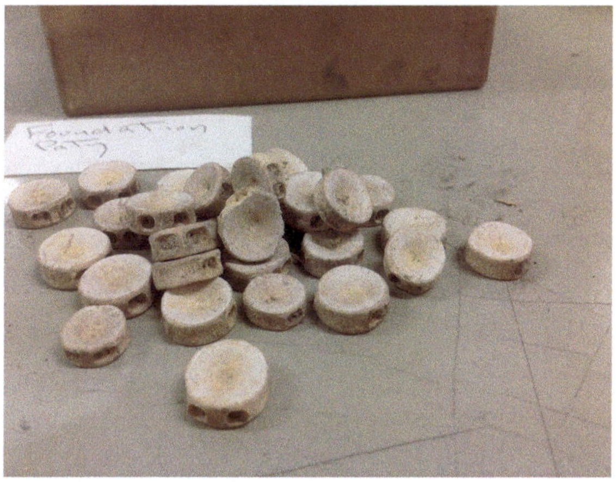

Fig. 13.4 Shark centra from an Early Dynastic private grave (144) uncovered along a foundation path within the Royal Cemetery at Ur. NHMUK ARC.1970.3090. (Photo by author at NHM, London. Courtesy of NHM)

In the past, I had the opportunity to analyse the zooarchaeological material from the Royal Cemetery at the ancient site of Ur stored at The British Museum and The Natural History Museum, London. Within the specially selected organic remains, I was struck by the presence of fish remains that Woolley chose to bring to London in the 1920s and 30s. Even though it was a very small assemblage (and only a tiny fraction of what was once buried in the cemeteries and private graves), he thought it was important enough to curate at The British Museum (Woolley 1934). Amidst the finds within the burials were the *centra* of shark vertebrae (Figs. 13.4 and 13.5), and a large assemblage of various *Pisces* sp. bones buried as grave goods. It also highlighted how the exotic species from the oceans and seas were collected and buried as symbols of prestige and power. The presence of aquatic beings in funerary contexts is the perfect example of conspicuous consumption. I firmly believe it is

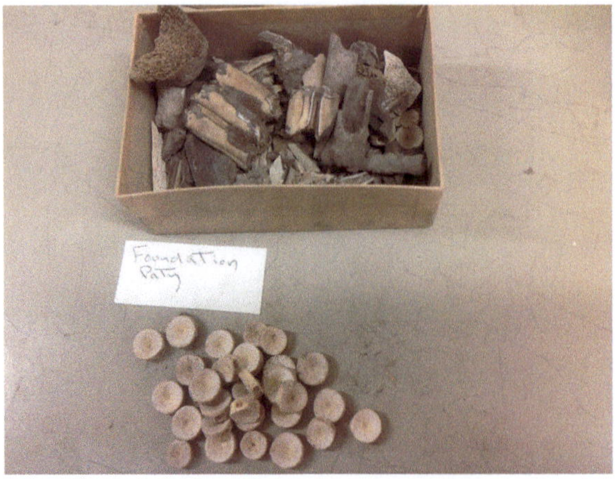

Fig. 13.5 Faunal assemblage from an Early Dynastic private grave (144) in the Royal Cemetery of Ur. NHMUK ARC.1970.3090. (Photo by author at NHM, London. Courtesy of NHM)

imperative to not only conduct new excavations, but to study old collections in museums. It is an excellent way to gather new information from sites excavated long ago when less standardised techniques were implemented, giving the opportunity to investigate long forgotten assemblages, and to apply holistic and multidisciplinary approaches to reassess the significance of such impossible to replicate archaeological data. We can never re-excavate the Royal Cemetery of Ur, but we can continue to apply new scientific techniques to the older collections that will enhance our limited understanding of the past, especially in regard to the lesser studied aquatic creatures and their by-products that were highly utilised in antiquity.

13.2.3 Entanglement and Interconnectedness

Entanglement and interconnectedness are embedded within the framework of most of the studies in the volume. This is highlighted in the study of the relationship between worship, trade, body and material decoration in the Aegean and its near and far neighbours (Schlegel; Patrevita). Urban and rural life are interconnected. One cannot survive well without the other. This was recognised almost 5000 years ago in the following Sumerian proverb "Things may be traded in the city but it is the fisherman who brings in the food supply" (Collection XI at The Electronic Text Corpus of Sumerian Literature).

Understanding agency and how the presence of aquatic creatures affects societies both in antiquity and in modern times is to fully grasp the impact aquatic life as beings (see Yamasaki; Stamatopoulou on coastal and inland fishing). There is both continuity and discontinuity over time and space that is reflected in the interconnectedness of life. We see this in Michael's study of the complex and multidimensional fishing strategies beginning in the Neolithic and extending into the early Christian periods on the island of Cyprus. According to the 2021 UN (UNESCO) World Water Development Report, "In some, water is seen as a gift for humans to care for, whilst others embrace a view that accentuates water's importance for the environment and wildlife. The connection between water and place, often categorized as 'relational values', can be strong in many indigenous cultures." (UN Reports 2021).

The physical environment dictates to only a certain extent the presence or absence of aquatic beings on archaeological sites. Trade networks highlight the interconnectedness of distant bodies of water. For example, the identification of shark teeth at the Early Bronze Age city of Titris Hoyuk in southeast Turkey on a tributary of the Euphrates illustrates that aquatic beings from distant seas may be found far inland. While this statement is obvious to anyone working in the Aegean or Near East (and beyond), it is necessary to reiterate this here. The archaeological record is biased in what is discovered—and this is especially true when studying or identifying creatures that are deposited in sacred contexts in houses and temples, thrown into middens, and/or tossed onto the surface of streets and alleyways.

13.2.4 Power—Aquatic Imagery as Tool for the Symbolic Display of Power

While it has long been recognised that these deposits only reflect part of the archaeological record, historical documentation and iconographic images are also biased and reveal only a part of the narrative. In particular, texts and iconography do not fully reflect the reality of the majority of the populations as they are made for the elite or administrative purposes. Such elite or state sponsored narratives and iconographic images often ignore the needs of the larger population, as well as their relationships and nature of interaction with the aquatic world. Hence, one cannot only use texts, iconographic images, figurines and wall reliefs, or the various forms of archaeology to gain a fuller understanding of human-aquatic world interactions. It is imperative to use all these data collectively, whether it be via concentrated holistic approaches including text and iconography (Yamasaki), biochemistry (Mylona and Allhouse) or excavation protocols (Stamatopoulou) to name just a few in this volume. Such approaches allow us a deeper understanding of such complex relationships.

The archaeological sciences (specifically related bioarchaeology and environmental studies) have in the last decade afforded the opportunity to delve deep into our data to "flesh" out more nuggets of information than previously thought possible. This is especially true in the Aegean and Eastern Mediterranean littoral where such studies have long been at the forefront of the discipline (Davies and Schofield

1995; Arnold et al. 2016). However, not all analyses need to employ cutting-edge scientific methods to enhance the study of human-environment relationships.

Part of the task of zooarchaeology is to reconstruct behaviour from animal bone assemblages excavated at archaeological sites. Much like the investigations of molluscs (Mylona and Allhouse; Eleftheriou et al), carved wall reliefs/ paintings, ceramics and shells (Schlegel; Lekka; and Ftaimi), fishing ornaments (Michael; Stamatopoulou), burial chambers and ritual deposits (Recht & Tsouparopoulou; Eleftheriou et al), jewellery (Naseri et al.), and other material remains to understand the role of aquatic beings in relation to human behaviour, zooarchaeologists attempt to squeeze out as much information as the bones will allow (Hesse and Wapnish 1985; Reitz and Wing 1994; Gifford-Gonzalez 2018). Traditional zoological identifications and zooarchaeological analyses of these aquatic creatures are now supplemented with exceptionally cutting-edge scientific techniques (e.g. stable and radiogenic isotopes, ZooMS, DietaryMS, aDNA, pigment analyses, etc.) (see Greenfield et al. 2022; Tomilin et al. 2024 and Mylona and Allhouse; Eleftheriou et al. this volume).

There has been a revolution in knowledge about the contribution of aquatic beings to the archaeological record and the history of ancient lifeways in the Aegean and Mediterranean worlds and their neighbours (Egypt, Southwestern Asia, Mesopotamia) since the 1960s, with the introduction of very simple methodological improvement in excavation and collection procedures, such as sieving, flotation, and the gridding of primary spaces to capture every piece of evidence (French 1971; Payne 1972, 1975, 1992; Clason and Prummel 1977; Diamant 1979; Gordon 1993; O'Conner 2001; Zohar and Belmaker 2005; Bar-Yosef Mayer 2008; Greenfield 2014; Brown and Greenfield 2017; Sapir-Hen et al. 2017; Brown et al. 2021). However, we have a long way to go in instilling these best practices especially for collecting microfauna (the bulk of which is often aquatic creatures—or the remains of them). Otherwise, we are biasing our picture and knowledge of the past and any reconstructions remain skewed. The archaeological record is limited, and we need to maximise recovery of it.

13.2.5 Sacred and Mundane

In the spring of 2023, I participated in the new excavations at the ancient Sumerian city of Tello (ancient Girsu), more specifically in Tell K (*Maison des Fruits*). This is an area previously excavated by a French team in the late nineteenth to early twentieth century. This is the first time in 90 years that systematic archaeological work was conducted in this part of the city. The early excavations by Ernest de Sarzec and Gaston Cros uncovered temples and several deposits of fish (carp (*Cyprinidae* sp.), tuna (*Thunnus* sp.), and Pharaoh cuttlefish (*Sepia pharaonis*)) preserved in oil between layers of reed mats in the sacred precinct. Cros made special mention of some larger kinds of preserved fish: for example, tuna and the 'fish of Tobias' (as he calls them, referring to the story of the huge fish caught in the Tigris by Tobias that is found in the Apocryphal Book of Tobit), which might have been carp (Cros 1910: 76). The skeletons of both sea and river fish species were present in multiple stacked layers between reed mats and bitumen (Tsouparopoulou 2024). Based on cuneiform texts, these species were likely used as food for mortals, as well as offerings to the gods (Rey 2021). Their flesh was consumed, but they could also be used for fermented fish oil (Cros 1910: 76).

The discovery of the remains of pressed fish, with their skeletons and parts of their skin and scales still discernible, suggested that the remains were the result of rations that were not consumed but rather preserved for a later date. Indeed, these remains could have been preserved via salt or sun dried or pressed for their oil. It is also quite possible that they were preserved simply to make the journey to the temple for preservation (Ellison 1983: 148; Rey 2021; Rey 2023 pers. communication).

The acquisition of the different fish present in the deposits speaks to not only the link between fish in sacred ceremonies found across southern Mesopotamia during the third-first millennium BCE, but also to the "deep" connection (again, pun intended!) to the environment and economic relations necessary to receive these creatures from near and far. As mentioned above, the Sumerians recognised the importance of the fisherman in supplying food for the urban population. Textual documentation (i.e. CDLI P220954) attests to each family having their own fish merchant to provide them both with food for the family, but also for their portion of temple offering (Fig. 13.6).

Further from Girsu, the remains of burnt fish vertebrae were also identified from deposits found within the *tannurs* (ovens) of a kitchen connected directly with the Early Dynastic temple on the Mound of the Palace (Fig. 13.7) (Rey 2016; Jongsma-Greenfield et al. 2024; Rey et al. Forthcoming). Similar to Girsu, temples of the south during the third-first millennia BCE were repositories for thousands of fish remains (i.e. Nippur). Representations of gods dressed as fish are a constant in Mesopotamia iconography (Fig. 13.2).

Aquatic beings were embedded (literally) in the structure of and fabric of everyday lives from cultures across the Aegean to Iran, and Turkey to Egypt, in the third to first millennium BCE—see Fig. 13.8 of an incredible aquatic find that was identified embedded within a mudbrick!. While we cannot be certain of the use of the fish embedded in the brick (e.g. ritual or utilitarian—see Morales 1993 for the use of fish bones for brick mortar in the Mediterranean), the continuity and use of fish in both religious ceremonies and daily life in Sumer and today is clear (as Recht & Tsouparopoulou

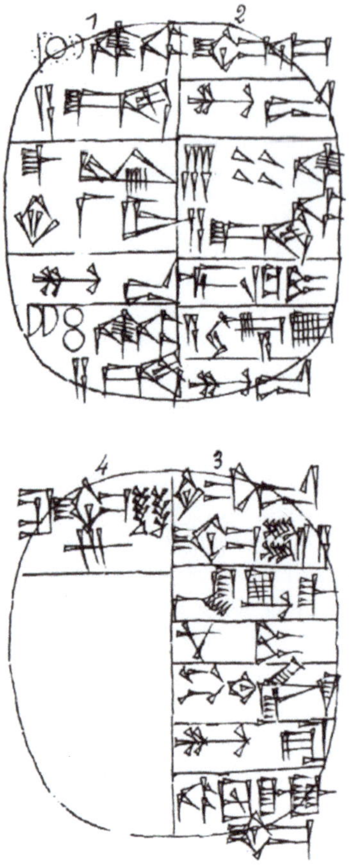

Fig. 13.6 Line drawing of cuneiform tablet from Tello/Girsu (Early Dynastic III) describing the role of each household in securing a man to provide them with fish to be sent to the temple as sacrifice. DP 304 CDLI P220954 image from https://cdli.mpiwg-berlin.mpg.de/ (accessed 27 June, 2024), after Englund, 1990

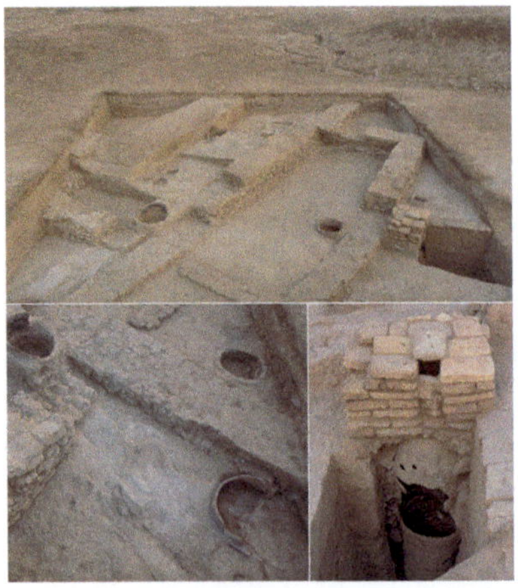

Fig. 13.7 Early Dynastic temple kitchen with *tannurs* filled with burnt animal and fish bones, with permission from S. Rey ©Tello/Girsu Project, Iraq Scheme, The British Museum

suggest) as even now in southern Mesopotamia fish are consumed as part of the regular diet, and also prepared into fish oil. A hundred years after the first excavations at Girsu,

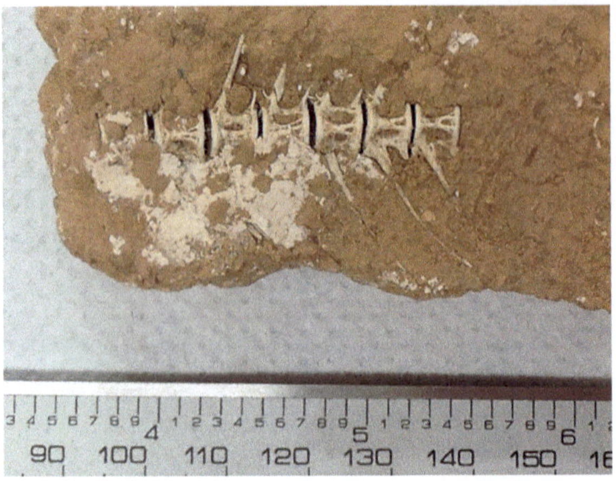

Fig. 13.8 Vertebral column of Pisces sp. embedded in mudbrick from Ur. No context number. Royal Cemetery of Ur. NHMUK ARC. (Photo by author at NHM, London)

fish remains continue to make an appearance both in archaeological deposits and as part of everyday and holiday meals by the modern inhabitants. This highlights both the sacred and mundane uses of aquatic creatures from the third millennium BCE onwards in ancient Sumer.

While the cuneiform textual data from most Sumerian sites is focused on domestic livestock, aquatic lifeforms (including molluscs and bivalves) can play a significant role in providing base-line data for evaluating isotopic information. These data have allowed for a fuller evaluation of the significance of radiogenic strontium analyses from domestic livestock from Ur and Abu Salabikh for understanding mobility of animals. Analysis has showed that the animals were raised and grazed locally and that there was no separation of royal and non-royal herds in terms of management strategies. They revealed that the historical texts do not always elucidate the reality of how animals are managed (Greenfield et al. 2022). Aquatic creatures are positioned to help not only modern peoples, but were also central to ancient humans. In modern times, they continue by way of archaeological sciences to help interpret past behaviours in every culture.

13.3 Conclusion

This book provides not only a fascinating picture of humanity's constant and complex relationship with all the creatures found in aquatic environments, it also provides a model moving forward to understanding the motivations behind our behaviour towards aquatic beings and how they have always been a part of our narrative. Through the diligent and comprehensive excavation and examination of these data, a greater picture of the interconnected world of aquatic creatures and humanity will rise to the surface to further educate us all.

Bibliography

Arnold, E.R., Hartman, G., Greenfield, H.J., Shai, I., Babcock, L.E. & Maeir, A.M. 2016. Isotopic Evidence for Early Trade in Animals between Old Kingdom Egypt and Canaan. *PLoS One,* 11(6): e0157650. https://doi.org/10.1371/journal.pone.0157650

Bar-Yosef Mayer, D.E. 2008. Dentalium Shells Used by Hunter-gatherers and Pastoralists in the Levant. *Archaeofauna,* 17: 103–10.

Bertrand, R. 1947. *A History of Western Philosophy*. New York: Simon & Schuster.

Brown, A. & Greenfield, H.J. 2017. Microdebris Analysis from the Early Bronze Age Levels at Tell eş-□âfi/Gath. *Near East Archaeology,* 80(4): 259–60.

Brown, A., Greenfield, H.J. & Maeir, A. 2021. Sweating the Small Stuff: Micro-debris Analysis at Tell eş-□âfi/Gath, Israel. In: J. Daujat, A. Hadjikoumis, R. Berthon, J. Chahoud, L. Kassianidou & J.D. Vigne, eds. *Archaeozoology of SW Asia and Neighboring Areas XIII: Proceedings of the 10th ICAZ-ASWA Conference*. Atlanta: Lockwood Press, pp. 81–91.

Clason, A.T. & Prummel, W. 1977. Collecting, Sieving and Archaeozoological Research. *Journal of Archaeological Science,* 4: 171–75.

Cowie, R.H., Bouchet, P. & Fontaine, B. 2022. The Sixth Mass Extinction: Fact, Fiction or Speculation? *Biological Reviews of the Cambridge Philosophical Society,* 97: 640–663. https://doi.org/10.1111/brv.12816

Cros, G. 1910. *Nouvelles fouilles de Tello*. Paris: Leroux.

Davies, W.V. & Schofield, L. eds. 1995. *Egypt, the Aegean and the Levant: Interconnections in the Second Millennium BC*. London: British Museum Press.

Davis, S.J.M., Gonçalves, M.J. & Gabriel, S. 2008. Animal Remains from a Moslem Period (12th/13th Century AD) Lixeira (Garbage Dump) in Silves, Algarve, Portugal. *Revista Portuguesa de Arqueologia,* 11(1): 183–258.

de Sarzec, E. & Heuzey, L. 1884–1912. *Découvertes en Chaldée*. Paris: Leroux.

Diamant, S. 1979. A Short History of Archaeological Sieving at Franchthi Cave, Greece. *Journal of Field Archaeology,* 6(2): 203–17.

Ellison, R. (1983). Some Thoughts on the Diet of Mesopotamia from c. 3000–600 B.C. *Iraq, 45*(1): 146–150. https://doi.org/10.2307/4200193

Englund, R.K. 1990. *Organisation und Verwaltung ser Ur III-Fischerei*. Berliner Beiträge zum Vorderen Orient 10. Berlin: Dietrich Reimer.

French, D.H. 1971. An Experiment in Water-sieving. *Anatolian Studies,* 21: 59–64.

Friesen, T. M. & Arnold, C.D. 1995. Zooarchaeology of a Focal Resource: Dietary Importance of Beluga Whales to the Precontact Mackenzie Inuit. *Arctic,* 48(1): 22–30.

Gibbard, P., Walker, M., Bauer, A., Edgeworth, M., Edwards, L., Ellis, E., Finney, S., Gill, J., Maslin, M., Merritts, D. & Ruddiman, W. 2022. The Anthropocene as an Event, Not an Epoch. *Journal of Quaternary Science,* 37. https://doi.org/10.1002/jqs.3416

Gifford-Gonzalez, D.P. 2018. *An Introduction to Zooarchaeology*. Berlin/Heidelberg: Springer.

Gordon, E.A. 1993. Screen Size and Differential Faunal Recovery: A Hawaiian Example. *Journal of Field Archaeology,* 20(4): 453–60.

Greenfield, H.J. 2014. The Faunal Remains from the Old Excavations. In: M.B. Cosmopoulos, *The Sanctuary of Demeter at Eleusis: The Bronze Age II*. Athens: The Archaeological Society at Athens, pp. 217–27.

Greenfield, T.L., McMahon, A.M., O'Connell, T.C., Reade, H., Holmden, C., Fletcher, A., Zettler, R.L. & Petrie, C. 2022. Were There Royal Herds? Understanding Herd Management and Mobility Using Isotopic Characterizations of Cattle Tooth Enamel from Early Dynastic Ur. *PloS ONE,* 17(6): e0265170. https://doi.org/10.1371/journal.pone.0265170

Hesse, B. & Wapnish, P. 1985. *Animal Bone Archaeology: From Objectives to Analysis*. Washington, DC: Taraxacum Press.

Jacobsen, T. & Adams, R.M. 1958. Salt and Silt in Ancient Mesopotamian Agriculture: Progressive Changes in Soil Salinity and Sedimentation Contributed to the Breakup of Past Civilizations. *Science,* 128(3334): 1251–58. https://doi.org/10.1126/science.128.3334.1251.

Jongsma-Greenfield, T., Di Michele, A., Husain, F., & Rey, S. (2024). Sacred Space and Ritual Behaviour in Ancient Mesopotamia: A View from Tello/Girsu. *Humans, 4*(3), 239–63. https://doi.org/10.3390/humans4030015?

Layard, A.H. 1853. *A Second Series of the Monuments of Nineveh*. London: John Murray.

Leeming, D.A. & Leeming, M.A. eds. 1994. *Encyclopedia of Creation Myths.* 2nd ed. Oxford: ABC-CLIO.

Leeming, D.A. & Leeming, M.A. eds. 2009. *A Dictionary of Creation Myths.* Oxford Reference Online ed. Oxford: Oxford University Press.

Monks, G.G. 2001. Quit Blubbering: An Examination of Nuu'chah'nulth (Nootkan) Whale Butchery. *International Journal of Osteoarchaeology,* 11: 136–49.

Monks, G.G. ed. 2017. *Climate Change and Human Responses: A Zooarchaeological Perspective*. Dordrecht: Springer.

Morales Muñiz, A. 1993. Where are the Tunas? Ancient Iberian Fishing Industries from an Archaeozoological Perspective. In: A. Clason, S. Payne & H.P. Uerpmann, eds. *Skeletons in Her Cupboard, Festschrift for Juliet Clutton-Brock.* Oxford: Oxbow Books, pp. 135–41.

O'Conner, T.P. 2001. Collecting, Sieving and Animal Bone Quantification. In: W. Prummel & H. Buitenhuis, eds. *Animals and Man in the Past: Essays in Honour of Dr. A.T. Clason Emeritus Professor of Archaeozoology*. Groningen: Rijksuniversiteit Groningen, pp. 7–16.

Payne, S. 1972. Partial Recovery and Sample Bias: The Results of Some Sieving Experiments. In: E.S. Higgs, ed. *Papers in Economic Prehistory*. Cambridge: Cambridge University Press, pp. 49–64.

Payne, S. 1975. Partial Recovery and Sample Bias. In: A.T. Clason, ed. *Archaeozoological Studies, Papers of the Archaeozoological Conference 1974, Held at the Biologisch-Archaeologisch Instituut of the State University of Groningen*. Amsterdam: North Holland, pp. 7–17.

Payne, S. 1992. *Some Notes on Sampling and Sieving for Animal Bones*. London: Historic Buildings and Monuments Commission for England.

Reitz, E.J. & Wing, E.S. 1994. *Zooarchaeology*, ed. G. Barker. 2nd edition, Reprint. Cambridge Manuals in Archaeology. Cambridge: Cambridge University Press.

Rey, S. 2016. *For the Gods of Girsu: City-state Formation in Ancient Sumer*. Oxford: Archaeopress.

Rey, S. 2021. Divine Cults in the Sacred Precinct of Girsu. *Near Eastern Archaeology,* 84(2): 130–39.

Rey, S., Di Michele, A., Husain, F., Charles, M., Faiers, C., Douche, C., & Jongsma-Greenfield, T. Forthcoming. For the Glory of the Gods: An Early Dynastic Temple Kitchen at Tello/Girsu.

Sapir-Hen, L., Sharon, I., Gilboa, A. & Dayan, T. 2017. Wet Sieving a Complex Tell: Implications for Retrieval Protocols and Studies of Animal Economy in Historical Periods. *Journal of Archaeological Science,* 82: 72–79.

Tomilin, M., Greenfield, T. Chumala, P. & Katselis, G. (2024). Use of Proteomics for Dietary Reconstruction: A Case Study Using Animal Teeth from Ancient Mesopotamia. *Journal of Proteome Research.* https://doi.org/10.1021/acs.jproteome.4c00446

Tsouparopoulou, C. 2024. The Early Dynastic 'Maison des Fruits' at Tell K in Tello (Ĝirsu). *Bulletin of the American Schools of Overseas Research,* 391: 191–225. https://doi.org/10.1086/730298

Woolley, C. L. (1934). *Ur Excavation. Vol. II: The Royal Cemetery.* British Museum and the Museum of the University of Pennsylvania to Mesopotamia. New York: Trustees of the two Museums.

Zohar, I. & Belmaker, M. 2005. Size Does Matter: Methodological Comments on Sieve Size and Species Richness in Fishbone Assemblages. *Journal of Archaeological Science,* 32: 635–41.

Additional Sources

Sumerian proverbs, Collection XI at *The Electronic Text Corpus of Sumerian Literature*, 3rd millennium BCE. https://www.unesco.org/reports/wwdr/2021/en/cultural-values water

UNESCO World Water Assessment Programme. (2021). The United Nations World Water Development Report 2021: Valuing Water. UNESCO.

Index

C. Tsouparopoulou, L. Recht (eds.), *Human and Aquatic Beings: Interactions in and beyond the Eastern Mediterranean (3rd–1st Millennia BCE)*, Themes in Contemporary Archaeology, https://doi.org/10.1007/978-3-031-73643-8

MIX
Papier aus verantwortungsvollen Quellen
Paper from responsible sources
FSC® C105338

If you have any concerns about our products,
you can contact us on
ProductSafety@springernature.com

In case Publisher is established outside the EU,
the EU authorized representative is:
Springer Nature Customer Service Center GmbH
Europaplatz 3, 69115 Heidelberg, Germany

Printed by Libri Plureos GmbH
in Hamburg, Germany